RIVERSIDE COMMUNITY COLLEGE
1916

This is the third edition of a well established and highly successful undergraduate text. The content of the second edition has been reworked and added to where necessary, and completely new material has also been included. There are new sections on amorphous solids and liquid crystals, and completely new chapters on colloids and polymers.

Using unsophisticated mathematics and simple models, Professor Tabor leads the reader skilfully and systematically from the basic physics of interatomic and intermolecular forces, temperature, heat and thermodynamics, to a coherent understanding of the bulk properties of gases, liquids and solids. The introductory material on intermolecular forces and on heat and thermodynamics is followed by several chapters dealing with the properties of ideal and real gases, both at an elementary and at a more sophisticated level. The mechanical, thermal and electrical properties of solids are considered next, before an examination of the liquid state. The author continues with chapters on colloids and polymers, and ends with a discussion of the dielectric and magnetic properties of matter in terms of simple atomic models. The abiding theme is that all these macroscopic material properties can be understood as resulting from the competition between thermal energy and intermolecular or interatomic forces.

This is a lucid textbook which will continue to provide students of physics and chemistry with a comprehensive and integrated view of the properties of matter in all its many fascinating forms.

Gases, liquids and solids

and other states of matter

Gases, liquids and solids

and other states of matter

D. TABOR

Emeritus Professor of Physics
Cavendish Laboratory, University of Cambridge

THIRD EDITION

CAMBRIDGE UNIVERSITY PRESS

Cambridge

New York Port Chester
Melbourne Sydney

e University of Cambridge
, Cambridge CB2 1RP
0011-4211, USA
3166, Australia

First published by Penguin Books Ltd 1969
Second edition published by Cambridge University Press 1979
Third edition published by Cambridge University Press 1991

Printed in Great Britain by J. W. Arrowsmith Ltd., Bristol

British Library cataloguing in publication data

Tabor, David
Gases, liquids and solids
1. States of matter
I. Title
530.4

Library of Congress cataloguing in publication data

Tabor, David
Gases, liquids, and solids: and other states of matter/D. Tabor.—3rd ed.
p. cm.
Includes index.
ISBN 0-521-40488-6. — ISBN 0-521-40667-6 (pbk.)
1. Matter. 2. Matter—Properties. I. Title.
QC173.T25 1991
530.4—dc20 91–9214 CIP

ISBN 0 521 40488 6 hardback
ISBN 0 521 40667 6 paperback

AS

To Hanna

Contents

Preface to the third edition

I am most grateful to the Syndics of the Cambridge University Press for inviting me to prepare a third edition of *Gases, Liquids and Solids*. The basic structure is unchanged. The main theme is that the three primary states of matter are the result of a competition between thermal energy and intermolecular forces. The second motif is that a whole range of properties e.g. the specific heat capacity of solids, the thermal conductivity of nonmetals, the elastic modulus of rubber, thermal expansion, surface tension, the viscosity of gases and liquids, osmotic pressure, the adiabaticity of sound waves in air, the dielectric properties of gases, liquids and solids, van der Waals forces between solid bodies, the hardness of metals, may be understood in terms of simple models and unsophisticated mathematics.

Few changes have been made in the early chapters on the properties of gases. In dealing with solids I have added short sections on the structure of surfaces and the phenomenon of surface melting and have extended the treatment of the elastic properties of crystalline solids and of rubber. In liquids there are further elaborations of the theory of viscosity and its application to the behaviour of lubricating oils.

But the main change in the book is the recognition that there are other states of matter which are of great scientific interest and which impinge on many aspects of everyday life. These include the amorphous state and liquid crystals which are dealt with rather briefly and two new chapters. One deals with the colloidal state which overlaps the interests of physicists, chemists and chemical engineers. The second deals with the polymeric state where the main physical properties and characteristics of thermo-plastics are described and discussed. Both these fields are the scene of active research in many university departments and industrial laboratories.

I am very grateful to various members of the Cavendish who have made helpful comments on the whole range of the added material. As to the two new chapters they could not have been finalized without the advice, criticism and suggestions of Robin Ball, Mike Cates, Athene Donald, Richard Jones, Paley Johnson, Chris Toprakcioglu and Mark Warner of the Cavendish and Malcolm Mackley of the Chemical Engineering Department. I am greatly indebted to them for their friendly and supportive attitude, though naturally I accept full responsibility for any errors or inadequacies. I should also like to acknowledge the generous help I received from Gina Murrell in typing and retyping parts of the manuscript.

Last but not least I wish to thank my wife for her unfailing encouragement and forbearance.

Cambridge 1990

Preface to the first edition

The subdivision of physics into mechanics; heat, light and sound; magnetism and electricity; properties of matter; and kinetic theory, goes back to the early days of classical physics. During the last twenty or thirty years there has been a discernible trend towards a regrouping of subjects, partly to allow for new knowledge and partly to allow for new methods of approach. Few of the areas have received such an impulse as those which, in classical days, were grouped together as kinetic theory and properties of matter. The new approach is to emphasize the atomic structure of matter and to show that, by assuming the existence of attractive and repulsive forces between atoms and molecules and the presence of thermal energy, it is possible to explain nearly all the bulk properties of gases, liquids and solids in terms of relatively simple models.

This book, which attempts to do this, is based on the lecture course given to first-year students at Cambridge. Its treatment is relatively simple and contains very little quantum physics or wave mechanics. It represents an attempt to bridge the gap between sixth-form physics and physical chemistry, and the more advanced courses which follow in later years of specialization. If it has any merit at all this is largely due to the devotion of many generations of Cavendish teachers who have hammered out the type of treatment and approach I have given here. I am particularly indebted to Dr T. E. Faber who first explained the Cavendish syllabus to me, and to Dr D. Shoenberg who has run a parallel course of lectures with me for the last few years and who has, on several occasions, saved me from perpetuating mistaken ideas. I also wish to thank the members of the Cavendish who have made comments on various parts of the manuscript, and Miss Shirley V. King of Birkbeck College for her comments on the first part of Chapter 10. Any mistakes, however, either of fact or of opinion are my own and I accept responsibility for them.

Dr Leo Baeck, in his spirited critique of romantic religion, referred to orthodoxy in the following terms: 'that bent of mind which would cultivate respect for the answer but, in the process, often loses respect for the problem'. A textbook such as this must naturally pay great respect to the answer. The student quite rightly expects to find the correct answer crisply and clearly given. This I have attempted to do. But it sometimes happens in physics, as in other walks of life, that the simple crisp answer is not the whole truth. Where it seems profitable I have discussed this as openly as possible. I hope it does not confuse the student, for this is not my intention. I recognize that, so long as a student's ability is judged by examinations, both student and examiner will persist in cultivating great respect for the answer. On the other hand, the progress of physics, in the long run, may well depend far more on our according greater respect to the problem.

Cambridge 1969

Preface to the second edition

The main change in this edition is the conversion to SI units. I have tried to be as consistent as possible and have been greatly aided in this task by the unflagging and meticulous care of the publishers. I have occasionally retained the °C particularly when referring to historical experiments and to the properties of water or ice near the melting point; 0 °C seems a more sensible point of reference than 273 K.

I have included some fresh material which fills out certain omissions in the first edition. I have also added some new themes which extend the coverage of the book; in particular the nature of van der Waals forces and the basic ideas of colloid stability. In some places, where it seemed relevant, I have added sections at a level rather more advanced than that required by the average readers; they may at some later stage find it useful to refer back to these items. Finally I have, as is inevitable in moving from c.g.s. to SI units, completely rewritten the introductory parts of the chapters dealing with the dielectric and magnetic properties of matter.

It remains for me to thank various colleagues and numerous correspondents who at various times have written to me proposing corrections and improvements. Though I have not been able to implement all their suggestions I am most grateful to them. For if this new edition represents any improvement on the old it is due in no small measure to their involvement. The faults are all my own.

Cambridge 1979

Note added at 1985 reprinting

In this reprint several typographical errors have been corrected and one or two misleading statements rectified. I am grateful to various corespondents who have written to me suggesting such improvements. In particlar I wish to thank Dr H. Mykura of Warwick University for a number of detailed and perceptive comments.

Cambridge 1985

Introduction

The three primary states of matter are the result of a competition between thermal energy and intermolecular forces. In this book we show that many of the bulk properties of a given substance in its gaseous, liquid and solid state may be correlated in terms of intermolecular forces thus emphasizing the fact that the three states are linked by a common property. Simple models are introduced and the mathematics is generally straightforward and unsophisticated.

Chapter 1 is a simple account of the nature of atoms and molecules and the forces between them, and includes a rather more detailed explanation of van der Waals forces. The second chapter deals with temperature and the concept of heat. These are the introductory chapters.

The next three chapters describe the properties of gases. This is a well-worn area where the molecular approach has long been used to describe the bulk behaviour. Chapter 3 gives the simple kinetic theory of ideal gases, while Chapter 4 deals with a more sophisticated treatment which includes a simplified approach to the Boltzmann distribution, and a discussion of the equipartition of energy. Chapter 5 deals with imperfect gases where perfect gas behaviour is modified to allow for the finite size of the molecules and for the forces between them.

We turn at once from gases to solids. In gases the molecules are virtually free, in solids they are bound to particular sites and their main thermal exercise consists in vibrating about their equilibrium positions. However, the molecular forces that cause the real gas to deviate from ideal gas behaviour are the same forces that are responsible for the existence of the solid state. In Chapter 6 we show that these forces can explain the heat of sublimation of solids, their thermal expansion, their elastic properties and the existence of the surface energy of solids. Some of these themes are analysed in greater detail in subsequent chapters. The greater part of Chapter 6 deals with solids in the crystalline form. To provide a fuller picture of solids a short section has been added on the amorphous state.

In Chapter 7 the elastic properties of solids are described in terms of intermolecular forces. This is applied quantitatively to ionic and van der Waals solids, to metals and tentatively to covalent solids. An account is also given of the elasticity of rubber in terms of the entropy of a single long-chain molecule and a simple model is then developed for deducing the elastic modulus of rubber in the bulk.

Chapter 8 deals with the plastic and brittle properties of solids and shows that these may be understood in molecular or atomic terms. For both types of strength property it is shown that the real strength is generally much less than the theoretical because of the presence of imperfections. Chapter 9 describes the thermal and electrical properties of solids. In this chapter we give a brief account of the specific heat capacities of solids in terms of atomic vibrations (the Einstein and Debye theories); we also show how thermal expansion arises as the amplitude of the vibration increases with increasing temperature. Finally we describe thermal and electrical conductivity; with metals these are both attributed to the transport of energy by the free electrons. The classical theory which treats the electrons as a gas gives the right answers for the wrong reason and it is shown how the theory must be modified to allow for the discrete energy levels of the electrons.

The next two chapters deal with the liquid state which still remains the Cinderella of modern physics. The main difficulty here is that, although in some ways the intermolecular forces dominate (a liquid occupies a specified volume unlike a gas), in other ways thermal motion dominates (a liquid shows mobility, unlike a solid). The molecules are both bound and free so that over small regions they appear to be highly ordered as in a solid whilst over large regions such order does not exist. These feaures and the way they have been tackled in Bernal's packing model and by Alder in his pioneering development of molecular dynamics are given in Chapter 10. This includes a molecular theory of vapour pressure, osmotic pressure and surface tension. The chapter concludes with a section on liquid crystals which have acquired new interest because of their use in electronic display devices and, surprisingly, in some types of polymer fibres.

In Chapter 11 we discuss the flow of liquids and using a simple form of Eyring's theory, show that viscosity may be understood in terms of intermolecular forces. These results are then applied to the behaviour of lubricating oils at high pressure and shear rates where it is shown that the shear properties of the oil resemble those of a solid wax. Finally because intermolecular forces are involved in both surface tension and viscosity a simple connection between them is made.

Chapter 12 deals with the colloidal state, in particular the factors that determine the stability of colloids. Colloidal particles in suspension are attracted by van der Waals forces and would if unimpeded coalesce, coagulate and fall out of suspension. It is shown that by introducing short range repulsive forces either of an electrical or entropic nature various levels of colloid stability may be achieved.

Chapter 13 provides a review of some of the main physical properties of polymers. Starting with the random configuration of a long hydrocarbon chain it is shown how the deformation of such a chain is dependent on both temperature and deformation rate, a low temperature being equivalent to a high rate of deformation. This basic property is carried right through the behaviour of bulk polymer in the liquid phase and, with some qualifications, into the solid glassy phase. The flow, deformation and fracture of polymers are described and finally a brief reference is made to the action of reinforcing fibres.

The last two chapters deal with the dielectric and magnetic properties of matter. It is shown that these properties arise from the polarization of atoms and molecules produced by the applied electrostatic or magnetic field. The dielectric behaviour, described in Chapter 14, receives a very satisfactory explanation in terms of simple atomic and molecular models. In addition, the treatment provides a direct correlation between dielectric properties, optical properties and van der Waals forces. The explanation of magnetic properties is more difficult since quantum effects are of major importance. The classical explanation of diamagnetism which we reproduce in Chaper 15 yields the right answer but, at a more fundamental level, is quite wrong. Paramagnetism is more easily explained and, assuming the existence of strong interactions, a satisfactory and satisfying model for ferromagnetism is found.

Some of the chapters lend themselves to brief summaries and where this has been possible they have been introduced at or near the end of the relevant chapter.

1

Atoms, molecules and the forces between them

1.1 Atoms and molecules

1.1.1 *The evidence for atoms and molecules*

Matter is not a continuum of uniform density, but consists of discrete particles or, if one wishes to be more up to date, of localized regions of very high density separated by regions of almost zero density. The particulate nature of matter has been known to us since the time of the Greeks and the idea of atoms (units which could not be further cut or divided) is generally attributed to Democritus (*c*. 460–370 BC). Many of his concepts have a surprisingly modern ring and constitute a tribute to the immense power, as well as the originality, of the Greek approach to logical inference and abstract reasoning. Much of his work was, indeed, the result of thought rather than of direct experiment. One must not, however, read too much modern science into these early ideas. For example in a passage quoted by Theophrastus, *De Sensu*, 61–2, Democritus states, 'Hard is what is dense, and soft what is rare . . . Hard and soft as well as heavy and light are differentiated by the position and arrangement of the voids. Therefore iron is harder and lead heavier.' But it would be wrong from this to attribute to Democritus a knowledge of dislocations and point defects.

The first real attempt to get to grips with the basic 'atoms' of matter had to wait until more quantitative measurements and generalizations had been made. The law of multiple proportions, largely due to the work of J. Dalton (1766–1844) was one of these. It stated that if two elementary substances combined chemically to form more than one compound the weights of one which combine with a fixed weight of the other are in a simple ratio to one another. For example nitrogen and oxygen combine to form nitrous oxide, nitric oxide, nitrogen sesquioxide and nitrogen dioxide; three of these oxides were, in fact, amongst the compounds which Dalton first studied in the course of his investigation. In these compounds 14 unit masses of nitrogen combine respectively with 8, 16, 24 and 32 unit masses of oxygen: i.e. the ratio is 1:2:3:4. This fits in naturally with the concept of unit masses which can combine in simple multiples with one another.

We consider a simpler case, the reaction of hydrogen and oxygen to form water. Quantitatively we find 1 unit mass of hydrogen combines with 8 unit masses of oxygen to form 9 unit masses of water. According to Dalton's atomic theory (Dalton 1808) we should write

$$1 \text{ atom hydrogen} + 1 \text{ atom oxygen} \rightarrow 1 \text{ atom water}$$

or

$$H + O \rightarrow HO.$$

When these reactions are studied in the gaseous state and the volumes are measured at a standard temperature and pressure, the equation becomes

$$2 \text{ volumes hydrogen} + 1 \text{ volume oxygen} \rightarrow 2 \text{ volumes water vapour.}$$

The next step in the argument depends on the hypothesis put forward by A. Avogadro in 1811 that at a specified temperature and pressure a given volume of any gas contains the same number of unit masses.

$$2 \text{ unit masses of hydrogen} + 1 \text{ unit mass of oxygen} \rightarrow 2 \text{ unit masses of water.}$$

This statement makes sense only if the 'unit mass' of oxygen is itself divisible into two equal parts each of which goes into one of the unit masses of water. Arguments along these lines lead finally to chemical equations of the type

$$2H_2 + O_2 \rightarrow 2H_2O,$$

where H_2, the smallest quantity of hydrogen which can exist in the free state, is called the molecule whilst H, the smallest quantity of hydrogen to enter into chemical combination, is called the atom. Avogadro's 'unit masses' are therefore molecules.

1.1.2 *Avogadro's law*

This is still called a hypothesis but deserves to be promoted to the status of a law. It states that at a specified temperature and pressure a fixed volume of any gas or vapour contains the same number of unit masses, i.e. molecules or atoms if the free molecule is monatomic. If we redefine the atom of hydrogen as 1 unit mass, the molecule of hydrogen consists of 2 unit masses. In the older literature these quantities in grams were called the gram-atom or gram-molecule respectively. In SI units the standard is no longer hydrogen but carbon-12: the gram-atom and gram-molecule are replaced by the mole (and the unit is written 'mol'). It is defined as the 'amount of substance of a system which contains as many elementary particles as there are atoms in 0.012 kg of carbon-12. We can only regard this definition in terms of 12 grams of carbon as a failure of nerve on the part of physical chemists. The logic of SI units demands the use of 12 kg of carbon-12 as the unit and indeed some physicists use the kilomole in this way. We shall not, however, introduce such units into this book. One mole of any gas at a pressure of 760 mm Hg (101 325 Pa) and a temperature of 273 K fills approximately 0.0224 m^3. The number of molecules in a mole of any substance is called Avogadro's number (N_A) and has the value of 6.02×10^{23}. For example (N_A) molecules of hydrogen weigh approximately 2 g: these constitute the mole-mass of these materials to the nearest integer.

1.1.3 *First experimental deduction of the size of molecules*

Pioneer work by Fraülein Pockels in 1891 showed that certain organic molecules with polar end groups such as oleic acid which are insoluble in water in bulk will spread over the surface of clean water.

Figure 1.1. Schematic diagram showing the surface tension of water as small quantities of surface-active material are added. The surface tension remains fairly high until a critical amount, corresponding approximately to a complete monolayer of surfactant, has been added; it then falls rapidly to a constant lower value.

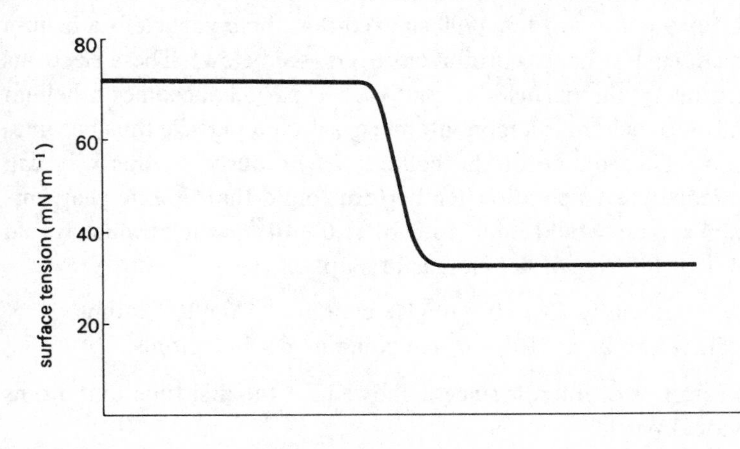

Lord Rayleigh experimented further with this and showed in 1899 that if the amount of acid added to the water surface is less than a certain amount there is practically no reduction in surface tension of the water. If, however, the amount added exceeds a certain critical value the surface tension falls rapidly. This is shown schematically in figure 1.1. Rayleigh assumed that this change occurred when there was a continuous monomolecular layer of fatty acid on the surface. He was thus able to determine the weight of acid per m^2 of surface at which the monolayer was formed. Assuming the monolayer to have the same density as the acid in bulk, the thickness of the monolayer was calculated and found to be about 10^{-9} m, i.e. 10 Å. This was the first direct indication that atomic and molecular dimensions are of the order of angstroms.

1.1.4 *Determination of Avogadro's number* N_A

We shall mention three very different ways of determining N_A. They are all of historic interest but only the last two provide reasonably accurate values.

1.1.4.1 *Radioactive disintegration*. Radium disintegrates to give off α particles and leave radon gas. Radon disintegrates to give off α particles and leave behind polonium. Further stages in the disintegration are not, in the present context, of great importance. The α particles emitted can be counted by allowing the particles emitted within a small known solid angle to strike a fluorescent screen. Each particle may be counted since it produces an individual scintillation on the screen. From the size of the solid angle chosen the total emission over a solid angle of 4π can be deduced. Lord Rutherford in 1909 showed that the α particle is a helium atom which has lost its two orbital electrons (see below). These electrons are picked up by the particles so that each α particle becomes a helium atom, and since helium is a monoatomic gas each α particle thus becomes a gas atom. The volume of the helium gas produced in this way can be measured. By extrapolation Rutherford found that, in one year, one kilogram of radium would emit a total of 11.6×10^{20} particles which would turn into 43×10^{-6} m^3 of helium gas at s.t.p.

Consequently 43×10^{-6} m^3 He contains 11.6×10^{20} atoms.
Therefore 22.4×10^{-3} m^3 He contains 6×10^{23} atoms.

This derivation is of interest since it showed for the first time that atoms are individual particles.

1.1.4.2 *Electrolysis.* If a quantity of electricity known as a faraday is used to electrolyse a dilute acid, exactly 1 gram of hydrogen is liberated. Since the faraday is equal to 96 500 coulombs, this means that 96 500 coulombs are involved in converting N_A ions of hydrogen into N_A atoms of hydrogen gas. The charge on the hydrogen ion is due to the loss of a single electron, hence

$$96\,500 \text{ coulomb} = \text{charge of } N_A \text{ electrons} = N_A e.$$

The next step is to determine e. This was carried through successfully by R. Millikan in 1909. He produced very fine droplets of oil between the plates of a parallel plate condenser, the plates being set in a horizontal plane. The particles gradually fall because of gravity. If, however, the air surrounding them is subjected to ionizing radiation the particles will pick up charge of say q. It then becomes possible to apply an electrostatic field E which is able just to hold the drop in equilibrium in a stationary position. In this state

$$Eq = mg.$$

Therefore as m, g and E were known, q was calculated and found always to be an integral number of some fundamental unit charge. This was assumed to be the charge e of the electron. The current value of e found in this way is very nearly

$$e = 1.6 \times 10^{-19} \text{ coulomb}.$$

Hence

$$N_A = \frac{96\,500}{1.6 \times 10^{-19}} = 6.03 \times 10^{23}.$$

1.1.4.3 *X-rays.* X-rays are electromagnetic waves of short wavelength (about 10^{-10} m). M von Laue was the first to suggest, in 1912, that since the spacing of crystal planes in a simple inorganic crystal is of this order, it ought to be possible to use the crystal as a diffraction grating for X-rays. The wavelengths λ of the X-rays may be determined accurately by diffracting them with a conventional ruled grating, at glancing incidence. The X-rays may then be used to find the lattice spacing in the crystal. If, for example, a collimated beam of X-rays strikes a crystal plane at an angle θ_1, it is first refracted through an angle θ_2, but since the refractive index for X-rays is almost unity one may take $\theta_1 = \theta_2 = \theta$. The condition for reinforcement is then that $2d \sin \theta = n\lambda$, where d is the distance between scattering planes and n is an integer. By observing the angles for which strong diffracted beams are obtained, the various scattering planes and their separation may be found. In this way the crystal structure and the lattice spacings may be deduced.

As an example, consider the sodium chloride crystal. This has a cubic structure with sodium and chloride ions at alternate sites (see figure 1.2). The distance a between each site is found by X-ray analysis to be 0.2818 nm. Consider one of the unit cubes of side a. Each of these cubes has 4 sodium and 4 chloride ions, i.e. 4 NaCl molecules. But each corner of the cube is shared by eight contiguous cubes. Consequently each volume a^3 contains one-eighth of four molecules of NaCl $=\frac{1}{2}$ molecule of NaCl. If M = molecular weight, ρ = density, the volume occupied by the half-molecule is $\frac{1}{2}(M/N_A)(1/\rho)$. Consequently

$$a^3 = \frac{1}{2}\frac{M}{N_A}\frac{1}{\rho}.$$

Inserting $M = 0.0585$ kg, $\rho = 2170$ kg m^{-3}, $a = 0.2818$ nm, $N_A = 6.02 \times 10^{23}$.

1.1.5 *Energy states in atoms: the earliest evidence for quantization*

By the turn of the century the work of J. J. Thomson and others had shown that the atom consists of a positively charged massive nucleus and several very light negatively charged particles – the electrons. The problem that perplexed physicists for the next 25 years was the way in which the electrons were linked to the nucleus and the factors that determined the energies of the electrons.

Figure 1.2. Structure of sodium chloride crystal. The sodium and chloride ions are arranged alternately on a cubic lattice, the separation between ions being 0.2818 nm. The sodium fluoride structure is identical except that $a = 0.231$ nm. The unit cell size is $2a$.

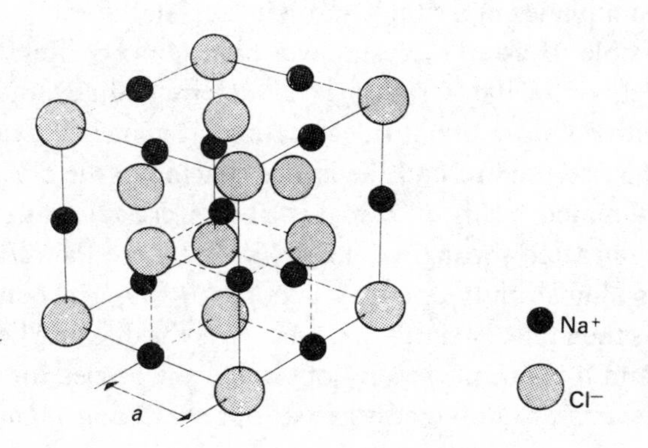

The following argument concerning the energy of electrons within the atom has been given by Sir Neville Mott,† and brings out very simply the need for quantization. Consider the thermal properties of a monatomic gas such as neon. The atom has 10 electrons interacting in some way with the nucleus. Careful calorimetric measurements show that the thermal energy per atom of neon is exactly $\frac{3}{2}kT$ where k is Boltzmann's constant (1.380×10^{-23} J K^{-1}) and T is the absolute temperature. If we increase the temperature of the gas by 1 degree the energy increase per atom (at constant volume) is exactly $\frac{3}{2}k$. Now the kinetic theory of gases shows that $\frac{3}{2}kT$ is the kinetic energy of a single particle in equilibrium with its surroundings at temperature T. This means that the only energy which is changed when we heat a neon atom is its kinetic energy as a single entity. If anything within the atom is moving it is not speeded up by the addition of heat, i.e. no energy has been transferred to the constituent parts of the atom.

This has an important consequence for if the electronic energies were continuous they would have to share in the thermal energy of the atom as a whole. They do not do so. This implies two things. First, the electronic energy states are discontinuous: they are at certain specific levels and no energies between these levels can exist. Secondly the gap between these energy levels must be very large compared with normal thermal energies: for this reason electrons cannot be promoted to higher levels at ordinary temperatures.‡ In fact the gaps between the electron energy levels are of order 5 eV which is very much greater than the thermal energy per atom, about $\frac{1}{40}$ eV at room temperature.

This argument provides, with some hindsight, the simplest and most basic evidence for the existence of quantized energy states within the atom.

1.1.6 *Basic structure of the atom*

It is not the purpose of this book to provide a full account of atomic structure. This would involve the wave properties of matter and other concepts used in describing electronic configurations. We shall very briefly fall back on the simple Bohr model: for although this has been superseded by the ideas of orbitals and the uncertainty principle it provides a very convenient model on which to base many order-of-magnitude calculations.

† N. Mott, *Contemporary Physics*, vol. 5 (1964), p. 401.

‡ This has a bearing not only on the specific heat of a gas but also on the Langevin treatment of diamagnetism described on pp. 387–90.

If the nucleus has a charge Ze and a specified electron of mass m, charge $-e$, describes a circular orbit of radius r at a speed v, we may balance the centrifugal force mv^2/r with the Coulombic attraction between electron and nucleus. Assuming that there is no shielding of charge by other electrons we may write

$$\frac{mv^2}{r}=\frac{Ze^2}{r^2}\frac{1}{4\pi\varepsilon_0}, \tag{1.1}$$

where ε_0 is the permittivity of free space. The kinetic energy of the electron is

$$\tfrac{1}{2}mv^2=\frac{1}{2}\frac{Ze^2}{r}\frac{1}{4\pi\varepsilon_0}. \tag{1.2}$$

The potential energy is the work done in carrying the electron from infinity to a distance r from the nucleus. This is given by

$$\int_\infty^r \frac{Ze^2}{r^2}\frac{1}{4\pi\varepsilon_0}\,\mathrm{d}r=-\frac{Ze^2}{r}\frac{1}{4\pi\varepsilon_0}. \tag{1.3}$$

Note that attractive forces always give negative potential energies. The total energy W of the electron, which is the sum of the potential and kinetic energies, is the sum of equations (1.2) and (1.3) so that

$$W=-\frac{1}{2}\frac{Ze^2}{r}\frac{1}{4\pi\varepsilon_0}. \tag{1.4}$$

The major defects of this model are twofold. First, an electron rotating about a nucleus should, if treated classically, radiate electromagnetic energy and spiral into the nucleus. Secondly, even if such orbits were stable all possible values of W should be possible corresponding to all possible values of r.

When the wave nature of the electron is incorporated into this model both these difficulties disappear. The idea of a crisp orbit is now replaced by a somewhat fuzzy orbital describing the probability of locating the electron. Such an orbital is found to correspond to a stable energy state and no energy is radiated by the electron. Further, a whole series of stable orbitals are permitted as determined by certain (integral) quantum numbers. For example the angular momentum, which in the simple Bohr model has a value mvr, must be quantized to be $n(h/2\pi)$, where n is an integer and h is Planck's constant (6.626×10^{-34} J s). We may pursue this a little further. We write

$$mvr = n(h/2\pi) \tag{1.5}$$

so that

$$mv^2 = \frac{v}{r}\,mvr = nh(v/2\pi r). \tag{1.6}$$

But $2\pi r$ is the circumference of the orbit (or the orbital) so that $v/2\pi r$ is the frequency ν with which the electron orbits the nucleus. Comparing equations (1.4) and (1.2) we see that the total energy W is equal to $-\frac{1}{2}mv^2$. We thus obtain

$$W = -\tfrac{1}{2}nh\nu = -\frac{1}{2}\frac{nZe^2}{r}\frac{1}{4\pi\varepsilon_0}. \tag{1.7}$$

For the hydrogen atom, where $Z=1$, the frequency ν becomes

$$h\nu = \frac{e^2}{r}\frac{1}{4\pi\varepsilon_0}, \tag{1.8}$$

a result we use later in this chaper in deriving van der Waals forces.

Of course equations (1.7) and (1.8) still contain r. For each integral value of n in equation (1.5) there will be an associated radius r and if this is fed into equation (1.4) we shall obtain a whole series of permitted energy levels. The result is

$$W = -\frac{1}{n^2}\frac{Z^2e^4m}{2(4\pi\varepsilon_0)^2}\left(\frac{2\pi}{h}\right)^2. \tag{1.9}$$

Using known values of e, m and h we find that for the hydrogen atom ($Z=1$) the deepest energy level ($n=1$) has a value of -24×10^{-19} J or -15 eV; the next level ($n=2$) has a value of -3.75 eV so that there is a gap of over 11 eV. With larger atoms containing more electrons, higher values of n are involved for the lowest energy states but Z is also larger, even if we allow for the effect of screening of the nuclear charge by some of the orbital electrons. In general, the gap between the lowest energy level and the next highest level is of order 5 eV.

Quantum mechanics gives a slightly more complicated picture. The energy levels remain quantized and are still given by equation (1.9), where n is the principal quantum number and can have any value from 1 upwards. The angular momentum, however, is no longer given by $n(h/2\pi)$: it has a value $l^{\frac{1}{2}}(l+1)^{\frac{1}{2}}(h/2\pi)$, where l is an integer which can be either positive or negative but numerically cannot be greater than $n-1$.

This has an unexpected consequence. The lowest energy state is $n=1$: to this corresponds $l=0$. This means that the lowest energy state (the S orbital) has zero angular momentum – the electron does not move in a circle, it moves in and out radially. There is no simple classical analogue although one might regard the smeared electron cloud as pulsing in and out.

An orbiting electron behaves like a circulating current and its interaction with an applied magnetic field determines the magnetic properties of the atom or molecule. For this reason the quantization of angular momentum is of great importance in explaining atomic magnetism. The detailed behaviour is very complicated since the angular momentum must itself follow further quantization rules when exposed to a magnetic field. However, as we shall see in a later chapter, many of the basic magnetic properties of atoms may be understood in terms of the simplest models.

1.1.7 *Charge, mass and configuration of atoms: the periodic table*

We may now summarize the main features of atomic structure and establish the basis on which the periodic table of elements may be constructed.

1. An atom consists of a minute nucleus, which is positively charged and contains the main mass of the atom.

2. The nucleus consists of protons and neutrons. These are of equal mass but each proton carries a positive charge whilst the neutron carries no charge. The resultant positive charge (referred to the proton as unit charge) is called the atomic number Z, i.e. Z is equal to the number of protons in the nucleus.

3. As a crude approximation, valid especially for smaller atoms of atomic number less than 20, there are about equal numbers of protons and neutrons so that the mass of the nucleus, referred to the proton as unit mass, is about twice the atomic number. This is known as the mass number and in practice is the integer nearest to the mass of the atom, referred to the carbon-12 atom as 12.000. In the older literature the mass number is sometimes loosely referred to as the atomic weight.

4. The nucleus is surrounded by electrons. These have a mass only $\frac{1}{1840}$ of the mass of the H atom (i.e. of a proton or neutron) and they have a unit negative charge.

5. Since the atom is electrically neutral the number of orbital electrons is equal to the atomic number.

6. The chemical properties of the atom are determined by the configuration of the electrons, especially of the outermost ones.

7. There are certain stable arrangements, the electrons falling into well-defined energy levels or shells, defined by quantum conditions. Each energy level can hold no more than two electrons and these must have opposite spins (see 8 below). This is known as the Pauli exclusion principle.

8. The lowest energy shell, with principal quantum number 1, is the K shell. This ground state is called the 1s orbital. Here the electron is distributed with spherical symmetry about the nucleus and has zero angular momentum. However, the electrons can have a spin quantum number which can be described as being either parallel or antiparallel to some specified direction, for example ↑ or ↓. If the K shell contains 1 electron, as in hydrogen, it has a single spin; if it contains two electrons, as in helium, the spins are paired in opposite directions and the K shell is now complete. The atom is thus inert.

9. The next higher energy shell has principal quantum number 2 and is known as the L shell. The lowest level is called the 2s state (again spherically symmetrical) and this can hold 2 electrons of opposite spin. There is a higher energy orbital known as the 2p. This has angular momentum, is dumb-bell in shape, and there are three directions available. Each can hold 2 electrons of paired spins so that altogether this state can hold 6 electrons. If only two or three electrons have to go into the 2p state the energy is lowest if they each have parallel spins (this is part of a more general principle, due to F. Hund, which says that when several levels within a given state are available the lowest energy corresponds to parallel spins): they can do this by each going into the three available orbitals. These electrons can readily play a part in chemical bonding and other processes. A fourth electron will have to pair its spin with one of the existing three. The remaining two unpaired electrons are then readily available for chemical reaction. When the 2p state is filled with its six electrons the atom – neon – is chemically non-reactive (but see below).

Table 1.1. *Shells and stable sub-groups*

Shell	Name	Number of electrons (maximum)	Number of electrons in stable sub-groups
1st	K	2	2
2nd	L	8	2, 6
3rd	M	18	2, 6, 10
4th	N	32	2, 6, 10, 14

10. The next higher level has principal quantum number 3 and is the M shell. This can hold 2 electrons in the 3s orbital, 6 in the 2p and 10 in the 3d. (Again, like the p orbitals, the d orbitals are directed and have angular momentum.) The next higher energy level is the N shell which contains 2 electrons in the 4s, 6 in the 4p, 10 in the 4d and 14 in the 4f states. However, it happens that the lowest state in the N shell (4s) generally has lower energy than the 3d states in the M shell: electrons will therefore go into the 4s state before they start filling the 3d orbitals. The arrangement of shells and stable sub-groups are summarized in Table 1.1.

As mentioned above if an atom contains a complete shell or sub-group it is relatively stable and may even be inert. In Table 1.2 we construct the

Table 1.2. *The beginning of the periodic table of elements*

Principal quantum numbers		1	2		3			4
Shell		K	L		M			N
Electronic states		1s	2s	2p	3s	3p	3d	4s 4p 4d 4f
At. No.	Atom							
1	H	↑						
2	He	⇅	inert					
3	Li	filled	↑					
4	Be	,,	⇅					
5	B	,,	filled	↑				
6	C	,,	,,	↑, ↑				
7	N	,,	,,	↑, ↑, ↑				
8	O	,,	,,	⇅, ↑, ↑				
9	F	,,	,,	⇅, ⇅, ↑				
10	Ne	,,	,,	⇅, ⇅, ⇅	inert			
11	Na	,,	,,	filled	↑			
12	Mg	,,	,,	,,	⇅			
13	Al	,,	,,	,,	filled	↑		
18	Ar	,,	,,	,,	,,	⇅, ⇅, ⇅	inert	
19	K	,,	,,	,,	,,	filled	empty	↑
20	Ca	,,	,,	,,	,,	,,	empty	⇅
21	Sc	,,	,,	,,	,,	,,	↑	⇅
29	Cu	,,	,,	,,	,,	,,	filled	↑
Maximum occupancy number		2	2	6	2	6	10	2 6 10 14
Total		2	8		18			32

first part of the periodic table based on the shell-model described above.

We see that the outermost electrons, especially those with unpaired spins, determine the chemical properties of the atoms. As a simple approximation we may say that the valency reflects the way in which the atom loses, gains or shares electrons in an attempt to form stable shells or sub-shells. When the atom already contains such a configuration (He, Ne, and Ar above) it is already stable and is chemically inert. This needs some qualification since in recent years it has been shown that the larger inert atoms such as xenon can, in fact, be ionized by the presence of a powerful electronegative ion such as the fluoride ion. As a result it is possible to form a stable covalent compound of xenon hexafluoride.

1.1.8 *Size of atoms*

The simplest way of forming some quantitative idea of the size of atoms is to measure the distance between atoms in the solid state. The result, of course, depends on the crystal structure. Some typical results are given in Tables 1.3 and 1.4.

In Table 1.3 the outermost configuration is unchanged. As we fit in additional shells or sub-shells there is an increase in *d*. (The inner orbitals contract because of the increase in nuclear charge.)

In Table 1.4 the main process is one of filling up an unfilled shell whilst the outer configuration scarcely changes. The increasing nuclear charge causes shrinkage in the atomic diameter.

1.1.9 *Size of ions*

Some results are given in Table 1.5 for monovalent atoms which either lose an electron to become a positive ion, or gain an electron to become a negative ion. It is seen that the effect of removing one electron

Table 1.3. *Atomic diameters for vertical columns in periodic tables*

		Electron shells						
Atom	Z	K	L	M	N	O	P	d (nm)
H	1	1						—
Li	3	2	1					0.3
Na	11	2	8	1				0.37
K	19	2	8	8	1			0.46
Rb	37	2	8	18	8	1		0.49
Cs	55	2	8	18	18	8	1	0.52

from the atom is to reduce d by 0.18–0.2 nm; the effect of adding an electron is to increase d by about the same amount.

Table 1.4. *Atomic diameters for horizontal columns in periodic table*

Atom	Z	Electrons shells K	L	M	N	d (nm)
Ca	20	2	8	8; 0	2	0.4
Sc	21	2	8	8; 1	2	0.35
Ti	22	2	8	8; 2	2	0.30
V	23	2	8	8; 3	2	0.26
Cr	24	2	8	8; 5	1	0.26
Mn	25	2	8	8; 5	2	0.26
Fe	26	2	8	8; 6	2	0.25
Co	27	2	8	8; 7	2	0.25
Ni	28	2	8	8; 8	2	0.25
Cu	29	2	8	8; 10	1	0.26

Note that Ir ($Z=77$) has smaller diameter than Al ($Z=13$); d (Ir) = 0.27 nm, d (Al) = 0.28 nm.

One obtains a better idea of what is involved by comparing Na^+, Ne and F^-. These have the same final electron configuration, i.e. are isoelectronic but the nuclear charge is 11, 10, 9 respectively. Consequently one would expect the diameter of Na^+ to be smaller than Ne and the diameter of Ne to be smaller than that of F^-. This is indeed the case, the values being 0.20, 0.25 and 0.27 nm for Na^+, Ne and F^- respectively.

Table 1.5. *Effect of ionization on size*
(size obtained from spacing in appropriate compounds)

Metals			Non-Metals		
	Diameter d (nm)			Diameter d (nm)	
Element	Atom	Ion	Element	Atom	Ion
Li	0.3	0.136	F	0.13	0.27
Na	0.37	0.196	Cl	0.20	0.36
K	0.46	0.296	Br	0.23	0.39
Rb	0.49	0.296	I	0.27	0.44
Cs	0.52	0.330			

1.1.10 *Density of solid elements*

An increase in atomic number usually implies an increase in atomic weight. There is also some increase or decrease in atomic diameter (see Tables 1.3 and 1.4) but it is not as large a change as the change in atomic weight. Consequently although there is generally an increase in density of the solid element as the atomic number increases it is not necessarily a monotonic relation.

It is interesting to note that if we compare the atomic weight of hydrogen with that of uranium (1 : 238) it is not very different from the ratio of their densities (70 : 18 400 kg m^{-3}).

1.2 **The forces between atoms and molecules**

1.2.1 *Forces due to the ionic bond*

The simplest forces between atoms are those which arise as a result of electron transfer. A simple example is that of, say, sodium fluoride. The sodium atom has a nuclear charge of +11, with 2 electrons in the K shell, 8 in the L shell and 1 in the M shell. The fluorine atom has a nuclear charge of 9 with 2 electrons in the K shell, and 7 in the L shell. The outermost electron in the sodium atom may transfer readily to the fluorine atom; both atoms then have a complete shell but the sodium now has a net charge of +1 and the fluorine a net charge of −1. These ions therefore attract one another by direct coulombic interaction. The force between them is strong – it varies as x^{-2}, where x is the distance between the ions, and it acts in the direction of the line joining the ions. Furthermore it is unsaturated – one positive ion can attract several negative ions around it and the force exerted by the positive ion on each negative ion is not affected by the presence of other negative ions. Of course the negative ions will also repel one another.

1.2.2 *Forces due to the covalent bond*

These are sometimes referred to as valency forces, for they are the type of force which accounts for the binding together of atoms in those molecules where the primary attraction is not ionic. For example the force binding two hydrogen atoms together to form a hydrogen molecule is of this nature. Covalent bonding always involves the sharing of electrons. One way of describing this is to say that in the hydrogen molecule the single electron of each atom interacts with both nuclei, so that each atom regards itself as 'possessing' two electrons which provide a complete, stable 'molecular' K shell. However, the binding cannot be described in classical terms. It arises from the fact that the electrons e_1

and e_2 are indistinguishable, so that if e_1 and e_2 are interchanged this does not involve a new configuration, but leads to an 'exchange energy' which provides the force binding the atoms together. Descriptively we may say that there is an appreciable concentration of electrons (negative charge) between the two nuclei (positive charge), and this binds the atoms together. This is indicated schematically in figure 1.3.

Covalent forces are strong, they fall off rapidly with separation; they act in specified directions (valency bonds) and they are saturated.

1.2.3 *Van der Waals forces*

Van der Waals forces exist between all atoms and molecules whatever other forces may also be involved. We may divide van der Waals forces into 3 groups.

(*a*) *Dipole–dipole forces* (Keesom forces). If we consider a substance such as HCl, where the bonding is largely ionic, the individual molecule will consist of a positive and a negative charge separated by a small distance of the order of 0.1 nm, so that it constitutes a minute electric dipole. This will interact with a neighbouring HCl molecule and produce a net attractive force. The force (see (*c*) below) will be proportional to x^{-7}, where x is the distance between the molecules.

(*b*) *Dipole-induced dipole forces* (Debye forces). Consider the interaction between say an HCl molecule and a neon atom. The latter has no natural dipole. However, the HCl dipole will induce a dipole in the neon atom (see (*c*) below) and there will then be an attractive force between them again proportional to x^{-7}.

(*c*) *Dispersion forces*. The most 'mysterious' type of force is that which exists between two non-polar atoms such as neon. We know that these atoms must attract one another, for otherwise we could never obtain

Figure 1.3. Electron concentration in a pair of hydrogen atoms; (*a*) unstable, the molecule dissociates into two atoms, (*b*) stable, the electron concentration between the nuclei serves to bond the nuclei together as the H_2 molecule.

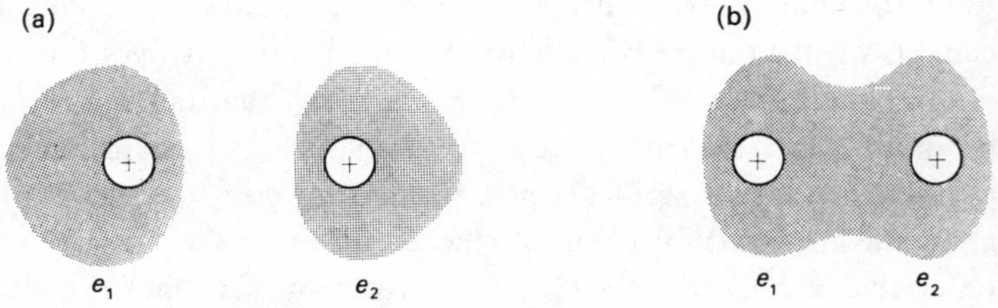

(a) e_1 e_2 (b) e_1 e_2

liquid or solid neon simply by cooling the material and so reducing the thermal energy of the atoms below some critical level. The explanation is as follows. Although the neon atom has a symmetrical distribution of electrons around the nucleus so that the atom has no dipole, this statement is true only as a time average. At any instant there may be some asymmetry in the distribution of the electrons around the nucleus. The neon atom at this instant, therefore, behaves as an electric dipole. This produces a field which distorts the electron distribution in a neighbouring atom (polarizes it) so that the neighbouring atom itself acquires a dipole. The two dipoles will attract. This is the basis of the 'dispersion' van der Waals force.

We can easily derive the law of force in the following way. Suppose that at a given moment the first atom has a dipole moment μ (figure 1.4). At a distance x along the axis of the dipole the electrostatic field E is given by

$$E=\frac{2\mu}{x^3}\frac{1}{4\pi\varepsilon_0}, \tag{1.10}$$

where ε_0 is the permittivity of space. (This equation is only valid if x is reasonably large compared with the length of the dipole in figure 1.4. For closer approach, a more complicated equation would be needed.) If there is another dipole at this point it will be polarized and acquire a dipole μ'. In order to describe this we invoke a property discussed in Chapter 14: that the dipole produced by a field is approximately proportional to the strength of the field. We write

$$\mu'=\alpha E, \tag{1.11}$$

where α is termed the atomic polarizability of the atom.

Now a dipole μ' in a field V has a potential energy V given by

$$\begin{aligned} V &= -\mu' E \\ &= -\alpha E^2 = \frac{-4\alpha\mu^2}{x^6}\left(\frac{1}{4\pi\varepsilon_0}\right)^2. \end{aligned} \tag{1.12}$$

Figure 1.4. A dipole produces an electric field at a distance x along its axis of amount $E=(2\mu/x^3)(1/4\pi\varepsilon_0)$. This field can polarize another molecule such that mutual attraction occurs.

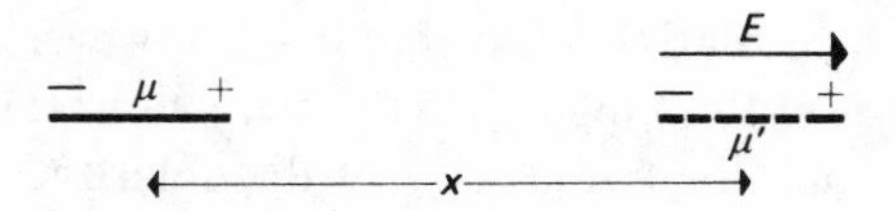

This implies a force between the atoms of magnitude

$$F=\frac{\partial V}{\partial x}=\frac{\text{constant}}{x^7}. \tag{1.13}$$

We see that the force is always attractive and falls off as x^{-7}.

We may take this two stages further. For simplicity let us consider the attraction between two hydrogen *atoms*, where only one electron can be involved in each atom. If the effective radius of the atomic orbital is r, the maximum value of the instantaneous dipole is re. Substituting in equation (1.12) we have

$$V=-\frac{4\alpha r^2e^2}{x^6}\left(\frac{1}{4\pi\varepsilon_0}\right)^2. \tag{1.14}$$

We now make use of a derivation given in Chapter 14. For a simple atom, or for a small molecule which is virtually spherically symmetrical, the polarizability is given by

$$\alpha=r^3(4\pi\varepsilon_0), \tag{1.15}$$

where r is the radius of the atom or molecule. For reasons which will soon become apparent we select from this relation the value of r^2

$$r^2=\frac{\alpha}{r}\,\frac{1}{4\pi\varepsilon_0}. \tag{1.16}$$

Insertion in equation (1.14) gives

$$V=\frac{-4\alpha}{x^6}\,\frac{\alpha}{(4\pi\varepsilon_0)^2}\,\frac{e^2}{r(4\pi\varepsilon_0)}. \tag{1.17}$$

By comparison with equation (1.8) we see that the last term on the right is the quantized energy level $h\nu$ of the electron when its frequency is ν. We are left with

$$V=-\frac{4\alpha^2}{x^6}h\nu\left(\frac{1}{4\pi\varepsilon_0}\right)^2. \tag{1.18}$$

This derivation is, of course, rather crude since it assumes that the dipoles are always oriented end-on and we have ignored the fact that the second atom must itself influence the polarization of the first. However, it contains all the essential features of the van der Waals interaction: the polarizability α, the separation x and the frequency ν of the polarized orbital.

The rigorous derivation by F. London, in 1930, considered the perturbation produced in the solution of the Schrödinger equation. His result is

$$V=-\frac{3}{4}\frac{\alpha^2}{x^6}h\nu\left(\frac{1}{4\pi\varepsilon_0}\right)^2 \tag{1.19}$$

and is exactly the same as our simple derivation except for the numerical factor. As we shall see in Chapter 14 the frequency ν is the same as that which accounts for the dependence of refractive index on the frequency of the incident light, i.e. it is the frequency involved in optical dispersion. For this reason those van der Waals forces which arise by mutual polarization of the atoms or molecules are referred to as dispersion forces.

Van der Waals forces are much weaker than ionic and covalent forces. They are central and, like ionic forces, they are unsaturated: also to a first approximation they are additive (see below).

1.2.4 *Retardation effects in van der Waals forces*

The derivation of the London relation is valid only if the atoms are less than a few hundred angstroms, i.e. a few tens of nm, apart for the following reason. For large separations the electric field from one atom takes an appreciable time to reach its neighbour: by the time it has done so and polarized it, it itself will have acquired a different electron configuration. There is a lag in the interaction and this leads to what is known as retarded van der Waals forces. In different words we may say that when the atoms are close together the electron fluctuations between neighbours are in close correlation: when they are far apart this correlation gets poorer and poorer, and the potential now falls off as $1/x^7$. We can form some estimate at which non-retarded forces go over to retarded forces by noting that an orbiting Bohr electron takes about 3×10^{-16} s to rotate about the nucleus so that an electric field, travelling at the velocity of light (3×10^8 m s^{-1}) will cover a distance of about 10^{-7} m or 100 nm in this time. Thus when atoms are more than this distance apart we should expect the correlation to be lost and the interaction to be almost completely retarded. Experiments described below show that, in fact, the transition from non-retarded to retarded forces occurs at a separation of the order of 10 to 50 nm.

There is no simple derivation of retarded van der Waals forces but the following argument provides some insight into their origin. The first point to emphasize is that retarded van der Waals forces are not basically a new type of interaction. All that really happens is that the atoms select from the fluctuations in the electron distribution those which occur at a low enough frequency to give good correlation between neighbouring atoms. We replace ν in equation (1.19) by c/λ, where c is the velocity of light and λ the wavelength of the fluctuation. We now specify that if λ is of the same order as the separation x (or greater) fluctuations of this wavelength in one atom will be in good correlation with fluctuations in another atom at distance x.† All shorter wavelengths (i.e. higher frequencies) will give such poor correlation that their contributions to the interaction energy can be neglected. We thus convert equation (1.19) into

$$V = -\frac{3}{4}\frac{\alpha^2}{x^6}\frac{hc}{\lambda}\left(\frac{1}{4\pi\varepsilon_0}\right)^2,$$

where λ is of order x. This gives as a final result

$$V \approx -\frac{3}{4}\frac{\alpha^2}{x^7}hc\left(\frac{1}{4\pi\varepsilon_0}\right)^2, \tag{1.20}$$

so that the attractive force now falls off as $1/x^8$. Equation (1.20) is very close to the correct answer obtained by extremely difficult theory (Casimir gives $23/8\pi^2$ instead of $\frac{3}{4}$ in equation (1.20)).

van der Waals forces exert their influence in the behaviour of individual atoms or molecules in the gaseous state (see Chapter 5). The magnitude of these forces can then be deduced by comparing the behaviour of a real gas where these forces operate with that of an ideal gas where intermolecular forces are negligible. A more direct way of determining the magnitude of these forces is by measuring the forces between macroscopic bodies. We assume that the forces are roughly additive and carry out a pair-wise addition for each atom in one body with every atom in the other body. Additivity is not strictly true but is a reasonably good first approximation. The detailed calculations are given in Chapter 12. For a sphere of radius R at a distance x of *nearest* distance to a flat surface (or for crossed cylinders each of radius R) the force is

$$F_{\text{non-retarded}} = AR/6x^2, \tag{1.21a}$$

$$F_{\text{retarded}} = 2\pi BR/3x^3, \tag{1.21b}$$

where A and B are known as the Hamaker constants for non-retarded and retarded forces respectively.

† Sophisticated theory shows that the loss in correlation occurs at distances greater than $\lambda/2\pi$ rather than λ: but this does not change the type of argument given here.

Direct measurements of van der Waals forces have been made using mica as the model surface. Mica may be cleaved over fairly large areas to be molecularly smooth on both faces: the faces are therefore molecularly parallel. The back surface is silvered to be highly reflecting and glued to a cylindrical glass surface. This provides a rigid molecularly smooth surface of cylindrical shape. Two such surfaces may be mounted at right angles and the force between them measured by suitable devices. White light is shone through the gap and multiple reflected beams from the silver backing give extremely high optical resolution (1–2 Å). A spectrometer separates the different wavelengths. The interferograms provide a measure of both the separation and the geometry of the surfaces so that the effective radius R of the system may be determined.† The set-up is sketched in figure 1.5.

Figure 1.5. Arrangements for determining the variation of surface forces with separation using molecularly smooth mica surfaces.

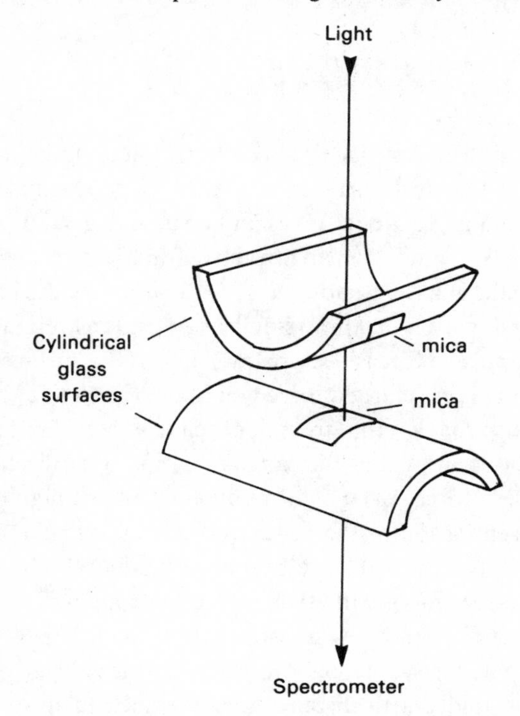

† References: D. Tabor and R. H. S. Winterton (1969) *Proc. Roy. Soc. Lond.* **A312**, 435–50. J. N. Israelachvili and D. Tabor (1972) *Proc. Roy. Soc. Lond.* **A331**, 19–38. J. N. Israelachvili *Intermolecular and Surface Forces*, Wiley (1985).

Some typical experimental results for crossed cylinders of mica are shown in figure 1.6. It is seen that the power law of x changes from 2 to 3 over the range of 10 to about 50 nm. The results also show that for mica the constant A has a value of about 1.3×10^{-19} J and B a value of about 10^{-28} J m. Retarded forces are not of great importance since they are small and play little part in most physical processes. By contrast the non-retarded forces can be relatively large, particularly if the bodies come into atomic contact. Values of A range from about 10^{-19} J for inorganic solids to about 3×10^{-19} J for metals. If the bodies are immersed in a liquid the value of A is less, often by about an order of magnitude.

For extended parallel surfaces the forces are naturally proportional to the area of the surfaces. The force f per unit area (see Chapter 12) has the value

$$f_{\text{non-retarded}} = \frac{A}{6\pi x^3} \qquad (1.22a)$$

and

$$f_{\text{retarded}} = \frac{B}{x^4}. \qquad (1.22b)$$

Consider two smooth polished surfaces placed in contact and kept apart by a few asperities or a few dust particles so that the separation is of the order 1 μm. Then, assuming retarded forces and a value of $B \approx 10^{-28}$ J m, the force f is of order 10^{-4} N m^{-2}, i.e. an attractive force per square metre about equal to the weight of a fly. If, however, the separation were reduced to near atomic dimensions, say 0.4 nm (smaller separations might bring us into the range of repulsive forces – see below), and if the surfaces were molecularly smooth, the non-retarded force would be of order 10^8 Nm^{-2}, i.e. 10 kg mm^{-2}: this approaches the strength of real van der Waals solids. These simple calculations show that the range of action of surface forces is small: in theory these forces extend to infinity but their magnitude is appreciable only for separation of the order of a few atomic diameters.

The consequences of this may at once be seen. Even in the absence of ionic or valency forces, surfaces will stick together strongly if they can come into close contact (within a few angstroms) of one another. Again gases will generally be adsorbed strongly for the first monolayer but the attraction falls off so rapidly with distance that second and third layers are far less likely. Roughly speaking the formation of adsorbed layers depends on whether the potential energy drops by a greater amount than the mean thermal energy. For a molecule, the latter is of the order of kT, where k is the Boltzmann constant and T the absolute temperature (see Chapter 3). This is not the whole story since, of course, the latent heat of condensation is also a factor encouraging further adsorption, whilst the entropy decrease associated with an 'ordered' adsorbed film (see Chapter 2) tends to oppose adsorption. The net result is that polymolecular adsorption is unlikely unless the gas or vapour is very near saturation.

With large particles, such as those in colloidal systems or Brownian suspensions, the position is different. The thermal energy of the individual particle is still given by kT but the van der Waals energy must be integrated for the millions of atoms in the particles. In such systems the van der Waals forces can in fact play a basic part in determining whether particles will stick together or not. This will be discussed in greater detail at the end of Chapter 12.

1.2.5 *Repulsion forces*

If atoms were subjected only to attractive forces all atoms would coalesce. Thus it is clear that for very short distances of separation some

Figure 1.6. Variation of power-law index n of the van der Waals law of force between crossed mica cylinders with distance of separation x. Below 10 nm the forces are non-retarded and the law of force is $F = AR/6x^n$ with $n = 2$, where R is the radius of the cylinders. Above 50 nm the forces are retarded and the law of force is $F = 2\pi BR/3x^n$, where $n = 3$.

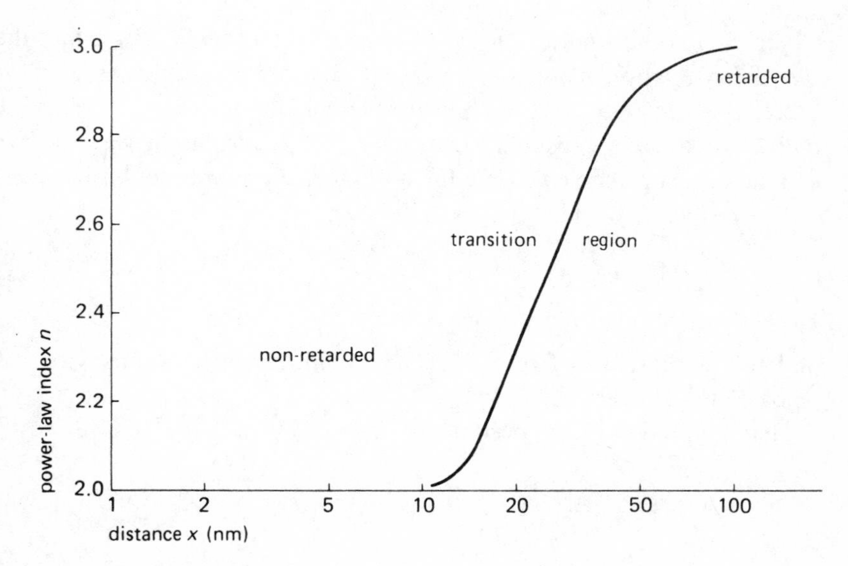

type of repulsive force must operate. In ionic systems the attractive force arises from the electrostatic charge and the repulsive force from the electron shells, which act as a sort of tough elastic sphere resisting further compression. The repulsion may be described as arising from two effects. First, the penetration of one electron shell by the other, which means that the nuclear charges are no longer completely screened and therefore tend to repel one another. The other effect arises from the Pauli exclusion principle which states that two electrons of the same energy cannot occupy the same element of space. For them to be in the same space (overlapping) the energy of one must be increased – this is equivalent to a force of repulsion.

With covalent and van der Waals forces the repulsion and attraction are all part of a single mechanism and it is not correct to consider them as arising from separate mechanisms; however, for the purpose of simple calculation it is very convenient to do so. We therefore describe the long-range attraction by a term of the form

$$F=\frac{A}{x^m}, \tag{1.23}$$

where A and m are suitable constants, and the repulsion by a relation of the form

$$F=\frac{B}{x^n}, \tag{1.24}$$

where B is a constant and n has a value of the order 8–10. In fact the power-law relation is not very good and there are theoretical reasons for preferring a relation for the repulsive force of the form $B\,\mathrm{e}^{-x/s}$ where B and s are suitable constants. In the following calculation we shall for simplicity retain the power law for both attraction and repulsion forces.

The resultant force is then

$$F=\frac{A}{x^m}-\frac{B}{x^n} \tag{1.25}$$

and this is plotted in figure 1.7. The equilibrium separation x_0 occurs when $F=0$. This gives $B=Ax_0^{n-m}$.

Hence equation (1.25) becomes

$$F=A\left(\frac{1}{x^m}-\frac{x_0^{n-m}}{x^n}\right). \tag{1.26}$$

We may at once note that for small displacements from the equilibrium position the restoring force is

$$dF = A\left(\frac{n-m}{x_0^{m+1}}\right)dx. \tag{1.27}$$

Since the force is proportional to the displacement, the motion of the displaced particle will be simple harmonic. This conclusion is of course not restricted to power-law relations. Any relation of the form $F = f(x)$ will lead to a similar conclusion.

1.2.6 *Potential energy*

The potential energy V involved in bringing one atom from infinity to a distance x from another atom is given by

$$V = \int (\text{external force})\, dx = -\int (\text{internal force})\, dx = \int F\, dx, \tag{1.28}$$

since F acts in the opposite direction to dx.

Figure 1.7. (*a*) Force and (*b*) potential energy curves for two atoms (or molecules) as a function of separation. There is a long-range attractive force and a short-range repulsive force which operates only when the molecules come close together. The equilibrium separation occurs when the net force is zero (at *A*) or when the potential energy is a minimum.

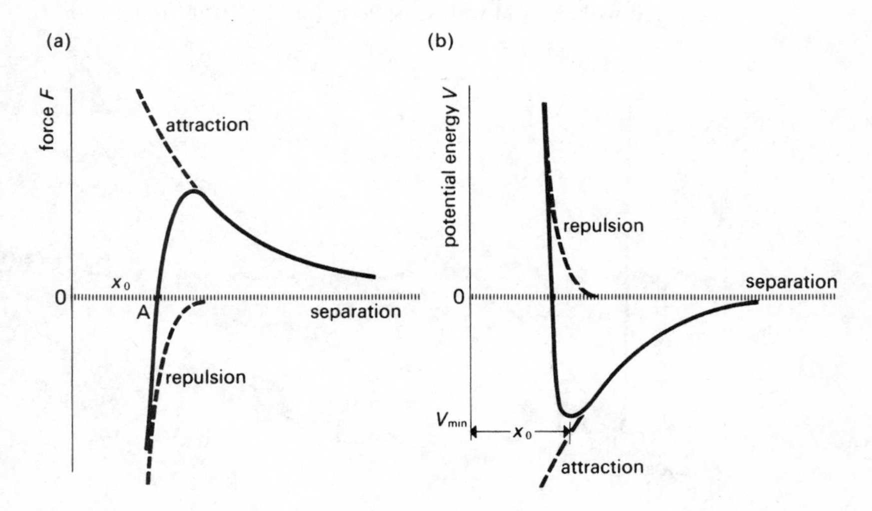

From equation (1.26) this gives

$$V = A\left(-\underbrace{\frac{1}{m-1}\frac{1}{x^{m-1}}}_{\text{attractive term}} + \underbrace{\frac{1}{n-1}\frac{x_0^{n-m}}{x^{n-1}}}_{\text{repulsive term}}\right).$$

A potential energy equation of this form was first proposed by Mie in 1903. Note that the attractive term gives a negative potential energy, the repulsive term a positive potential energy (see figure 1.8).

At the equilibrium position, where $x = x_0$, the potential energy has a minimum value

$$V_{\min} = -A\frac{1}{x_0^{m-1}}\left(\frac{1}{m-1} - \frac{1}{n-1}\right).$$

The interatomic forces are summarized in Table 1.6.

1.2.7 *Intermolecular forces*

Intermolecular forces are of the same nature as those occurring between atoms. For ionic crystals such as rock-salt one cannot distinguish individual molecules of NaCl, as the crystal is an array of positive (Na^+) and negative (Cl^-) ions held together by coulombic forces. In covalent

Figure 1.8. Stylized figure of potential energy curve of one molecule in relation to a single neighbour emphasizing the depth of the potential energy trough $\Delta\varepsilon$, the extremely steep repulsive part of the curve, and the effective diameter σ of the molecule. The zero-point energy, $\frac{1}{2}h\nu_0$, where ν_0 is the fundamental frequency of the pair, is very small compared with $\Delta\varepsilon$ and will be neglected in the rest of this treatment.

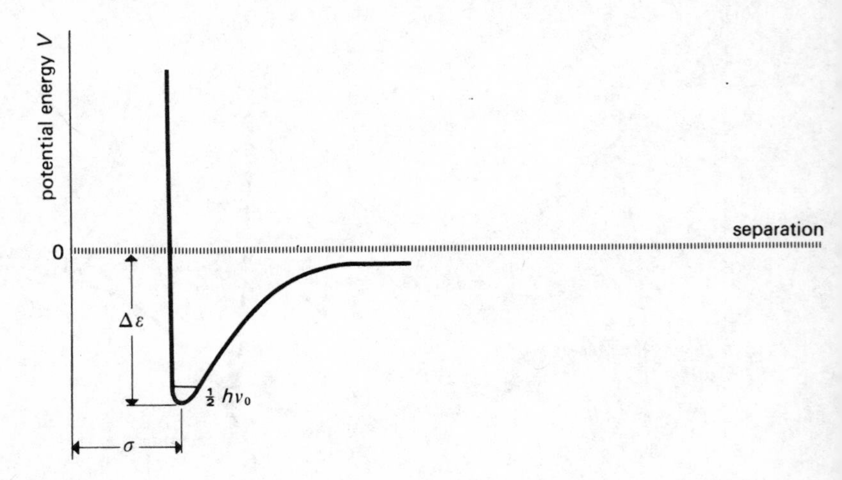

Table 1.6. *Interatomic forces*

Type	Nature	A	m	Energy to separate 2 atoms or ions
Ionic	Central unsaturated	Very large	2	$Na^+ + F^-$; 10^{-18} J
Covalent	Directed saturated	Very large	[a]	H–H; 7×10^{-19} J
van der Waals	Central unsaturated	Small to very small	7	Ne: Ne; 5×10^{-22} J

[a] The attractive force between two hydrogen atoms in the hydrogen molecule is roughly proportional to $1/x^3$. This only applies for small displacements; it cannot be applied over large displacements nor can it be considered as being valid for covalent forces in general. See Chapter 7, section 7.3.6, on the bulk modulus of covalent solids.

solids such as diamond the individual carbon atoms are joined, by valency bonds, to form a strongly-linked crystal. In a solid such as paraffin wax, or solid hydrogen, the molecules are held together by weak van der Waals forces. However, in a polymer these may be augmented by the attraction between polar groups or by chemical bonds. There is an additional type of bond which occurs only in the aggregate; this is the metallic bond. Individual metal atoms exist as such in the vapour phase, where a small number of diatomic molecules may occur, but there is little tendency for the outermost valency electrons to join forces and produce polyatomic molecules in the vapour phase. There is, however, strong bonding in the condensed solid state. The atoms become ionized and exist as positive ions in a sea of free valency electrons; they are held together by the strong attraction between the ions and the electrons. In effect, energy has to be provided to ionize the metal atoms, but this is more than compensated for by the binding energy between the ions and the electron sea.

The simplest measure of the intermolecular forces is the heat of sublimation, which is the amount of energy needed to separate all the molecules from all their neighbours. This is highest for ionic and covalent solids, high for metals and least for van der Waals solids. We may make this more quantitative in the following way. If the heat of sublimation E_s is given in Joules per mole, we divide it by 6×10^{23} to obtain the energy in Joules per single molecule. This conversion factor is 1.66×10^{-24}. The result is the energy E_s required to break bonds between one molecule and all its neighbours. If, therefore, the coordination number (number of nearest neighbours) is n and only nearest-neighbour interaction is involved, we may write for the binding energy ε between one molecule and one neighbour

$$E_s = \tfrac{1}{2}n\varepsilon.$$

The $\frac{1}{2}$ is introduced since otherwise every bond would be counted twice. We then have

$$E_s \times 1.66 \times 10^{-24} = \frac{n}{2}\varepsilon \quad \text{or} \quad \varepsilon \approx \frac{3.3 \times 10^{-24}}{n} E_s.$$

As an example we quote the simplest type of solid, solid neon, which forms a close-packed face-centred cubic (f.c.c.) structure of the van der Waals type. A sketch illustrating this structure is given in Chapter 7, figure 7.5. The latent heat of sublimation is found by experiment to be approximately 1.88 kJ per mole (see Table 7.2 where E_s is written as U_0) per mole. The number of nearest neighbours, n, is 12.

Hence

$$\varepsilon \approx 5 \times 10^{-22} \text{ J}.$$

This agrees well with the theoretical interaction between one neon atom and its neighbour.

Another example, which is less direct, is worth quoting because it provides a check on our deduced value of ε by a completely different method. In sodium fluoride the energy to convert 1 mole of NaF to Na^+ and F^- free ions is 899 kJ (see Table 7.1). This is approximately 15×10^{-19} J per molecule of NaF or 7.5×10^{-19} J per ion of Na^+ or F^-. In the crystal lattice each ion is surrounded by 6 ions of opposite sign (i.e. $n = 6$). The energy per unit bond (it is of course negative) is therefore

$$\varepsilon = \frac{7.5 \times 10^{-19}}{3} = 2.5 \times 10^{-19} \text{ J}.$$

We may now consider the ionic interaction between a Na^+ and a F^- as they are separated from their equilibrium position in the lattice (separation $x_0 = 0.231$ nm) to infinity. Ignoring the repulsion term, we have

$$\varepsilon = -\frac{1}{4\pi\varepsilon_0}\int_{x_0}^{\infty} \frac{e^2}{x^2}\,\mathrm{d}x = -\frac{1}{4\pi\varepsilon_0}\frac{e^2}{x_0} = -10^{-18} \text{ J}.$$

This assumes only nearest-neighbour interaction. There is, in fact, a very large interaction between the next nearest neighbours. For example the potential energy due to the attraction between one positive ion and the negative ions surrounding it is greatly reduced by the coulombic repulsion between the positive ion and the nearest positive ions. These are only a little further away than the nearest negative charges and as coulombic potentials fall off only as $1/x$, the resultant potential energy is greatly reduced. Detailed calculations such as those indicated in Chapter 9 show that the energy is reduced from $-(4\pi\varepsilon_0)^{-1}e^2/x_0$ to 0.2905 times this value. This gives a value for ε of about 2.9×10^{-19} J which agrees well with the value quoted on p. 28.

It should be emphasized that the calculation of intermolecular forces from the experimentally determined heat of sublimation can only be applied in a direct way to van der Waals solids. In this case the sublimed material consists of the same chemical units as those existing in the solid. With ionic solids such as NaF sublimation breaks up the solid into a vapour of NaF molecules. Thus the energy involved in separating all the Na^+ and F^- ions to infinity is greater than the heat of sublimation by the dissociation energy of NaF. It is this quantity which was used in the calculation on p. 28. (We ignore the kinetic energy of the vapour molecules or ions.) Again when a metal is sublimed the vapour consists primarily of free atoms: consequently the energy involved in separating the metal ions and the binding electrons is greater than the sublimation energy by, approximately, the ionization potential of the metal. Some of these issues are discussed in greater detail in Chapter 7, where it is shown that the elastic constants of solids can be calculated in terms of intermolecular forces. At this stage we merely summarize the main features of intermolecular forces in Table 1.7.

Table 1.7. *Intermolecular forces*

Type	Nature	Example	Mode of separation	Energy per bond $J \times 10^{21}$
Ionic	Identical with interatomic forces	NaF crystal	Break up into Na^+ and F^- ions	300
Covalent	Giant molecule held by covalent forces	Diamond	Break up into C atoms	600
van der Waals	As for atoms	Solid methane CH_4	Break up into CH_4 gas	3
Metal	Metal ions in sea of free valency electrons	K	Vaporize to metal ions + electrons	500

1.2.8 *Comparison of electrical and gravitational attraction*

Consider the interaction between, say, two helium atoms, where only the weakest electrical attraction exists as a result of the van der Waals mechanism. If the atoms are packed together as close as possible (say in liquid state) the separation x_0 is about 0.2 nm. The potential energy V_W due to van der Waals for this separation turns out to be about -10^{-22} J. Now consider the potential energy V_g due to gravity.

$$V_g = \int_{x_0}^{\infty} G\frac{m_1 m_2}{x^2}\,dx = -G\frac{m_1^2}{x_0} = -\frac{6.7\times 10^{-11}(0.7\times 10^{-26})^2}{2\times 10^{-10}},$$

i.e.

$$V_g = -10^{-53}\ \text{J},$$

whereas

$$V_W = -10^{-22}\ \text{J}.$$

We see that even when only van der Waals forces are involved the gravitational energy is trivial compared with the electrical.

Table 1.8

Molecule	σ (nm)	$\Delta\varepsilon$ (J) (approx.)
He	0.22	1×10^{-22}
H_2	0.27	4×10^{-22}
Ar	0.32	16.5×10^{-22}
N_2	0.37	13×10^{-22}
CO_2	0.45	40×10^{-22}

1.2.9 *The simplified potential energy curve*

In this section we shall treat molecules as though they are hard elastic balls of diameter σ which attract each other according to some suitable law of force, and then experience a very powerful repulsion force for separations less than σ.

The resulting potential energy curve is then as shown in figure 1.7 and the two most important factors are σ and the depth $\Delta\varepsilon$ of the potential energy trough. Some typical values are given in Table 1.8.

In later chapters we shall be able to explain, in terms of such a potential energy curve, a large number of phenomena. These include critical temperature of a gas, surface tension and viscosity of a liquid, heat of sublimation of a solid, elastic constants and the thermal expansion of a solid.

2

Temperature, heat and the laws of thermodynamics

In this chapter we consider briefly some of the basic concepts of heat and temperature. The treatment is simple and restricted to those aspects which will be of use in later parts of the book.

2.1 **Temperature**

2.1.1 *The concept of temperature*

The concept of temperature originally arose from a sensory feeling of hot or cold. Probably all physical concepts arise in a similar way; the next step is to turn the idea into something more general and, in particular, more objective. One such approach is as follows. It is found that a given mass of gas is completely specified by its volume V and its pressure P. Suppose we start with 1 kg of a gas and subject it to any changes we wish, whether by compressing it, cooling it, allowing it to expand, heating it; we then take another 1 kg specimen of the same gas and carry out a completely different series of operations. If we end up with the same values of V and P for the two samples, the final states will be found identical in every way – colour, warmth, viscosity, and every objective and subjective test to which we can subject the sample.

If we end up with different P, V values there is something different about the samples and if they are placed in contact changes take place until they reach identical P, V values (see section 2.1.3 below).

The attribute which produces this difference is the temperature, and the changes brought about when the specimens are placed in contact result in their eventually reaching the same temperature. For gases and fluids there is a unique function of (P, V) which determines their temperature θ if the mass is specified: $f(P, V) = \theta$. For solids one needs more parameters.

2.1.2 *Temperature scales*

There are, of course, sophisticated means of deriving an absolute thermodynamic scale of temperature (see section 2.3.6 below). In practice one can use the $P-V$ properties of a gas. One may also use less direct standards such as the expansion of a metal rod, the expansion of a liquid column (a conventional thermometer), or the e.m.f. produced by a thermocouple.

2.1.3 *Zeroth law of thermodynamics*

In the development of thermodynamics the first two laws were developed and so named, before it was realized that a preliminary law was required to satisfy a rigorous formulation of thermodynamic laws. This was therefore called the zeroth law. (It has nothing to do with Professor Zero.) The law simply states that if two bodies A and B are individually in thermal equilibrium with a body C, then A and B are in thermal equilibrium with one another.

If C is a thermometer this is clearly a matter of common experience.

2.2 **Heat**

2.2.1 *The concept of heat*

Suppose we have a body of mass M_1 at temperature θ_1, which is placed in thermal contact with another body of mass M_2 and temperature θ_2, where θ_2 is greater than θ_1. Equilibrium will be reached at a common temperature θ, $\theta_2 > \theta > \theta_1$. We say that heat has flowed from θ_2 to θ_1.

Without specifying what heat is, we can define its units. In the earlier work the heat required to raise the temperature of 1 gram of water at 15 °C by 1 K was defined as the calorie. If we assume that in any experiment involving the flow of heat from one body to another heat is conserved we may define the specific heat capacity of a given substance in terms of the specific heat capacity of, say, water. The specific heat capacity s would then be defined as the number of calories required to raise the temperature of 1 gram of the substance by 1 K.

We can now carry out experiments with different masses of different materials at various temperatures using only a centigrade scale thermometer. Common experience then shows that all the results are consistent with our original assumption that heat is always exactly conserved. We have to qualify this by adding that this is true only if any volume changes which occur do a negligible amount of work.

So far we have said nothing about the nature of heat, but this does not affect the validity of the calorimetric conservation principle which we have just described. The next step is to consider briefly J. P. Joule's classical experiments which show that heat is, in fact, a form of energy.

2.2.2 *Joule's experiments*

J. P. Joule (1818–89) began his famous series of experiments on the relation between heat and energy in 1840 and summarized his major conclusions in a massive *mémoire* published in the *Philosophical Transactions* in 1850. His first experiments (1843) involved the combined effects of mechanical and electrical energy, while his second experiments (1845) were concerned with the heat liberated when a gas is compressed. His more famous experiment in which he directly compared the mechanical work expended in stirring water with the heat evolved did not appear until later in 1845, and a greatly improved version was described by him in 1878. The principle was to stir water in a copper vessel with paddles and to calculate the mechanical work done W. Using a very sensitive thermometer he observed the temperature rise Δt of the water, paddles and vessel, and estimated heat losses by conduction, radiation and (if relevant in the mechanical arrangements of the driving system) the losses due to friction in the bearings. From subsidiary calorimetric experiments he was able to determine the specific heat capacities of the vessel and the paddles in terms of the unit heat capacity of water. In this way he could calculate the total amount of heat H in a purely calorimetric experiment required to produce the same temperature rise as in the paddle-stirring experiment, viz:

$$H = \Delta t[\text{mass of water} + m_1 s_1 \text{ (vessel)} + m_2 s_2 \text{ (paddles)}] + \text{losses}. \qquad (2.1)$$

He then found that H was closely proportional to the mechanical work W.

We may summarize some of his other experiments,

(*a*) He repeated the stirring experiments with mercury as the liquid.
(*b*) He rubbed iron rings together against their interfacial friction.
(*c*) He passed an electric current through a resistance wire immersed in a liquid.
(*d*) He compressed air in an insulated cylinder.

In all cases he found that the observed temperature rises could be achieved simply by adding heat of an amount proportional to the mechanical or electrical work done. Further, the factor of proportionality was a constant

$$W = JH, \qquad (2.2)$$

where J is Joule's constant or the mechanical equivalent of heat and has the value $J = 4.187\ \mathrm{J\ cal^{-1}}$.

Joule's constant is of immense importance in all mechanical–thermal operations, but in effect this reduces to translating specific heat capacity data into energy data. It is clearly not a fundamental unit of physics. Indeed if calorimetric studies (involving pounds or grams of water and degrees centigrade or fahrenheit) had not been so attractive as a convenient means of describing thermal energy, it is possible that our basic unit would have been the Joule; we should then have had a calorimetric heat-unit defined as the heat required to raise the temperature of one gram of water by 0.238 K. Indeed tables now quote specific heat capacities in Joules per g or per kg per degree K.

2.3 The laws of thermodynamics

2.3.1 *Internal energy and the first law of thermodynamics*

If we carry out an experiment in a thermally isolated calorimeter, such that no heat can flow into or out of the system, we can change the state of the system by performing mechanical or electrical work on it. Joule's experiments show that for the same amount of work we shall always arrive at the same final state. We may now define the difference in internal energy U of the system as the work done on the system in an adiabatic calorimeter.

Since we can change from state 1 to state 2 by an infinite number of combinations of heat and work, if we wish to define internal energy as a function of state only, we must include both work and heat in our definition of energy. The following example illustrates this in a more concrete way.

Suppose we take a mass of water and perform mechanical and/or electrical work on it; provided we do the same amount of work W, we produce the same change in temperature. Alternatively we can apply a quantity of heat $H = W/J$ and produce the same change. If we start with water at 0 °C and end with water at 100 °C, then the water ends with the same energy state whatever the path adopted. For example we could add heat to raise the temperature to 30 °C and then stir paddles to supply the remaining 70 °C rise. Clearly an infinite number of paths is available. Yet the change in internal energy $U_{100} - U_0$ (ignoring any work done by the liquid on expanding) must be determined only by the initial and final temperature, not by the path. If this were not true we should be able to devise heating and cooling cycles whereby we could raise the temperature to 100 °C by one path and cool it back to 0 °C by another, but still obtain energy from the system. This is contrary to Joule's experiments.

We are left with the conclusion that a change in U is a function only of the initial and final states. If we do external work ΔW and add heat ΔQ we have

$$\Delta U = \Delta W + \Delta Q, \tag{2.3}$$

where ΔU is uniquely defined, but ΔW and ΔQ can have any values provided their sum is equal to ΔU.

For a specified change in state

$$W_A + Q_A \text{ over path A} = W_B + Q_B \text{ over path B.} \tag{2.4}$$

Thus work plus heat are conserved. We may regard this as an extension of the conservation of energy which is well established in *mechanics* and conclude that:

energy is conserved if we include heat as one of the forms of energy.

This is one way of formulating the first law of thermodynamics.

In the simple example given above we have considered U to consist of heat and mechanical or electrical energy. A little consideration shows that the internal energy can embrace all forms of energy, thermal, mechanical, electrical, gravitational, electromagnetic, etc. It may again be shown by similar arguments that, however many types of energy are involved, any change in the internal energy does not depend on the path; it is a function solely of the initial and final states.

2.3.2 *Internal energy of fluids*

For a fluid, U is found to be a unique function of only two variables; either (P, V), (P, T) or (V, T), where P is the pressure, V the volume and T the temperature. Since U is independent of path, changes in it can always be expressed in terms of an exact differential. This also means that, in principle, we can integrate $\mathrm{d}U$ between initial and final states and always end up with the same answer. Thus we can always write for $U(T, V)$

$$\mathrm{d}U = \left(\frac{\partial U}{\partial T}\right)_V \mathrm{d}T + \left(\frac{\partial U}{\partial V}\right)_T \mathrm{d}V, \tag{2.5a}$$

for $U(P, V)$

$$\mathrm{d}U = \left(\frac{\partial U}{\partial P}\right)_V \mathrm{d}P + \left(\frac{\partial U}{\partial V}\right)_P \mathrm{d}V. \tag{2.5b}$$

In our example of raising the temperature of water from 0 °C to 100 °C we concentrated on the variation of U with temperature; $(\partial U/\partial T)_V$ in equation (2.5*a*). That is why we made the proviso that for this to be strictly true any changes in volume would have to be ignored, i.e. $(\partial U/\partial V)_T = 0$.

2.3.3 *Reversible changes in a gas*

If we allow a gas to expand, it does work against its surroundings. The work done depends on the pressure of the gas, the volume change and the pressure outside the gas against which it does work.

Consider a cylinder of cross-sectional area A closed with a weightless piston (figure 2.1). The work done during an expansion can be defined uniquely only if (*a*) the piston is frictionless, (*b*) the expansion is slow and always in quasi-equilibrium, i.e. the pressure outside the piston is only infinitesimally smaller than P (i.e. $\mathrm{d}P \to 0$) so that, by a minute increase in the external pressure, we would be able to *compress* the gas. Under these conditions the gas pushes with a force PA against the surroundings and if the piston moves a distance $\mathrm{d}x$ the work done is $PA\,\mathrm{d}x = P\,\mathrm{d}V$, where $\mathrm{d}V$ is the volume increase. For such a *frictionless*, *reversible* expansion $P\,\mathrm{d}V$ is uniquely defined for the increment $\mathrm{d}V$.

Consider now what happens if we increase the internal energy of a gas by allowing it to do work w, and also add heat of amount q. We have

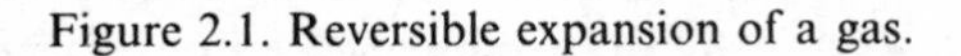

$$\mathrm{d}U = -w + q \,. \tag{2.6}$$

Here w is the work done *by* the gas and is therefore counted negative. For reasons which we described before, $\mathrm{d}U$ is determined uniquely by the

Figure 2.1. Reversible expansion of a gas.

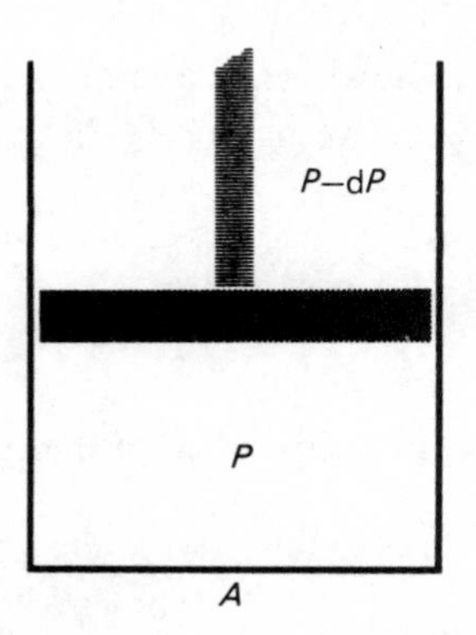

initial and final states. This is not generally true of $-w$ and q separately and for this reason they cannot be expressed as exact differentials.

If we now specify that the work w is a reversible frictionless expansion we have

$$\overset{\text{unique}}{\mathrm{d}U} = \overset{\text{unique}}{-P\,\mathrm{d}V} + q. \tag{2.7}$$

Consequently q must also be uniquely defined over this increment of change so that it can be written as $\mathrm{d}Q$.

Hence

$$\mathrm{d}Q = \mathrm{d}U + P\,\mathrm{d}V. \tag{2.8}$$

Each of the quantities in equation (2.8) is now in the form of a differential, so if the relevant paths are known, they can all be integrated exactly. We see now why frictionless reversible expansions or compressions are so important in thermodynamics. In an electrochemical system the term equivalent to $P\,\mathrm{d}V$ would be $E\,\mathrm{d}q$, where E is the potential difference and $\mathrm{d}q$ the amount of charge passing. Here again for an equation like (2.8) to be valid the current flow would have to take place against an opposing potential infinitesimally less than E. In a magnetization process, where a magnetic element of moment M is exposed to a magnetic field $\mathrm{d}H$, the equivalent term would be $M\,\mathrm{d}H$.

2.3.4 *Enthalpy*

We see from equation (2.8) that the internal energy may be written

$$\mathrm{d}U = \mathrm{d}Q - P\,\mathrm{d}V. \tag{2.8a}$$

Unfortunately we cannot integrate this unless we know the equation of state and the path followed by the expansion cycle. For example an adiabatic expansion would give a value of $\int P\,\mathrm{d}V$ which differs from that occurring in an isothermal expansion. If, however, we specify that the expansion is maintained at constant pressure we have

$$U_2 - U_1 = Q_2 - Q_1 - P(V_2 - V_1). \tag{2.9}$$

The heat absorbed at constant pressure is then given by

$$Q_2 - Q_1 = (U_2 + PV_2) - (U_1 + PV_1). \tag{2.10}$$

We now introduce a new thermodynamic quantity called enthalpy which is defined as

$$H = U + PV. \tag{2.11}$$

This is also a function solely of initial and final states. Thus for a small change

$$dH = dU + P\,dV + V\,dP = dQ + V\,dP. \tag{2.12}$$

Hence

$$\left(\frac{\partial H}{\partial T}\right)_P = \left(\frac{\partial Q}{\partial T}\right)_P. \tag{2.13}$$

As we shall see later, enthalpy is conserved in the steady flow of a gas through a throttle (Joule–Kelvin expansion).

2.3.5 *Specific heat capacities*

These may be defined as the amount of heat required to raise the temperature of a specified amount of material (1 mol, 1 g, 1 kg) by a unit degree of temperature:

$$C = \frac{dQ}{dT}. \tag{2.14}$$

In general we can specify two different conditions for the measurement of C.

Constant volume

$$C_V = \left(\frac{\partial Q}{\partial T}\right)_V = \left(\frac{dU}{\partial T}\right)_V \text{ from equation (2.8}a\text{).}$$

Constant pressure

$$C_P = \left(\frac{\partial Q}{\partial T}\right)_P = \left(\frac{\partial H}{\partial T}\right)_P. \tag{2.15}$$

For gases, as we shall see, there is an appreciable difference between C_P and C_V. For liquids and solids the difference is very much smaller. Some typical values of C for metals are given in Table 2.1. The last column

Table 2.1. *Specific heat capacities of some metals*

Metal	Specific heat capacity C (J kg^{-1} K^{-1})	Atomic mass (kg)	Product (J K^{-1})
Al	920	0.027	25
Cu	380	0.063	24
As	230	0.108	25
Pb	120	0.207	25

shows that the product of the specific heat capacity and the atomic weight (this is in effect the specific heat capacity per mole-atom) has a value of about 25 J K^{-1}. This conclusion is known as the law of Dulong and Petit and we shall discuss its basis and validity in a later chapter.

2.3.6 *The second law of thermodynamics*

It is not the purpose of this book to provide a complete account of thermodynamics (readers will find a fascinating and stimulating account in A. B. Pippard's *Elements of Classical Thermodynamics*†). Indeed, we shall be able to deal with practically all our needs using only the first law. There is, however, some point in emphasizing the essential difference between heat and other forms of energy. When a force does work, the work done is the product of force and distance moved, the force always being measured in the direction of the line along which it is acting. When an electric current does work the current flows in the same direction as the resultant e.m.f. Mechanical work and electrical work are in principle absolutely convertible into one another. A dynamo in principle produces an amount of electrical power exactly equal to the work done in driving it (apart from friction and electromagnetic losses). If this is converted into heat there is, again, an exact equivalence as Joule showed. If, however, we wish to convert heat into work this is *not* true, because thermal energy always has associated with it random movement, vibration, thermal agitation, so there is always part of the thermal energy which is not free for conversion into mechanical or electrical energy.

N. L. S. Carnot (1824) studied the efficiency of a heat engine working between a source at a higher temperature and a sink at a lower temperature, and defined the efficiency as the ratio of useful work done to the thermal energy absorbed from the higher temperature source. He then found that the engine would achieve its greatest efficiency if all the parts of the cycle were carried out reversibly. Under these conditions the efficiency was independent of the working material. For reversible cycles, if the heat taken in is Q_1 and the heat rejected to the sink is Q_2, the efficiency η is

$$\eta = \frac{Q_1 - Q_2}{Q_1}. \tag{2.16}$$

† A. B. Pippard, *Elements of Classical Thermodynamics* (Cambridge University Press, 1957).

Since this is independent of the working material it depends only on the temperature of the two sources. This provides a means of establishing a thermodynamic scale of temperature, which is achieved by defining the ratio of heat absorbed at the higher temperature to that rejected at the lower temperature as equal to the ratio of the higher to the lower temperature, i.e. $Q_1/Q_2 = T_1/T_2$. For such an absolute scale of temperature

$$\frac{Q_1 - Q_2}{Q_1} = \frac{T_1 - T_2}{T_1}. \tag{2.17}$$

We see that fundamentally the efficiency can never be unity unless T_2, the temperature of the sink, is zero. Under these conditions the heat, as it were, is acting completely in the direction of doing useful work.

For the reversible cycles we see that

$$\frac{Q_1}{T_1} = \frac{Q_2}{T_2}. \tag{2.18}$$

Hence travelling from the initial condition, through the cycle, and then back to the initial condition again

$$\frac{Q_1}{T_1} - \frac{Q_2}{T_2} = 0. \tag{2.19}$$

We may now define $\mathrm{d}Q/T$ as the entropy change $\mathrm{d}S$ in a reversible process. The word entropy was invented by R. Clausius; it comes from the Greek word *trope* meaning transformation and he deliberately added the prefix *en* to convey a connection with energy. For a reversible cycle (the sign $\oint$ indicates that the integral is taken over the complete cycle)

$$\oint \frac{\mathrm{d}Q}{T} = \oint \mathrm{d}S = 0. \tag{2.20}$$

This implies that $\mathrm{d}S$ is an exact differential; for any increment in a reversible process

$$\mathrm{d}S = \frac{\mathrm{d}Q}{T} \text{ or } \mathrm{d}Q = T\,\mathrm{d}S. \tag{2.21}$$

One of the most interesting properties of entropy is the following. Even for a reversible cycle the integral of $\mathrm{d}Q$ depends on the details of the paths followed. We saw this in our discussion of enthalpy in section 2.3.4. If, however, we divide the increment of $\mathrm{d}Q$ by the temperature at which the heat $\mathrm{d}Q$ is absorbed (or rejected), the quantity $\mathrm{d}Q/T$ in any reversible path can be integrated between the initial and final state. It is then found to be independent of the path followed.

Because the change in entropy is a function only of the initial and final states we can always determine the entropy change in a system even if we have brought about the change in state by irreversible or non-equilibrium processes. We merely connect the initial and final states (A and B respectively) by some series of reversible changes involving the addition or removal of heat and calculate the integral $\int_A^B \mathrm{d}Q/T$. This is then the relevant entropy change. We could imagine an extreme case in which we added no heat to state A but brought it to state B by some type of irreversible dissipative process such as friction. In that case of course the integral $\int \mathrm{d}Q/T$ would be zero – although the entropy change would still be that given by the reversible heating from state A to state B. Clearly in any irreversible change the entropy change is always greater than $\int \mathrm{d}Q/T$ if $\mathrm{d}Q$ refers to the heat absorbed in the irreversible processes.

There are two other thermodynamic functions which are independent of path, the Gibbs free energy G for isothermal changes at constant pressure and the Helmholtz free energy A for isothermal changes at constant volume. They are defined as follows

$$G = U + PV - TS = H - TS, \tag{2.22a}$$

$$A = U - TS. \tag{2.22b}$$

The following simple argument (based on A. H. Cottrell's *Theoretical Structural Metallurgy*†) gives some idea of how these functions arise.

We are almost always involved in systems which exchange energy with their surroundings. However, in most cases we want to derive properties belonging to the system itself, without reference to the surroundings, in order to define the equilibrium state of the system. Suppose the system can be specified in terms of its internal energy U, temperature T, volume V, pressure P and entropy S and that it can exchange energy with its surroundings. For an infinitesimal change, using the first law we have

$$\mathrm{d}U = \mathrm{d}Q - P\,\mathrm{d}V, \tag{2.8a}$$

where $\mathrm{d}Q$ is the heat taken by the system from its surroundings and $P\,\mathrm{d}V$ is the work it does on its surroundings. Let $\mathrm{d}S$ and $\mathrm{d}S_s$ be the entropy changes in the system and surroundings respectively.

The system and its surroundings constitute a closed system so that the total entropy change either increases or, in the limiting case of ideal reversibility, is zero.

$$\mathrm{d}S + \mathrm{d}S_s \geq 0. \tag{2.23}$$

Suppose the surroundings are at the same temperature T as the system.

† A. H. Cottrell, *Theoretical Structure Metallurgy* (Edward Arnold, 1948).

Then since the system has taken heat $\mathrm{d}Q$ from the surroundings the entropy change is

$$\mathrm{d}S_s = -\mathrm{d}Q/T. \tag{2.24}$$

Thus from equation (2.23) we have

$$\mathrm{d}S - \mathrm{d}Q/T \geq 0. \tag{2.25}$$

Substituting in equation (2.8*a*)

$$\mathrm{d}S - (\mathrm{d}U + P\,\mathrm{d}V)/T \geq 0$$

or

$$\mathrm{d}U + P\,\mathrm{d}V - T\,\mathrm{d}S \leq 0. \tag{2.26}$$

We now have a relation involving quantities belonging to the system itself. In equilibrium, when all changes are reversible,

$$\mathrm{d}U + P\,\mathrm{d}V + T\,\mathrm{d}S = 0. \tag{2.27}$$

At constant pressure and temperature equation (2.27) may be written

$$\mathrm{d}G = \mathrm{d}(U + PV - TS) = 0 \text{ for constant } P, T. \tag{2.28}$$

Thus the Gibbs free energy G is a minimum in the equilibrium state for changes at constant P and T.

At constant volume and temperature $P\,\mathrm{d}V = 0$; substituting in equation (2.26) the equilibrium condition reduces to

$$\mathrm{d}U - T\,\mathrm{d}S = 0$$

or

$$\mathrm{d}A = \mathrm{d}(U - TS) = 0 \text{ for constant } T. \tag{2.29}$$

Thus the Helmholtz free energy is a minimum in the equilibrium state for changes at constant V and T.

There is another way of looking at these relations. Consider a constant temperature, constant volume change

$$\mathrm{d}A = \mathrm{d}U - T\,\mathrm{d}S. \tag{2.30}$$

The change $\mathrm{d}A$ is less than the change $\mathrm{d}U$ by the quantity $T\,\mathrm{d}S$. This is the 'unavailable' energy arising from entropy changes in the system. Thus the energy that one can extract from the system is always less than the energy available from the internal energy (except at absolute zero when $T = 0$). That is why A is called the 'free' energy: part of U is always entangled with the entropy factor. Similarly, at constant temperature and constant pressure the maximum work that the system can perform, $\mathrm{d}G$, is the change in enthalpy $(U + PV)$ less the unavailable energy $T\,\mathrm{d}S$. For a non-reversible process the available energies in both cases will be less.

The unavailability of thermal energy is associated with the random nature of heat. It is not surprising therefore that entropy should in some way be connected with the probability or distribution-randomness of the system under consideration. What is surprising is that this concept can be made quantitative. If the probability of finding a system in a given state is W_1 and in another state is W_2 the entropy change may be written

$$S = k \ln \frac{W_2}{W_1}, \tag{2.31}$$

where k is the Boltzmann constant. Of course, the probability needs to be correctly defined and this forms the subject of statistical mechanics.

One of the most striking examples of 'entropy in action' is the occurrence of vacancies in a crystalline solid. A finite amount of energy is required to form a vacancy so that a solid with vacancies has a higher internal energy than one without. One would therefore expect to find that, in equilibrium, a solid would tend to become vacancy-free. This is not so. Indeed the vacancy concentration increases as the temperature increases. This is because the vacancies increase the disorder: the probability distribution of atoms and vacancies is increased and this increases the entropy of the crystal. Thus, although under equilibrium conditions we are tied to zero change in the free energy

$$\mathrm{d}A = 0 = \mathrm{d}U - T\,\mathrm{d}S,$$

the increase in U due to vacancies is compensated for by the associated increase in S (see Chapter 8). Another example of the importance of entropy is in the elasticity of rubber-like materials. This is discussed in Chapter 7, while Chapter 12 describes the role of entropy in the stabilization of colloids.

With gases the probability associated with the translational motion of gas molecules turns out to be proportional to the volume occupied by the gas; the larger the volume the greater the randomness. The entropy change in increasing the volume from V_1 to V_2 at constant temperature is

$$\Delta S = k \ln\left(\frac{V_2}{V_1}\right)^N,$$

where N is the number of molecules in the quantity of gas considered. We shall use this result in a later chapter of the book.

3

Perfect gases – bulk properties and simple theory

3.1 Bulk properties

3.1.1 *Summary of main bulk properties*

As we shall see in the course of this and the next two chapters, all gases will approximate to perfect or ideal gases if they are sufficiently dilute and if the intermolecular forces are negligible compared with the thermal energy. This condition applies reasonably well to oxygen, nitrogen and hydrogen at normal temperatures and pressures. For this reason the earliest studies of the 'springiness' of air were concerned with the behaviour of a perfect gas.

The main bulk properties were established long ago. We summarize the main conclusions:

1. Boyle's law (*c.* 1660). For a given mass of gas at a fixed temperature the product of pressure and volume is a constant.

$$PV = \text{constant.} \tag{3.1}$$

2. Charles' law (1787) or Gay-Lussac's law (1802). For a given mass of gas, if the pressure be kept constant the volume increases linearly with the temperature, t.

$$V = V_0(1 + \alpha t). \tag{3.2}$$

If the temperature is measured on the centigrade scale and V_0 is the volume at 0 °C, it is found that $\alpha = \frac{1}{273}$. We may therefore write equation (3.2) as

$$V = V_0\left(1 + \frac{t}{273}\right) = \frac{V_0}{273}(273 + t) \tag{3.3}$$

$$= \frac{V_0}{273}T,$$

where T is a scale of temperature, which has -273 °C as its zero point. This scale turns out to be the same as the thermodynamic scale.

3. Variation of pressure with temperature if the volume is kept constant.

$$P = P_0(1 + \beta t). \tag{3.4}$$

The constant β has the same value of $\frac{1}{273}$ so that we have

$$P = \frac{P_0}{273} T. \tag{3.5}$$

4. General gas equation. Rewriting equation (3.1) in the light of equations (3.3) and (3.5) we obtain a general equation of state

$$PV = RT, \tag{3.6}$$

where R depends only on the quantity of gas used.

5. Avogadro's law. For all perfect gases R has the same value if the same molecular quantity of gas is considered. For one mole,

$$R \approx 8.3 \text{ J K}^{-1} \text{ mol}^{-1}.$$

6. Dalton's law. This states that, at a fixed temperature, the pressure of a mixture of gases is equal to the sum of the pressures which would be exerted by each gas separately, if the other constituents were not there.

3.1.2 *Reversible isothermal expansion*

We allow a gas to expand in a frictionless cylinder against a pressure infinitesimally less than the gas pressure. The temperature T is maintained constant. The gas does work

$$dW = P\,dV. \tag{3.7}$$

For an ideal gas $PV = RT$ so that P in equation (3.7) can be replaced by RT/V. Then

$$dW = \frac{RT}{V} dV.$$

If the expansion occurs from volume V_1 and pressure P_1 to volume V_2 and pressure P_2, the work done is

$$W = RT \ln \frac{V_2}{V_1} = RT \ln \frac{P_1}{P_2}. \tag{3.8}$$

This work is provided by the heat absorbed from the constant temperature source, which must be included in the system if T is to be kept constant.

3.1.3 *Fast adiabatic expansion into a vacuum*

Consider the experimental arrangement shown in figure 3.1. A volume V of gas is maintained in the left-hand flask at pressure P and temperature T. The right-hand flask is evacuated. The whole system is thermally insulated so that heat cannot flow into or out of the flasks, then the tap is opened and the gas rushes into the empty flask. Some inequalities of temperature occur initially but a steady state is quickly reached. We write

$$\mathrm{d}U_{\text{total}}=\Delta Q+\Delta W. \tag{3.9}$$

For an adiabatic process there is no heat flow so that $\Delta Q=0$. The gas expanding into the right-hand flask does work in compressing the gas as it enters the flask: but there is no net work done on or by any external system. The overall work done is zero, i.e. $\Delta W=0$. Hence

$$\mathrm{d}U_{\text{total}}=0. \tag{3.10}$$

Now the internal energy consists partly of thermal energy (which depends on temperature) and partly on potential energy due to intermolecular forces. For dry air Joule's results showed that the final temperature is indistinguishable from the original temperature. The thermal part is therefore unchanged, i.e. $\mathrm{d}U_{\text{thermal}}=0$. But since $\mathrm{d}U_{\text{total}}=0$ as in equation (3.10) it follows that $\mathrm{d}U_{\text{volume}}$ also equals zero. This implies that there are negligible forces between the molecules.

Analytically we may write

$$U=f(V,T) \quad \text{or} \quad (P,T)$$

$$\mathrm{d}U=\left(\frac{\partial U}{\partial V}\right)_T \mathrm{d}V+\left(\frac{\partial U}{\partial T}\right)_V \mathrm{d}T.$$

Figure 3.1. Fast adiabatic expansion of a gas into a vacuum. Joule found that for dry air there was no net temperature change.

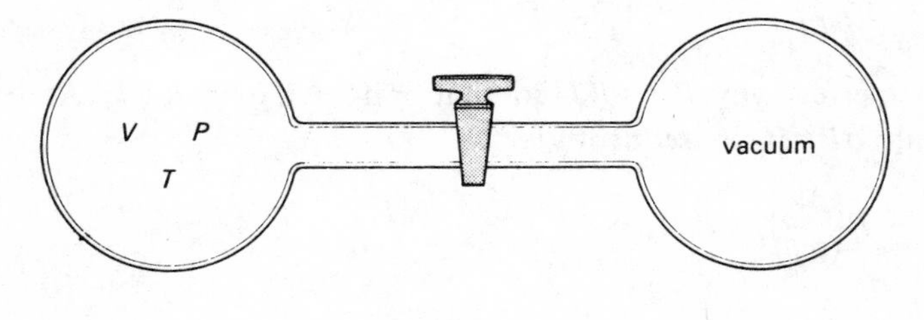

Since there is no temperature rise, $\mathrm{d}T=0$; also from equation (3.10), $\mathrm{d}U=0$, so consequently

$$\left(\frac{\partial U}{\partial V}\right)_T=0, \quad \text{and similarly} \left(\frac{\partial U}{\partial P}\right)_T=0. \tag{3.11}$$

Thus for a perfect gas, U is a function of T only and neither P nor V is involved.

3.1.4 *Specific heat capacity of one mole of a perfect gas*

Consider a small equilibrium change in the gas if we increase its temperature by $\mathrm{d}T$. The basic equation is

$$\mathrm{d}Q=\mathrm{d}U+P\,\mathrm{d}V$$

$$=\left(\frac{\partial U}{\partial V}\right)_T \mathrm{d}V+\left(\frac{\partial U}{\partial T}\right)_V \mathrm{d}T+P\,\mathrm{d}V.$$

For a perfect gas the first term on the RHS is zero. Thus we have

$$\mathrm{d}Q=\left(\frac{\partial U}{\partial T}\right)_V \mathrm{d}T+P\,\mathrm{d}V. \tag{3.12}$$

Consider now two different conditions.

(*a*) *Constant volume*: $\mathrm{d}V=0$. We can rewrite equation (3.12) as

$$\mathrm{d}Q=\left(\frac{\partial U}{\partial T}\right)_V \mathrm{d}T.$$

Hence

$$\left(\frac{\partial Q}{\partial T}\right)_V=C_V=\left(\frac{\partial U}{\partial T}\right)_V. \tag{3.13}$$

(*b*) *Constant pressure*: the gas expands during heating. Equation (3.12) now becomes

$$\mathrm{d}Q=C_V\,\mathrm{d}T+P\,\mathrm{d}V.$$

Hence

$$\left(\frac{\partial Q}{\partial T}\right)_P=C_V+P\left(\frac{\partial V}{\partial T}\right)_P. \tag{3.14}$$

For a perfect gas $PV=RT$ so that $P\,\mathrm{d}V+V\,\mathrm{d}P=R\,\mathrm{d}T$. At constant pressure $\mathrm{d}P=0$, consequently

$$P\left(\frac{\partial V}{\partial T}\right)_P=R.$$

So equation (3.14) becomes

$$\left(\frac{\partial Q}{\partial T}\right)_P = C_P = C_V + R,$$

or

$$C_P - C_V = R. \tag{3.15}$$

3.1.5 *Reversible adiabatic expansion*

We start with our basic equation

$$\mathrm{d}Q = \mathrm{d}U + P\,\mathrm{d}V,$$

specifying that, for an adiabatic process, $\mathrm{d}Q = 0$. Our starting equation then becomes

$$0 = C_V\,\mathrm{d}T + P\,\mathrm{d}V. \tag{3.16}$$

For a perfect gas, the gas equation always applies:

$$P\,\mathrm{d}V + V\,\mathrm{d}P = R\,\mathrm{d}T.$$

Equation (3.16) becomes:

$$-C_V\,\mathrm{d}T = P\,\mathrm{d}V = R\,\mathrm{d}T - V\,\mathrm{d}P$$
$$= (C_P - C_V)\,\mathrm{d}T - V\,\mathrm{d}P.$$

Hence

$$C_P\,\mathrm{d}T = V\,\mathrm{d}P. \tag{3.17}$$

From (3.16)

$$C_V\,\mathrm{d}T = -P\,\mathrm{d}V. \tag{3.18}$$

Taking the ratio of equations (3.17) and (3.18)

$$\frac{C_P}{C_V} = \gamma = \frac{-V\,\mathrm{d}P}{P\,\mathrm{d}V},$$

or

$$\gamma\frac{\mathrm{d}V}{V} = -\frac{\mathrm{d}P}{P}.$$

Integrating from initial to final conditions

$$\gamma \ln\frac{V_2}{V_1} = -\ln\frac{P_2}{P_1}$$

or

$$\ln\left(\frac{V_2}{V_1}\right)^\gamma + \ln\left(\frac{P_2}{P_1}\right) = 0$$

or

$$\ln \frac{P_2 V_2^\gamma}{P_1 V_1^\gamma} = 0. \tag{3.19}$$

Since the antilogarithm of 0 is 1 this gives

$$P_1 V_1^\gamma = P_2 V_2^\gamma = \text{constant}. \tag{3.20}$$

If we plot the pressure against the volume we see (figure 3.2) that the adiabatic curve is steeper than the isothermal.

3.1.6 *Work done in reversible adiabatic expansion*

The work done by the gas in expanding from $P_1 V_1$ to $P_2 V_2$ is

$$\mathrm{d}W = P\,\mathrm{d}V. \tag{3.21}$$

At any stage we can write

$$PV^\gamma = P_1 V_1^\gamma$$

so that P in equation (3.21) can be replaced by $P_1 V_1^\gamma V^{-\gamma}$. The equation becomes

$$W = \int_{V_1}^{V_2} P_1 V_1^\gamma V^{-\gamma}\,\mathrm{d}V.$$

Therefore

$$W = \frac{P_1 V_1}{\gamma - 1}\left[1 - \left(\frac{V_1}{V_2}\right)^{\gamma - 1}\right] = \frac{P_1 V_1}{\gamma - 1}\left[1 - \left(\frac{P_2}{P_1}\right)^{(\gamma - 1)/\gamma}\right]. \tag{3.22}$$

Figure 3.2. Pressure–volume relations for a perfect gas. The adiabatic (lower) curve is steeper than the isothermal (upper) curve.

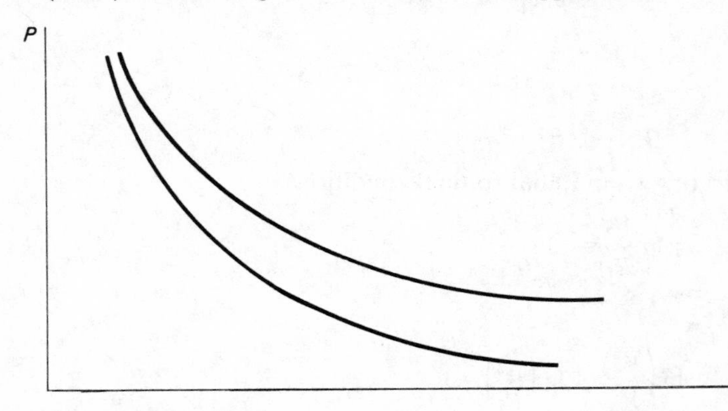

3.1.7 *Cooling as a result of adiabatic expansion*

The work done by the gas is done at the expense of the internal energy. Instead of calculating this *de novo* we merely combine the general gas equation:

$$P_1V_1^\gamma = P_2V_2^\gamma,$$

$$\frac{P_1V_1}{T_1} = \frac{P_2V_2}{T_2}. \tag{3.23}$$

Taking the ratio of the LHS of the two equations and equating it to the ratio of the RHS we obtain

$$\frac{P_1V_1^\gamma}{P_1V_1/T_1} = \frac{P_2V_2^\gamma}{P_2V_2/T_2}.$$

Hence

$$\frac{T_2}{T_1} = \left(\frac{V_1}{V_2}\right)^{\gamma-1} = \left(\frac{P_2}{P_1}\right)^{(\gamma-1)/\gamma}. \tag{3.24}$$

3.1.8 *Entropy changes on expansion*

When an ideal gas expands isothermally the molecules are provided with a greater volume without any other change. This implies a greater randomness or probability and this, in turn, implies an increase in entropy.

For an isothermal expansion the entropy change for one mole of gas is easy to calculate, though the proof given here cannot be considered very rigorous. If the initial volume is V_1 and the final volume V_2 we may divide the volume into a number of unit cells, each of which could just hold one molecule. The number of cells represents the number of ways in which we could place an individual molecule in the given volume (the distribution probability). Clearly in increasing the volume from V_1 to V_2 this distribution probability for one molecule will be increased in the ratio V_2/V_1. If there are N molecules to be considered, the distribution probability will have been increased in the ratio $(V_2/V_1)^N$. The change in entropy is then

$$\Delta S = k \ln \frac{\text{probability 2}}{\text{probability 1}} = k \ln \left(\frac{V_2}{V_1}\right)^N$$

$$= kN\left(\ln \frac{V_2}{V_1}\right) = R \ln \left(\frac{V_2}{V_1}\right), \tag{3.25}$$

if we are dealing with a mole of gas, i.e. if N = Avogadro's number, N_A.

For the adiabatic expansion we may divide the process into two independent reversible parts. We first allow the gas to expand isothermally from V_1 to V_2, giving the entropy increase derived above. We then cool the gas (at constant volume V_2) until its temperature is reduced from T_1 to the final (adiabatic) temperature T_2. The decrease in entropy for such a reversible cooling is

$$\Delta S = \int \frac{dq}{T} = \int_{T_1}^{T_2} \frac{C_V\,dT}{T} = C_V \ln \frac{T_2}{T_1}. \tag{3.26}$$

But, as we saw in equation (3.24), T_2/T_1 for the adiabatic expansion is equal to $(V_1/V_2)^{\gamma-1}$. The entropy loss on cooling is therefore

$$\begin{aligned} C_V \ln \frac{T_2}{T_1} &= C_V(\gamma-1)\left(\ln \frac{V_1}{V_2}\right) = C_V\left(\frac{C_P}{C_V}-1\right)\ln\frac{V_1}{V_2} \\ &= (C_P - C_V)\ln\left(\frac{V_1}{V_2}\right) \\ &= R\ln\frac{V_1}{V_2} = -R\ln\frac{V_2}{V_1}. \end{aligned} \tag{3.27}$$

If we add this to the entropy change associated with the isothermal expansion part of the process (equation (3.25)), we see that the total entropy change is zero. This is because the volume increase tends to increase the randomness whereas the temperature decrease tends to reduce it, the two effects exactly balancing out. This is, of course, what we should expect, since in the adiabatic process itself no heat is added or withdrawn from the gas, so $\Delta Q = 0$ and therefore $\Delta S = \int (dQ/T)$ must be zero.

3.1.9 *Adiabatic lapse rate*

The pressure of the atmosphere diminishes with height and in the next chapter we shall explain why. The atmospheric pressure drops by about a tenth part for every 1 km of ascent. If at ground level the pressure is P_0, the pressure 1 km up is about $\frac{9}{10}P_0$.

Consider now what happens if some disturbance initiates an upward movement of air. The air will expand and in the course of this it will cool. Assuming the cooling to be purely adiabatic, ground temperature to be T_0, and the temperature at the 1 km level to be T, we have from equation (3.24)

$$\frac{T}{T_0} = \left(\frac{P}{P_0}\right)^{(\gamma-1)/\gamma} = \left(\frac{P}{P_0}\right)^{(1.4-1)/1.4} = \left(\frac{1}{10}\right)^{2/7} = \left(1 - \frac{2}{70}\cdots\right), \tag{3.28}$$

since γ for air is approximately 1.4. If $T_0 = 300$ K,

$$T = 300(1 - \tfrac{2}{70}) \approx 300 - 9,$$

so that the temperature drop is 9 K. This is close to the observed temperature drop per km. In practice if there is an appreciable amount of moisture present in the air, cooling leads to condensation and this releases energy which lessens the cooling. A more representative value is about 6 K per km, or about 10 K per mile.

Figure 3.3(*a*) shows the variation of temperature with height. During the day the temperature drops from about 20 °C at ground level to −60 °C at a height of 10 to 12 km, where the troposphere ends. Often in the early morning the temperature gradient is reversed near the ground level due to the cooling of the earth overnight (figure 3.3(*b*)). Inversion temperatures of this type can be an important cause of fog and smog.

3.2 Elementary kinetic theory of ideal gases

The idea that gas molecules are in random motion was considered by the Greeks (as Lucretius' description of Brownian motion shows (see the end of Chapter 4)). It was revived in a surprisingly modern form by D. Bernoulli (1700–82) in 1738, but seems not to have been widely adopted until J. C. Maxwell (1831–79) took the matter in hand and provided practically the whole of our present theory of gases. We first present a very simple theory. Its basic assumptions are:

Figure 3.3. The decrease of temperature with height in air; (*a*) steady conditions, (*b*) early morning.

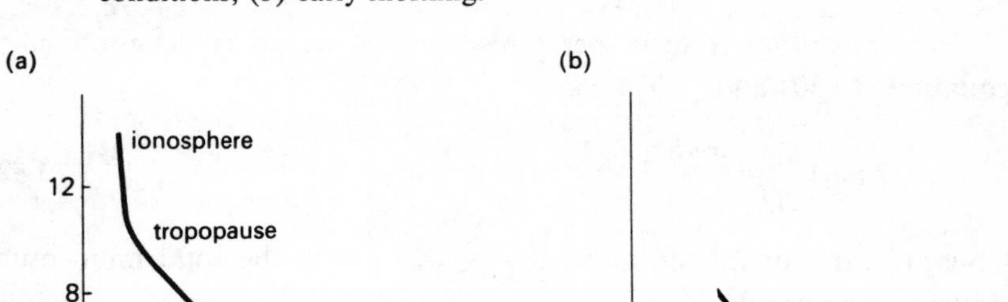

(*a*) the gas consists of identical molecules of mass m;
(*b*) they have zero size, and do not collide with one another;
(*c*) they exert no forces on one another;
(*d*) they undergo random motion within the gas and their collisions with the walls of the container are perfectly elastic.

We now consider the collisions of the gas molecules with the container walls and show that the pressure exerted by the gas on the walls is due to the momentum transfer accompanying the collisions.

Suppose there are n molecules per m^3. They all have varying velocities but because the numbers are generally so enormous we can subdivide these up into groups of $n_1, n_2, n_3 \ldots$ molecules per m^3 with velocity ranges c_1 to $c_1 + dc$; c_2 to $c_2 + dc$; c_3 to $c_3 + dc$, etc.

Let the gas be held in a cubic container of side l, and let us first consider the group with the velocity c_1 – there are $n_1 l^3$ of these in the container.

We resolve c_1 into three mutually orthogonal components u_1, v_1, w_1, parallel to the sides of the cube. Clearly

$$u_1^2 + v_1^2 + w_1^2 = c_1^2. \tag{3.29}$$

Along the u_1 direction the molecule has momentum mu_1 normal to the face of the cube before collision and momentum $-mu_1$ after collision. Thus for each collision

$$\text{momentum transfer to wall} = 2mu_1. \tag{3.30}$$

The molecule has to travel across the cube and back again (i.e. a distance of $2l$) to make the next collision. The time taken is $2l/u_1$.

$$\text{Hence the molecule makes } u_1/2l \text{ collisions per second.} \tag{3.31}$$

The momentum transfer per molecule per second is the product of equations (3.30) and (3.31), i.e.

$$2mu_1\left(\frac{u_1}{2l}\right) = \frac{mu_1^2}{l}. \tag{3.32}$$

Consequently for all the molecules of this group the total momentum transfer per second is

$$\sum \frac{m}{l} u_1^2 = \frac{m}{l} \sum u_1^2. \tag{3.33}$$

We now define the average value of u_1^2 for all this group as

$$\overline{u_1^2} = \frac{\sum u_1^2}{n_1 l^3}, \tag{3.34}$$

so that equation (3.33) becomes

$$\frac{m}{l} n_1 l^3 \overline{u_1^2} = m n_1 \overline{u_1^2} l^2. \tag{3.35}$$

The momentum transfer per second is the force and this is exerted on a face of area l^2. These molecules therefore exert a pressure p on the wall of amount

$$p = m n_1 \overline{u_1^2}. \tag{3.36}$$

We now make the following observation. For this group of molecules the velocity c_1 is the same but their paths are uniformly distributed in all directions; when averaged over the large number of molecules involved there can be no preferred direction so that

$$\overline{u_1^2} = \overline{v_1^2} = \overline{w_1^2} = \tfrac{1}{3} c_1^2. \tag{3.37}$$

Equation (3.36) becomes

$$p = \tfrac{1}{3} m n_1 c_1^2. \tag{3.38}$$

For the groups of molecules the total pressure becomes

$$\begin{aligned} p &= \sum \tfrac{1}{3} m n_i c_i^2 \\ &= \tfrac{2}{3} \sum \tfrac{1}{2} m n_i c_i^2 = \tfrac{2}{3}(\text{total kinetic energy per m}^3). \end{aligned} \tag{3.39}$$

For the total volume V we have

$$PV = \tfrac{2}{3}(\text{total kinetic energy of molecules in the volume } V). \tag{3.40}$$

We now define the mean square velocity of all the molecules in the gas as

$$\begin{aligned} \overline{c^2} &= \frac{n_1 c_1^2 + n_2 c_2^2 + n_3 c_3^2 + \cdots}{n_1 + n_2 + \cdots} \\ &= \frac{\sum n_i c_i^2}{n}. \end{aligned} \tag{3.41}$$

We can insert this in equation (3.39) and obtain

$$P = \tfrac{1}{3} \sum m n_i c_i^2 = \tfrac{1}{3} m n \overline{c^2}. \tag{3.42}$$

For a gas occupying volume V, where the number of molecules is N, where $N = nV$, we have

$$PV = \tfrac{1}{3} m N \overline{c^2}. \tag{3.43}$$

For one mole of a gas N is Avogadro's number, N_A, and

$$PV = \tfrac{1}{3} m N_A \overline{c^2} = RT. \tag{3.44}$$

Order of magnitude of $(\overline{c^2})^{\frac{1}{2}}$. The square root of $\overline{c^2}$ is called the root mean square velocity. As we shall see below it is a little less than the arithmetic mean of the velocities but not very different. From equation (3.44) $RT=\frac{1}{3}mN_A\overline{c^2}=\frac{1}{3}M\overline{c^2}$, where M is the molecular weight. Consequently

$$\overline{c^2}=\frac{3RT}{M}. \tag{3.45}$$

Typical values for $(\overline{c^2})^{\frac{1}{2}}$ are given in Table 3.1 for gases at 0 °C ($T=$ 273 K).

The kinetic energy of an individual molecule is $\frac{1}{2}mc^2$ and from (3.45) this is equal to $\frac{3}{2}(R/N_A)T$ or $\frac{3}{2}kT$, where k is the Boltzmann constant. At room temperature the kinetic energy has a value of about 6×10^{-21} J or 0.04 eV. It is evident that in a head-on collision between two molecules travelling in opposite directions, if all the energy could be absorbed it would be far too small to promote the internal energy of either of the molecules since the electron energies are discrete and the separations are of order 5 eV. Consequently within normal temperature ranges the kinetic energy is never converted into internal electronic energy. It is for this reason that, in the kinetic theory, we can treat the molecule as a hard sphere without any internal structure (see discussion in Chapter 1, p. 7).

3.3 The ether theory of the ideal gas

Before congratulating ourselves on the validity of the kinetic theory we may consider one of the theories which preceded it, and I am much indebted to Professor Eric Mendoza for pointing it out to me. This theory was based on the Newtonian view that a gas was like a solid. In a solid the molecules are held together by attractive forces. In a gas they are repelled by some agency which pushes them apart but still leaves them

Table 3.1. *Typical values of r.m.s. velocity*

Gas	r.m.s. velocity ($m\,s^{-1}$)
H_2	1840
He	1300
O_2	650
Ar	410
Benzene vapour	290
Mercury vapour	180
Electron gas	100 000

in a, more or less, uniform array. What pushes them apart? One view was that it was the swirling ether. If each molecule is surrounded by a shell of ether of radius r rotating at a very high speed, the centrifugal force exerts a pressure on the neighbouring sphere and so keeps the molecules apart. It is as though the molecules are surrounded by balloons which press against one another. If the ether has a mass m and the velocity is somehow in all directions, but confined to the great circles of the sphere, the net centrifugal force is mv^2/r and it is exerted over an area $4\pi r^2$. The pressure exerted on the balloon shell is therefore

$$p=\frac{mv^2/r}{4\pi r^2}=\frac{mv^2}{4\pi r^3}=\frac{\frac{1}{3}mn^2}{4\pi r^3/3}.$$

The numerator on the RHS is $\frac{2}{3}$ times the kinetic energy of the ether; the denominator is the volume occupied by the ethereal shell. We are left with the relation

$$p=\tfrac{2}{3}(\text{kinetic energy of ether per unit volume}).$$

This is very similar to equation (3.39) and shows that the right answer does not necessarily guarantee the validity of the physical model, a most salutary warning. There is, of course, more to this than coincidence. Any gas theory which attributes the gas pressure to the kinetic energy in the gas is bound to give a relation similar to that obtained in the kinetic theory. This follows because pressure has the dimensions of force/area and energy per unit volume has the dimensions of (force × distance)/ volume which is also force/area.

3.4 Some deductions from kinetic theory

3.4.1 *Thermal equilibrium between gas and container*

We now make use of a treatment described by J. Jeans in his *Kinetic Theory of Gases*† to analyse the conditions which determine the thermal equilibrium which must exist between the molecules of a gas and the molecules of the container. We start off by considering the collision between one gas molecule and one wall molecule. Let m, M be the mass of the gas molecule and wall molecule respectively, and let their velocity components in the x, y, z directions be u, v, w and U, V, W respectively. Suppose the collision is along the direction of u, U, so that v, w, and V, W, are unchanged. After collision, u and U become u' and U'.

† J. Jeans, *Introduction to the Kinetic Theory of Gases* (Cambridge University Press, 1940).

For an elastic collision the relative velocity is reversed so that

$$U' - u' = -(U - u). \tag{3.46}$$

The momentum is conserved, so that

$$mu + MU = mu' + MU'$$

or

$$m(u - u') = -M(U - U'). \tag{3.47}$$

Combining equations (3.46) and (3.47) we have

$$U' = \frac{1}{m+M}[(M-m)U + 2mu].$$

The gain in kinetic energy of the wall molecule as a result of the collision is

$$\tfrac{1}{2}M(U'^2 - U^2) = \frac{M}{2}(U' + U)(U' - U)$$
$$= \frac{2Mm}{(m+M)^2}[mu^2 - MU^2 + (M-m)uU]. \tag{3.48}$$

If we now consider the average behaviour of a large number of collisions we note that u must always be positive if collision is to occur, whereas U may be positive or negative and its average value must be zero. Consequently for a large number of collisions between gas molecules and wall molecules, the average value of uU is zero and the average gain of energy of each wall molecule becomes

$$\frac{2Mm}{(m+M)^2}(m\overline{u^2} - M\overline{U^2}), \tag{3.49}$$

where barred symbols refer to the mean square velocities of u and U. For thermal equilibrium there can be no net transfer of energy to the wall. Hence

$$m\overline{u^2} = M\overline{U^2}. \tag{3.50}$$

Similar arguments may be applied to the other components (v, V) and (w, W). The final conclusion is that for thermal equilibrium

$$m\overline{c^2} = M\overline{C^2}, \tag{3.51}$$

where $\overline{c^2}$ and $\overline{C^2}$ are the mean square resultant velocities of the gas and wall molecules respectively. This equation implies that for thermal equilibrium the colliding molecules must have the same average kinetic energy.

This treatment assumes that the wall molecule is completely free. It is, of course, bound to the neighbouring molecules in the wall so that any analysis should really include the 'spring constants' between the wall molecules. It turns out that this does not affect the final results so that the conclusions of equations (3.50) and (3.51) remain valid.

3.4.2 *Boyle's law*

For a given quantity of gas

$$PV = \tfrac{2}{3}\,(\text{total kinetic energy in volume } V).$$

Since the kinetic energy is constant if the temperature is constant this implies that

$$PV = \text{constant}.$$

3.4.3 *Charles' law*

If the pressure is kept constant the volume increases linearly with the kinetic energy of the molecules. Only a sophisticated thermodynamic treatment can show that the kinetic energy itself is proportional to the absolute temperature.

3.4.4 *Avogadro's law*

Two different gases at the same temperature must have the same mean molecular kinetic energy, since they are each individually in equilibrium with the walls of their container. But

$$\begin{aligned} PV &= \tfrac{2}{3}(\text{total kinetic energy in volume } V) \\ &= \tfrac{2}{3}(\text{average kinetic energy per molecule} \\ &\qquad \times \text{number of molecules in volume } V). \end{aligned}$$

Hence (at constant temperature) PV will have the same value for all gases provided they contain the same number of molecules.

3.4.5 *Dalton's law of partial pressure*

If we have two gases which separately have average molecular kinetic energies E_1 and E_2, and a concentration of n_1 and n_2 per m^3 respectively we may write,

$$P_1 = \tfrac{2}{3}(\text{kinetic energy per unit volume}),$$

$$P_1 = \tfrac{2}{3}(E_1 n_1).$$

Similarly

$$P_2 = \tfrac{2}{3}(E_2 n_2).$$

For gases at equal temperature $E_1 = E_2$, so by adding the gases together a new concentration $n_1 + n_2$ is obtained and the combined pressure is

$$P = \tfrac{2}{3}En = \tfrac{2}{3}E(n_1 + n_2) = \tfrac{2}{3}n_1E_1 + \tfrac{2}{3}n_2E_2$$
$$= P_1 + P_2. \tag{3.52}$$

3.4.6 *Mean free path*

We consider here the average distance between collisions of the gas molecules. Suppose σ is the effective diameter of a molecule, then any other molecule whose centre is within a distance σ from the molecule considered will touch it (see figure 3.4(*a*)). For our simplest model we assume all the other gas molecules to be instantaneously at rest and follow the fate of the single mobile molecule. If this molecule has, in a given instant of time, travelled a distance l, it will have swept out a volume $\pi\sigma^2 l$ within which any other molecule will touch it (see figure 3.4(*b*)). Thus if there are n molecules per m^3 the number of molecules with which our moving molecule will collide will be $\pi\sigma^2 ln$ molecules.

The mean distance between molecular collisions λ may now be defined as (distance travelled)/(number of collisions), i.e.

$$\lambda = \frac{l}{\pi\sigma^2 ln} = \frac{1}{\pi\sigma^2 n}. \tag{3.53}$$

This is, of course, the simplest approach possible. In practice all the molecules are not immobilized. If we allow for the true relative velocity between the molecules this reduces λ by a factor $1/\sqrt{2}$. Also, in a collision molecules will only rarely have equi-velocity head-on collisions and bounce away from one another; usually there will be some persistence of velocity in the original direction after collision and this will complicate

Figure 3.4. (*a*) The effective diameter σ of a molecule. (*b*) A single molecule is allowed to travel through gas molecules considered instantaneously at rest.

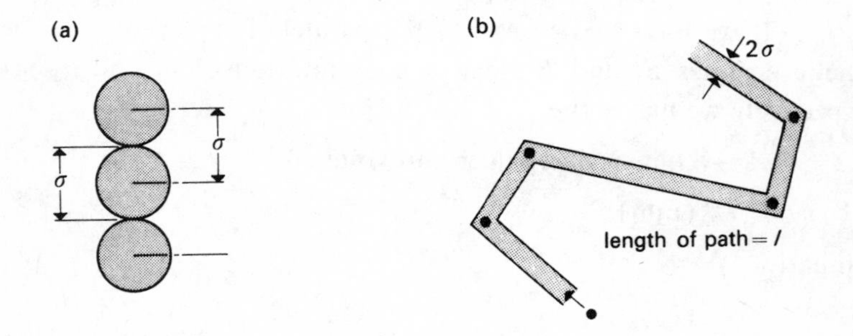

the analysis. Maxwell also considered the influence of intermolecular forces. All these refinements merely change the value of λ by a small numerical factor and we shall not consider this further. Some idea of the complexity of the problem may be gathered from Jeans' book on the kinetic theory of gases.

A more important question is: how far does the concept of molecular collisions and molecular diameter affect our simple kinetic theory, where the basic assumption made was that the molecules have no size, and do not collide with one another in travelling across the container from one wall to the other? The answer is that the finite size of the molecules does influence the simple theory since the 'free' volume available for molecular movement is reduced. This will be discussed in greater detail when we consider the behaviour of 'imperfect' gases in Chapter 5. The collisions between the molecules do not, however, alter the simple theory. As we shall see in the next chapter, collisions deflect molecules and remove them from the group under consideration, but these are constantly replaced (as a result of other collisions) by molecules of other groups. Thus the behaviour is the same as though we could divide the molecules into non-colliding groups with specified velocities and directions.

We see from equation (3.53) that λ varies as $1/n$, i.e. it varies as the reciprocal of the pressure P. Taking an average value of $\sigma=0.3$ nm, we find that for air at s.t.p. $\lambda\approx 100$ nm, whilst the mean distance between molecules is of the order $(1/n)^{\frac{1}{3}}\approx 3$ nm. These values, 0.3, 3, 100 nm, for diameter, separation and mean free path are typical of gases under normal or standard conditions of temperature and pressure (s.t.p.).

3.4.7 *Softening of molecules*

Molecules are not hard impenetrable spheres. If the molecular speed increases they will, during collision, penetrate further into the repulsion fields of their neighbours; but this effect is very small since the repulsive forces are very powerful.

A more important effect is that which arises from the attractive or repulsive forces themselves. These will deflect the molecular paths and produce a marked change in the effective collision diameter of the molecules. An analysis for repulsive forces, falling off as the fifth power of the separation, was first given in analytical form by Maxwell. Forces of this type are not representative of most gas molecules. A more realistic model is one which considers the attractive forces which molecules exert on others passing nearby. This is illustrated schematically in figure 3.5. If the attractive forces are zero (figure 3.5(a)), molecule B will just hit A if the separation is less than σ_0. If the attractive forces are zero and the separation is greater than σ_0, B will miss A (figure 3.5(*b*)). If, however, attractive forces are not negligible B may well hit A even though its original direction is such that it would miss it (figure 3.5(*b*), (*c*)). Hence, although there is no change in the size of the molecule, from the point of view of collision and transport phenomena, it behaves as though it had a much larger effective cross-section σ. Thus slow molecules will tend to collide even if they are very far apart, i.e. σ tends to infinity as T tends to 0. On the other hand, for very high velocities, i.e. very high temperatures, there will be practically no deflection of the molecular paths and collision will only occur when the separation is close to σ_0. A useful empirical relation for the effective molecular diameter σ at temperature T in terms of the 'geometric' diameter σ_0 is

$$\sigma^2 = \sigma_0^2\left(1 + \frac{C}{T}\right), \tag{3.54}$$

where C is known as Sutherland's constant for the gas. Some typical values of Sutherland's constant are given in Table 3.2, and Table 3.3 gives values of the effective cross-section of a number of gases at temperatures

Figure 3.5. The effect of attractive forces on the effective collision diameter.

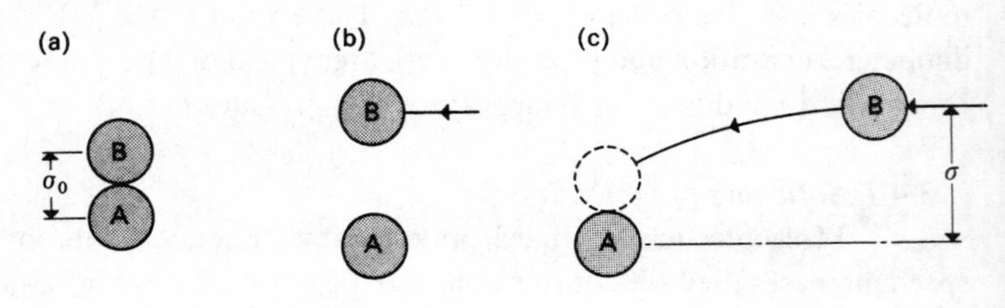

Figure 3.6. Collision of molecules travelling with velocity c_1 normal to a surface area A.

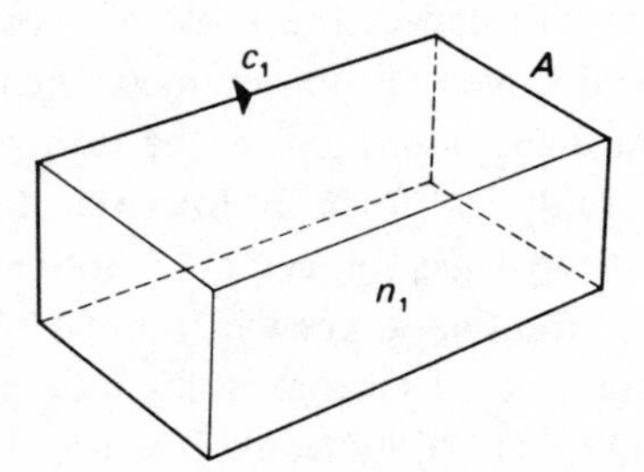

Table 3.2. *Typical values of Sutherland's constant C*

Gas	Ne	He	O_2	Cl_2	Steam
C (K)	56	72	125	350	650

Table 3.3. *Effective cross-section compared with cross-section at 0 °C*

Gas	−100 °C	0 °C	100 °C	200 °C	300 °C
Ne	1.1	1	0.94	0.92	0.9
H_2	1.13	1	0.94	0.92	0.9
O_2	1.18	1	0.9	0.87	0.81
Cl_2	1.32	1	0.85	0.76	0.70
H_2O	—	1	0.79	0.68	0.61

Based on values of C given in G. W. C. Kaye and T. H. Laby, *Tables of Physical and Chemical Constants*, 12th edn (Wiley, 1959).

from −100 °C to 300 °C. The effect is not large but is not negligible, especially for larger molecules.

3.4.8 *Number of molecular collisions per unit area per second*

Consider a surface of area A subjected to molecular collisions (figure 3.6). Divide the molecules into groups n_1 per m^3 with velocity c_1; n_2 per m^3 with velocity c_2, etc. Consider the first group n_1 and assume that they are all moving normal to the surface in question. Then in one second all the molecules in the volume c_1A will strike the surface, i.e. n_1c_1A molecules will strike. Hence the number of collisions per m^2 per second would be n_1c_1.

We now make use of the concept known as the Joule classification, which we used (without naming it) in our treatment of the kinetic theory. The velocity c_1 of a molecule may be described in terms of its u_1, v_1, w_1, components. Since for any one group all directions are equally possible, the number, on average, travelling in any *one* direction of given sign will be one-sixth of the total number.

Hence the collision rate $\mathrm{m}^{-2} = \frac{1}{6}n_1c_1$.

For all the groups we have

$$\text{collisions m}^{-2}\,\text{s}^{-1} = \tfrac{1}{6}\sum n_1c_1 = \tfrac{1}{6}n\bar{c} \tag{3.55}$$

where n is the total number per m^3 and $\bar{c}$ is the mean velocity.

3.5 Transport phenomena

There are three bulk properties of gases which can be explained satisfactorily in terms of a simple molecular model, involving the transport of properties through the gas. For this reason they are called transport phenomena. Viscosity can be explained in terms of transport of momentum, heat conduction in terms of transport of thermal energy, diffusion in terms of molecular transport produced by concentration gradients.

3.5.1 *Viscosity*

We consider the simplest arrangement where a plane surface YY is at rest and another parallel surface ZZ at a distance h away moves to the right with uniform velocity u. It is then found that a force F must be applied to the top plate to maintain the velocity u against the viscous drag of the fluid between the plates (figure 3.7). If the flow of the fluid between the surfaces is along lines parallel to the planes YY and ZZ (lamellar flow) it is found that F is proportional to the area A of the moving surface and the velocity gradient u/h. In the more general case where the velocity gradient is not uniform across the gap, F is proportional to A and to $\mathrm{d}u/\mathrm{d}y$. The factor of proportionality η is introduced such that

$$F = \eta A \frac{\mathrm{d}u}{\mathrm{d}y}, \tag{3.56}$$

where η is the viscosity of the liquid or gas between the surfaces.

We now show that the viscosity of a gas can be expressed in terms of momentum transported across the gas. Consider a plane XX within the gas parallel to the direction of the stream lines (figure 3.7(b)). Let the

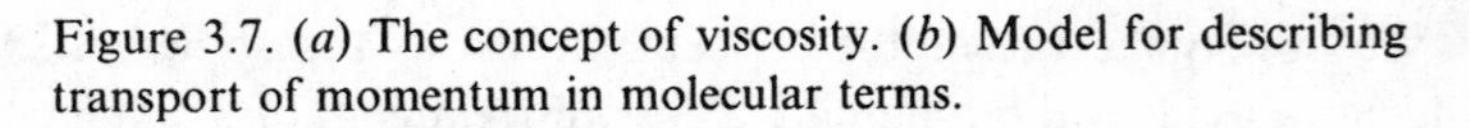

Figure 3.7. (a) The concept of viscosity. (b) Model for describing transport of momentum in molecular terms.

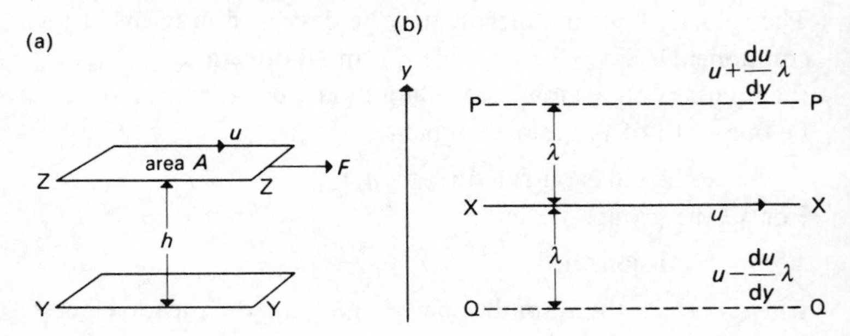

velocity of flow at XX be u and the velocity gradient $\mathrm{d}u/\mathrm{d}y$. We consider the molecules entering and leaving this plane. We assume

(*a*) that the flow velocity u is very small compared with the mean gas velocity c;

(*b*) that the only molecules reaching XX are those, which, on an average, have just made their collision at a distance λ from XX. Then at P the molecules have a flow velocity $u+(\mathrm{d}u/\mathrm{d}y)\lambda$ and at Q a flow velocity $u-(\mathrm{d}u/\mathrm{d}y)\lambda$.

From assumption (*a*) the number of molecules crossing an area A in one second is $\frac{1}{6}n\bar{c}A$. From Q these molecules bring to XX horizontal momentum

$$m\left(u-\frac{\mathrm{d}u}{\mathrm{d}y}\lambda\right)\frac{n\bar{c}}{6}A. \tag{3.57}$$

From P a similar flow brings horizontal momentum

$$m\left(u+\frac{\mathrm{d}u}{\mathrm{d}y}\lambda\right)\frac{n\bar{c}}{6}A. \tag{3.58}$$

The plane XX itself is also discharging $\frac{1}{6}ncA$ molecules per second towards both P and Q. The resultant behaviour is summarized in Table 3.4.

We see that there is no accumulation of molecules or momentum in plane XX. There is, however, a transport of momentum per second

Table 3.4. *Molecules entering and leaving an area A in plane* XX *per second*

	Number	Horizontal momentum
Entering from P	$\frac{1}{6}n\bar{c}A$	$m\left(u+\frac{\mathrm{d}u}{\mathrm{d}y}\lambda\right)\frac{n\bar{c}A}{6}$
Entering from Q	$\frac{1}{6}n\bar{c}A$	$m\left(u-\frac{\mathrm{d}u}{\mathrm{d}y}\lambda\right)\frac{n\bar{c}A}{6}$
Leaving from XX upwards	$\frac{1}{6}n\bar{c}A$	$mu\frac{n\bar{c}A}{6}$
Leaving from XX downwards	$\frac{1}{6}n\bar{c}A$	$mu\frac{n\bar{c}A}{6}$
Net addition to XX	0	0

through the plane given by the difference between equations (3.58) and (3.57). This amounts to

$$\frac{n\bar{c}A}{6}\left[m\left(u+\frac{\mathrm{d}u}{\mathrm{d}y}\lambda\right)-m\left(u-\frac{\mathrm{d}u}{\mathrm{d}y}\lambda\right)\right]=\frac{m\lambda n\bar{c}}{3}A\frac{\mathrm{d}u}{\mathrm{d}y}. \tag{3.59}$$

This horizontal momentum per second is a horizontal force which is transmitted through the gas to the top moving surface. As we saw before, in equation (3.56), the tangential force on the top plate is

$$F=\eta A\frac{\mathrm{d}u}{\mathrm{d}y}.$$

By comparison we see that we have derived a relation for the viscosity of the gas

$$\eta=\tfrac{1}{3}m\lambda n\bar{c} \tag{3.60a}$$

or

$$\eta=\frac{1}{3}\frac{m}{\pi\sigma^2}\bar{c}. \tag{3.60b}$$

We conclude that the viscosity of a gas is proportional to $\bar{c}$, i.e. to $T^{\frac{1}{2}}$, and is independent of n, i.e. of the pressure. Both these conclusions are well supported by experiment. If, however, the pressure is too low and the mean free path becomes comparable with the distance between the

Figure 3.8. To ensure continuity in the force transmitted through the gas we may assume that the velocity gradient within a distance λ from the walls, is doubled. This reduces the velocity gradient in the *bulk* of the gas from u/y to, approximately, $u/(y+2\lambda)$.

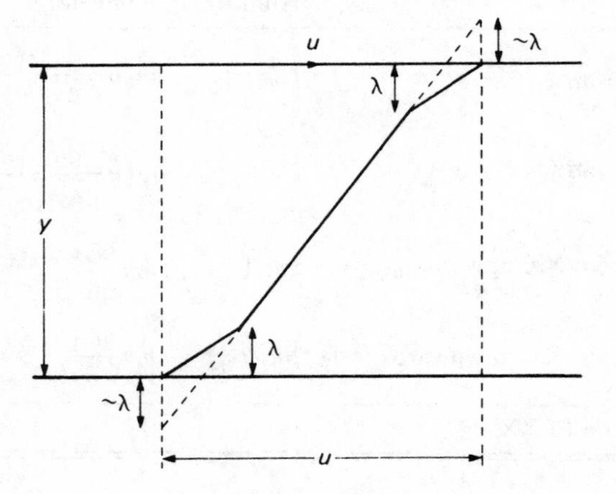

moving surfaces, equations (3.60) are no longer reliable. This may be seen in the following elementary way.

Consider the transport of momentum at a distance equal to or less than λ from the solid walls. Clearly one-half of the transport process is missing, since no molecules can pass through the walls. As there cannot be a discontinuity in the force acting across the gas we must find some way of doubling the momentum transfer. A convenient expedient is to assume that the effective velocity gradient, at a distance of about λ from the wall, is doubled. By a very simple geometric construction shown in figure 3.8 we see that the slope in the bulk of the gas is then not u/y but, approximately, $u/(y+2\lambda)$. Since our theory of viscosity is based on the behaviour in the bulk of the gas it is evident that the effective viscosity of the gas is reduced from the values given in equations (3.60*a*) and (3.60*b*) by a factor

$$\frac{y}{y+2\lambda}.$$

Of course a discontinuity in the velocity gradient within the gas is also inadmissible. In the more detailed analysis given below a better model is provided which explains why the factor $y/(y+2\lambda)$ is valid. The consequences are clear. If λ is appreciable compared with y the viscosity is greatly reduced. If λ is greater than y the whole analysis ceases to be valid. Finally if the molecules do not take up the tangential velocity of the wall after hitting it the transport of momentum will be less and the gas will behave as though there is slip at the walls. The more detailed analysis given below may be omitted on a first reading.

3.5.2 *Boundary conditions*

Suppose the distance between the plates is y, the lower plate is stationary and the upper plate has velocity u. If y is very large compared with the mean free path λ, the velocity gradient across the gas (see figure 3.9(*a*)) is

$$\frac{\mathrm{d}u}{\mathrm{d}y}=\frac{u}{y}. \tag{3.61}$$

Let us now consider in greater detail the situation at the solid plates themselves. It is clear from our derivation of η that at the solid walls only one half of the transport process is operative since molecules cannot pass *through* the walls. Some change in flow rates must occur close to the walls if the appropriate shear stress is to be communicated through the gas to the solid wall.

Consider first the stationary wall. Molecules arriving after their last collision at a mean distance λ away, bring with them horizontal velocity u_1. If all the molecules striking the wall are absorbed and then re-emitted with the tangential velocity of the (stationary) wall, this tangential velocity u_2 is zero (see figure 3.9(*b*)). Thus the average velocity of the molecules very close to the stationary wall is the average of u_1 and zero, i.e. $u_1/2$. This is equivalent to saying that the molecules slip over the walls with velocity $u_1/2$. A similar slip occurs at the moving surface. Between these two planes at which slip occurs there is full transport of molecules to and from the plates, so that a constant velocity gradient may be maintained throughout the bulk of the gas in accordance with equation (3.61). The velocity gradient is determined by the fact that between the stationary surface and the first mean free path there is a change in velocity of $u_1/2$ over a distance λ, i.e.

$$\frac{u_1}{2} = \lambda \frac{du}{dy}.$$

Figure 3.9. (*a*) Ignoring the problem of the solid boundaries the velocity gradient across the bulk of the gas is u/y. (*b*) If impinging molecules from the last mean-free-path separation acquire the tangential velocity of the solid wall there is an effective discontinuity in the mean tangential gas velocity close to the wall. This leads to a reduction in the velocity gradient in the bulk of the gas. (*c*) If impinging molecules rebound 'elastically' so that their horizontal velocity component after impact is the same as before, no horizontal momentum is communicated to the walls. There is complete slipping of the molecules over the walls and zero velocity gradient in the bulk of the gas.

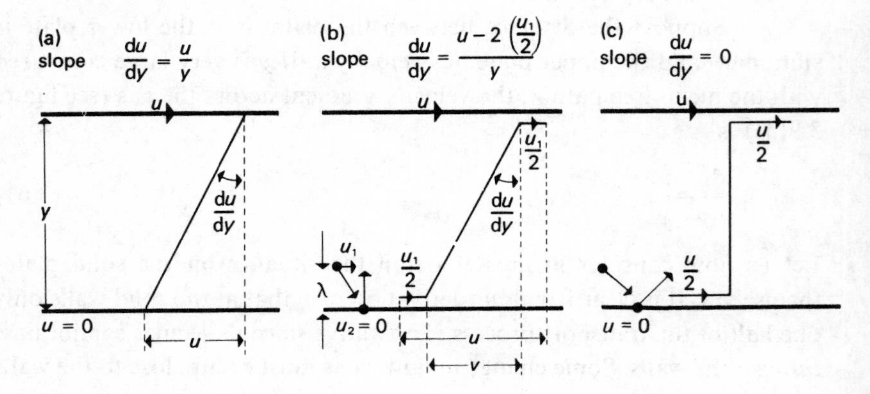

The velocity difference v between the gas near the stationary wall and the gas near the moving wall (see figure 3.9(*b*)) is clearly

$$v = u - \frac{2u_1}{2},$$

or

$$v = u - 2\lambda \frac{\mathrm{d}u}{\mathrm{d}y}.$$

Over this region there is a uniform velocity gradient of amount

$$\frac{v}{y} = \frac{(u - 2\lambda\,(\mathrm{d}u/\mathrm{d}y))}{y} = \frac{\mathrm{d}u}{\mathrm{d}y},$$

or

$$\frac{\mathrm{d}u}{\mathrm{d}y} = \frac{u}{y + 2\lambda}. \tag{3.62}$$

We now see that the shear stress across every section of the bulk of the gas is communicated to the solid walls by the velocity jump close to the solid surface. Comparing equation (3.62) with (3.61), the apparent viscosity of the gas is reduced by a factor

$$\frac{y}{y + 2\lambda}. \tag{3.63}$$

Clearly if the mean free path is an appreciable fraction of the separation there will be a large reduction in the apparent viscosity. In the limit, if λ is greater than y, none of the concepts used in this analysis is applicable.

There is yet a further factor involved. The molecules striking the surface may not all take up the tangential velocity of the surfaces before rebounding. In the extreme case, if the tangential velocities of the molecules leaving the surface are the same as those striking it, there will be no transfer of tangential momentum to the walls. Consequently there will be no resultant velocity gradient in the bulk of the gas, but a velocity discontinuity of $u/2$ at a short distance from each of the walls (figure 3.9(*c*)). This corresponds to perfect 'slipping' of the gas over the wall surfaces without the transfer of any shear stress. Under these conditions the gas would appear to have zero viscosity. In practice the situation is much closer to that involving adsorption and re-emission with the wall velocity. As a result the simple correction, given by equation (3.63), is valid.

Finally we may refer the reader to a more quantitative treatment of this problem given by S. Chapman and T. G. Cowling in *The Mathematical Theory of Non-Uniform Gases.*† If θ is the fraction of molecules absorbed and emitted with the tangential velocity of the wall, then the fraction reflected with the same tangential velocity as the incident molecules is $1-\theta$. The average tangential velocity of the 'reflected' molecules, for example from the stationary wall, is

$$u_2=(1-\theta)u. \tag{3.64}$$

The mean velocity at the wall is $\frac{1}{2}(u_1+u_2)$. This is the velocity of slip. Hence

$$u_1-\tfrac{1}{2}(u_1+u_2)=\lambda\frac{\mathrm{d}u}{\mathrm{d}y}. \tag{3.65}$$

Using equation (3.64) this gives for the slip velocity

$$\tfrac{1}{2}(u_1+u_2)=\frac{2-\theta}{\theta}\lambda\frac{\mathrm{d}u}{\mathrm{d}y}. \tag{3.66}$$

Because of similar slip at the moving wall the velocity difference across the bulk of the gas is

$$u-2\left(\frac{2-\theta}{\theta}\right)\lambda\frac{\mathrm{d}u}{\mathrm{d}y}. \tag{3.67}$$

The apparent viscosity of the gas is thus reduced by the factor

$$\frac{y}{y+2\lambda(2-\theta)/\theta} \tag{3.68}$$

For $\theta=1$ this reduces to equation (3.63); for $\theta=0$, the ratio is infinite. This corresponds to complete slip at the walls as shown in figure 3.9(c).

3.5.3 *Thermal conductivity*

Consider heat flow across the material lying between parallel surfaces YY and ZZ (figure 3.10(a)). If the temperature of the hotter surface is T_1 and the colder T_2, the rate of flow of heat is proportional to $(T_1-T_2)/h$ and to the area of the surfaces. More generally we write for the heat flow per second

$$Q=-KA\frac{\mathrm{d}T}{\mathrm{d}y}, \tag{3.69}$$

† S. Chapman and T. G. Cowling, *The Mathematical Theory of Non-Uniform Gases* (Cambridge University Press, 1939), Chapter 5. I am indebted to Dr N. J. Holloway and Dr B. Scruton for bringing this to my attention.

where K is the thermal conductivity of the material and $\mathrm{d}T/\mathrm{d}y$ the temperature gradient. The negative sign is because there is a flow of heat from the hotter to the colder region (positive Q for negative $\mathrm{d}T/\mathrm{d}y$).

We again assume that in a gas heat conduction is due to the transport of thermal energy. If c_V is the specific heat capacity at constant volume per molecule the rate of transport of thermal energy into plane XX from planes at a distance λ from it is (figure 3.10(*b*)):

from the upper plane:

$$\frac{n\bar{c}A}{6}c_V\left(T+\frac{\mathrm{d}T}{\mathrm{d}y}\lambda\right) \tag{3.70a}$$

from the lower plane:

$$\frac{n\bar{c}A}{6}c_V\left(T-\frac{\mathrm{d}T}{\mathrm{d}y}\lambda\right). \tag{3.70b}$$

Again we may show that there is no net accumulation of molecules or of thermal energy in plane XX. There is, however, a net transport of thermal energy per second of amount equal to the difference between equations (3.70*a*) and (3.70*b*).

Hence

$$Q=-\frac{n\bar{c}\lambda}{3}c_V\times A\frac{\mathrm{d}T}{\mathrm{d}y}.$$

By comparison with equation (3.69) we see that the thermal conductivity is given by

$$K=\frac{n\bar{c}\lambda}{3}c_V. \tag{3.71}$$

Figure 3.10. (*a*) The concept of thermal conductivity. (*b*) Model for describing transport of thermal energy in molecular terms.

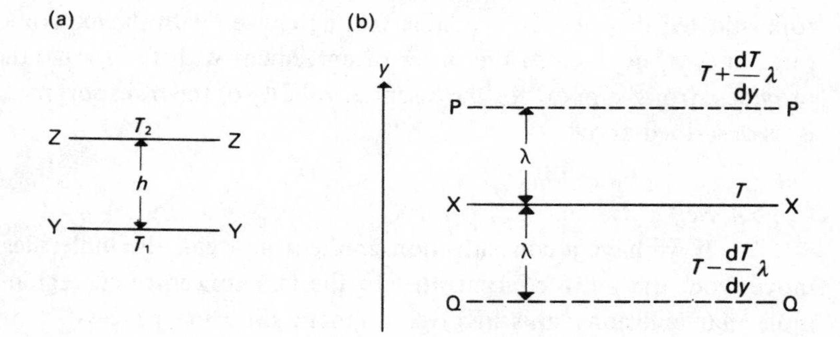

As in the preceding discussion of viscosity the temperature gradient near solid walls will depend on the degree of thermal equilibrium between the temperature of the molecules striking the walls, and the walls themselves. If the molecules acquire the temperature of the walls, the temperature very close to the wall will show a small discontinuity, and there will be a reduction in the temperature gradient through the bulk of the gas. If the collisions are perfectly elastic so that the molecules strike the wall and are reflected with their original thermal velocity, i.e. they ignore the temperature of the wall, no heat can be transferred to or from the walls. There will be a large discontinuity in temperature at the walls: the total temperature drop will occur in a very short distance comparable with λ and there will be a negligible temperature gradient across the mass of the gas. Consequently there will be no heat transfer. This is the limiting case in which the 'heat transfer coefficient' is zero.

Ratio of thermal conductivity and viscosity. From equations (3.71) and (3.60*a*) we have

$$\frac{K}{\eta}=\frac{c_V}{m}=\frac{c_V N_A}{m N_A}=\frac{C_V}{M}$$

where C_V is the specific heat capacity per mole and M the molecular weight. We therefore expect to find

$$\frac{K}{\eta}\frac{M}{C_V}=1. \tag{3.72}$$

The experimental value of this ratio lies between 1.4 and 2.5 for a very wide range of gases. The discrepancy is due partly to molecular repulsions which tend to reduce η. A more important factor is that, in thermal conduction, the transport of the translational kinetic energy takes place more efficiently than the transfer of the rotational and vibrational forms of thermal energy. This leads to an increase in K relative to η, and more sophisticated theories are available that agree well with the experimental ratio of $KM/\eta C_V$. Even the order of agreement with the simple theory provides strong support for the essential validity of the transport mechanisms described above.

3.5.4 *Self-diffusion*

If we have a concentration gradient in a gas, the molecules will move from the more concentrated to the less concentrated regions via molecular collisions; this involves a kinetic diffusion process. If over a distance dx the concentration increases by dn molecules per m^3 the concentration gradient is dn/dx. It is then found that the number of molecules per second crossing an area A normal to the gradient, i.e. the rate of diffusion of molecules per second, can be written

$$\frac{\mathrm{d}N}{\mathrm{d}t} = -D\frac{\mathrm{d}n}{\mathrm{d}x}A,$$

where D is called the coefficient of self-diffusion and the negative sign implies flow in the direction of smaller concentration.

Consider again a plane XX where the concentration is n and neighbouring planes P and Q distant from it where the concentrations are $n+(\mathrm{d}n/\mathrm{d}x)$ and $n-(\mathrm{d}n/\mathrm{d}x)$ respectively. The number of molecules per second crossing from P

$$= \frac{1}{6}\left(n + \frac{\mathrm{d}n}{\mathrm{d}x}\lambda\right)\bar{c}A \tag{3.73a}$$

and from Q

$$= \frac{1}{6}\left(n - \frac{\mathrm{d}n}{\mathrm{d}x}\lambda\right)\bar{c}A. \tag{3.73b}$$

There are also molecules leaving on each side of XX of amount $\frac{1}{6}n\bar{c}A$. Again there is no net accumulation of molecules in XX, but the net transfer is

$$-\frac{1}{3}\lambda\bar{c}A\frac{\mathrm{d}n}{\mathrm{d}x} = -D\frac{\mathrm{d}n}{\mathrm{d}x}A.$$

Consequently

$$D = \frac{\lambda\bar{c}}{3} \quad \text{or} \quad D = \frac{\bar{c}}{3\pi n\sigma^2}. \tag{3.74}$$

It is interesting to see the orders of magnitude involved in D. For air at s.t.p. we have:

$$\sigma = 0.25\ \mathrm{nm}; \quad \lambda = 100\ \mathrm{nm}; \quad \bar{c} = 450\ \mathrm{m\,s^{-1}}; \quad n \approx 3\times10^{25}\ \mathrm{m^{-3}}.$$

This gives a value of D of order 10^{-5} m^2 s^{-1}. Evidently D increases with the molecular velocity c but is reduced by molecular collisions as determined by n and the molecular cross-section σ^2.

There is only one direct way of studying D. This is to incorporate some radioactive molecules in one specimen of gas and place it in contact with an identical gas containing no radioactive molecules. The diffusion of the radioactive species may then be followed, say, using a Geiger counter.

Even here there is some difficulty since the radioactive species may not be the same size as the non-radioactive. For this reason most diffusion experiments involve the diffusion of one gas through another. If the concentration of gas 1 is n_1 per m^3 and that of gas 2 is n_2 per m^3; if c_1 and c_2 are the respective molecular velocities and σ_1 and σ_2 the respective molecular cross-sections, the diffusion coefficient of gas 1 into gas 2 may be shown to be given by

$$D_{1,2} = \frac{(c_1^2 + c_2^2)^{\frac{1}{2}}}{3\pi(n_1 + n_2)(\sigma_1 + \sigma_2)^2}. \tag{3.75}$$

A typical example of $D_{1,2}$ is the diffusion of a vapour through air. Consider a cylindrical vessel of cross-sectional area A (figure 3.11). It contains a small quantity of liquid at the bottom, and at the top open end is a cold surface on which the vapour immediately condenses. The vessel is full of air and diffusing vapour. Just above the liquid the vapour is at its saturation pressure implying a concentration n molecules per m^3. At the cold surface the vapour pressure is almost zero. In the steady state there can be no accumulation of vapour molecules in any section. This implies that the diffusion rate across every section is a constant. Consequently there must be a linear concentration drop from n at the surface of the liquid to zero at the cold surface, distance h. Then

$$\frac{dn}{dx} = \frac{n}{h}.$$

Figure 3.11. Diffusion of molecules from the surface of a liquid where vapour concentration corresponds to that of saturated vapour. The vapour condenses on a cold surface at the top of the vessel so that at this region the vapour concentration is virtually zero.

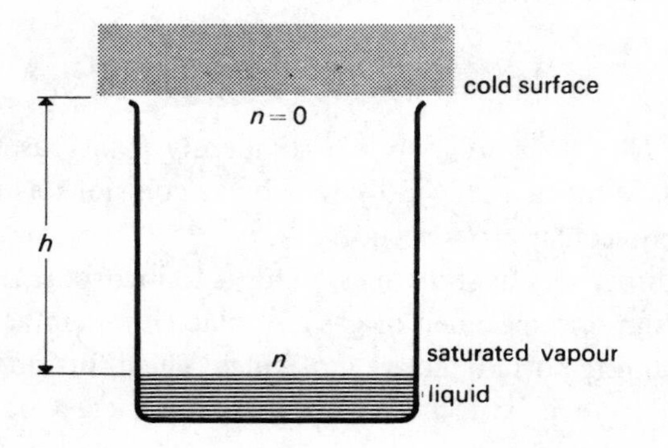

Hence the mass of vapour transferred per second and condensed on the cold surface is

$$w = D_{1,2} A \frac{\mathrm{d}n}{\mathrm{d}x} m = D_{1,2} A \frac{n}{h} m,$$

where m is the weight of the molecule. The product nm is the density ρ of the saturated vapour. If we, therefore, determine the weight w of liquid condensed on the cold surface (or lost from the liquid) per second, we have

$$D_{1,2} = \frac{wh}{A\rho}. \tag{3.76}$$

In this way it is possible to deduce values of $(\sigma_1 + \sigma_2)^2$. Some typical values of σ deduced from measurements of η, K and D are given in Table 3.5.

3.5.5 *Effects of high pressures on viscosity and thermal conductivity of a gas*

We have already seen that the viscosity and thermal conductivity of a gas are independent of the pressure. This is well supported by experiment over a wide range of pressures. If, however, the pressure is so small that the mean free path becomes comparable with the separation between

Table 3.5. *Molecular diameters* σ (10^{-10} m)

Gas	Deduced from		
	η	K	D
Ar	3.6	3.2	3.8
CO_2	4.5	3.8	4.5
CO	3.7	3.7	3.8
Cl_2	5.4	5.9	—
He	2.1	1.9	—
H_2	2.7	2.7	2.7
CH_4	4.1	4.3	4.0
N_2	3.7	3.6	3.7
O_2	3.6	3.5	3.6

Adapted from G. W. C. Kaye and T. H. Laby, *Tables of Physical and Chemical Constants*, 12th edn (Wiley, 1959).

the surfaces the behaviour changes as indicated in section 3.5.1, pp. 66–7. There is also a deviation from ideal behaviour at high pressures. For pressures above a few tens of atmospheres there is a steady increase in viscosity: for example with nitrogen at a pressure of 700 atmospheres the viscosity is doubled.

The explanation depends on the major mechanism involved. If intermolecular forces are important the closer packing of the molecules will clearly increase the energy of interaction. On the other hand, if intermolecular forces are of secondary importance the closer packing itself may play an important part in increasing the viscosity. This may be seen in simple terms in the following way.

In the simple estimation of the mean free path we did not specify whether λ was measured between centres or edges of the molecules because in general $\lambda \gg \sigma$. We can see that measuring λ between centres is misleading when the molecules are close together: for then molecular collisions transmit impulses over a distance $\lambda + \sigma$. If we assume that this is the main factor in influencing the viscosity we may write

$$\eta = \tfrac{1}{3}nm\bar{c}(\lambda + \sigma) = \tfrac{1}{3}nm\bar{c}\lambda + \tfrac{1}{3}nm\bar{c}\sigma. \tag{3.77}$$

We note that the first term is the normal viscosity η_0: for the second term we make use of the relation $P = \tfrac{1}{3}nm\overline{c^2}$. Hence

$$\eta = \eta_0 + P\left(\frac{\bar{c}}{\overline{c^2}}\sigma\right). \tag{3.78a}$$

Figure 3.12. The effect of high pressures on the viscosity of gaseous nitrogen. Below about 50 atmospheres there is little effect: above, the viscosity increases almost linearly with pressure.

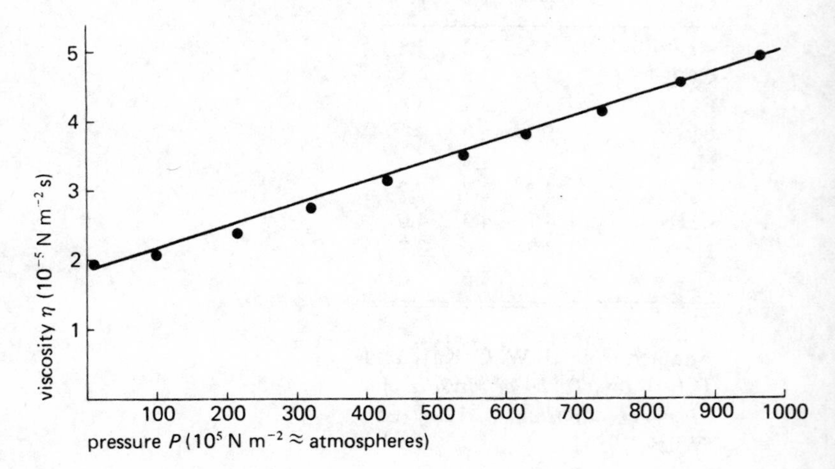

Ignoring the difference between $|\bar{c}|^2$ and $\overline{c^2}$ we may write

$$\eta = \eta_0 + P\left(\frac{\sigma}{\bar{c}}\right). \tag{3.78b}$$

Consequently a plot of η against pressure P should give a straight line of slope $\sigma/\bar{c}$. Data given by Jeans† for nitrogen have been plotted in figure 3.12. It is seen that except for a deviation between 1 and 200 atmospheres the points lie very close to a straight line of slope $3 \times 10^{-5}\ \mathrm{N\ m^{-2}\ s}/10^8\ \mathrm{N\ m^{-2}} = 3 \times 10^{-13}\ \mathrm{s^{-1}}$. This slope should equal $\sigma/\bar{c}$. For nitrogen at room temperature $\bar{c} \approx 400\ \mathrm{m\ s^{-1}}$. Hence

$$\begin{aligned}\sigma &= \bar{c} \times 3 \times 10^{-13} = 1.2 \times 10^{-10}\ \mathrm{m}\\ &= 0.12\ \mathrm{nm}.\end{aligned}$$

Considering the crudity of the assumptions this is a reasonable value, to be compared with the more direct value of about 0.36 nm.

3.6 Sound waves in a gas

3.6.1 *Velocity of sound in a gas*

All elastic media can transmit waves of well-defined velocity and frequency, and with solids the waves can involve shear as well as compression. But with gases the only waves that can be transmitted are those involving a succession of compressions and rarefactions. The velocity of such a longitudinal wave is given by

$$v = \left(\frac{\text{elastic modulus}}{\text{density}}\right)^{\frac{1}{2}}. \tag{3.79}$$

The elastic modulus is the bulk modulus defined as the pressure increment $\mathrm{d}P$ divided by the fractional volume increment $\mathrm{d}V/V$. Since a pressure increase produces a diminution in volume it is usual to define the modulus as

$$-\frac{\mathrm{d}P}{(\mathrm{d}V/V)}.$$

We shall soon show that sound waves are adiabatic and not isothermal. However, at this point we shall assume isothermal waves. If we write

$$PV = RT$$

† *Introduction to the Kinetic Theory of Gases*, p. 166.

then for an isothermal change $P\,\mathrm{d}V + V\,\mathrm{d}P = R\,\mathrm{d}T = 0$, so that the modulus

$$-\frac{\mathrm{d}P}{\mathrm{d}V/V} = P. \tag{3.80}$$

Hence from equation (3.79)

$$v = \left(\frac{P}{\rho}\right)^{\frac{1}{2}} = \left(\frac{nm\overline{c^2}}{3nm}\right)^{\frac{1}{2}} = \left(\frac{\overline{c^2}}{3}\right)^{\frac{1}{2}}, \tag{3.81}$$

since the density $\rho = nm$.

Therefore

$$v = (\overline{u^2})^{\frac{1}{2}}, \tag{3.82}$$

where $\overline{u^2}$ is the mean square molecular velocity in one direction. The value of $(\overline{u^2})^{\frac{1}{2}}$ is very nearly equal to u; this implies that the velocity of a sound wave is approximately equal to the mean molecular velocity in the direction of propagation. This at once tells us that the molecule is the messenger which carries the impulse through the gas. In this way we have linked the equation which describes wave propagation in terms of bulk properties (equation (3.79)), with a molecular process.

For an adiabatic wave the gas equation is

$$PV^{\gamma} = \text{constant}.$$

This leads to a value of the modulus

$$-\frac{\mathrm{d}P}{\mathrm{d}V/V} = \gamma P$$

and

$$v = \left(\frac{\gamma P}{\rho}\right)^{\frac{1}{2}}. \tag{3.83}$$

Consequently the adiabatic wave velocity is $\gamma^{\frac{1}{2}}$ times that given in equation (3.81). This increases v by about 20 per cent for sound in air, but does not change the basic conclusion that sound waves are propagated by molecules communicating an impulse to neighbouring molecules by collision with them.

This conclusion is derived from detailed consideration of the equations of wave propagation. It is interesting to note that a similar view was held long before these equations were derived:

> All sounds, whether articulate or inarticulate, are produced at the meeting of bodies with other bodies or of the air with bodies, not because the air assumes certain shapes, as some people think, but because it is set in motion . . . when it

clashes together by an impact from the breath or from the strings of musical instruments. For when the nearest portion of it is struck by the breath which comes in contact with it, the air is at once driven forcibly on, thrusting forward in like manner the adjoining air, so that the sound travels unaltered in quality as far as the disturbance of air manages to reach.
de Audibilibus; c. 350 BC†

3.6.2 *Why sound waves are adiabatic*

It has long been a tradition in textbooks of sound to assert that in practice sound waves are adiabatic, and not isothermal, because their frequencies are so high. There is a temperature rise at the compression and a cooling at the rarefaction, and this implies that the time taken for the wave to pass along (so reversing the position of compressions and rarefactions) is not sufficient for thermal equilibrium to be reached. Consequently the wave has not the time to become isothermal. This is true but we shall now show that this is because the *frequency is relatively low* and that high frequencies (in theory) would favour *isothermal* conditions. Basically this is because the time needed for temperature equilibration to occur turns out to be proportional to the square of the wavelength, whereas the time available for this to occur is simply proportional to the wavelength. Thus equilibration occurs only if the wavelength l is less than some critical length l_c as indicated schematically in figure 3.13(a). The following simple analysis shows how this arises.

Figure 3.13. (a) Time factors involved in achieving isothermal conditions in a sound wave; these are attainable only if the wavelength is less than l_c. (b) The compressions are rarefactions in a sound wave.

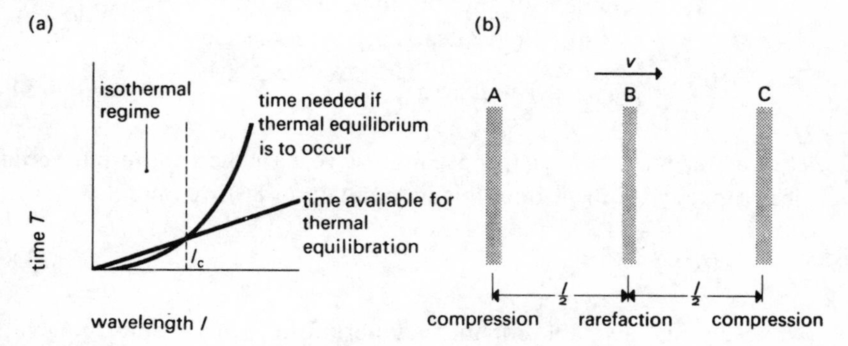

† From *The Works of Aristotle*, vol. 6, ed. W. D. Ross (Clarendon Press, 1913). *The Oxford Classical Dictionary* suggests that *de Audibilibus* is not the work of Aristotle but of Straton, a later contemporary.

Consider an instantaneous snapshot of a plane wave travelling from left to right with velocity v, frequency ν and wavelength l. There are compressions at A, C, and a rarefaction at B (see figure 3.13(b)). Suppose the heating at A and the cooling at B (under perfect adiabatic conditions) give a temperature difference between A and B of T. The temperature gradient is of course sinusoidal but for simplicity we shall assume that it is linear. Then we may write

$$\frac{\mathrm{d}T}{\mathrm{d}x}=-\frac{T}{\frac{1}{2}l}=-\frac{2T}{l}. \tag{3.84}$$

As a result heat flows from A to B tending to equalize the temperature, so reducing the temperature gradient. As we are only interested in an order of magnitude calculation we shall ignore such changes and assume that the gradient remains at $-2T/l$.

If K is the thermal conductivity of the gas the heat flow ΔQ per m^2 of section in time Δt is

$$\frac{\Delta Q}{\Delta t}=-K\frac{\mathrm{d}T}{\mathrm{d}x}=K\frac{2T}{l}. \tag{3.85}$$

Consequently the time Δt taken for a quantity of heat ΔQ to flow is

$$\Delta t=\frac{\Delta Q l}{2KT}. \tag{3.86}$$

Equalization of temperature between A and B demands that the heat ΔQ raises the temperature of the mass of gas between A and B by T, i.e. we need an amount of heat,

$$\begin{aligned}\Delta Q &= \text{volume} \times \text{number of molecules per m}^3 \\ &\quad \times \text{specific heat capacity per molecule} \times T \\ &= \frac{l}{2} n c_P T \text{ per unit area.}\end{aligned} \tag{3.87}$$

(For an instantaneous snapshot of the wave, c_P is the appropriate specific heat capacity.) Substituting this in equation (3.86) we obtain

$$\Delta t=\frac{l^2 n c_P}{4K}. \tag{3.88}$$

Now the time available for this equilibrium to occur cannot be greater than the time taken for the compression and rarefaction to change places, i.e. about $\tau/2$ where τ is the period of the wave. A more realistic estimate is $\tau/4=l/4v$. Consequently temperature equilibrium will occur, i.e. we shall approximate to isothermal conditions if

$$\frac{\tau}{4} > \Delta t, \quad \text{i.e.}\ \frac{l}{4v} > \frac{l^2 n c_P}{4K}.$$

This implies

$$l < \frac{K}{n c_P v} \quad \text{or} \quad \nu > \frac{n c_P v^2}{K}. \tag{3.89}$$

In a previous section we showed (see equation (3.71)) that for a perfect gas

$$K = \frac{n\bar{c}\lambda}{3}(c_V),$$

where λ is the mean free path. Hence isothermal conditions will be achieved if the frequency

$$\nu > \frac{3 n c_P v^2}{n \bar{c} \lambda c_V}, \quad \text{i.e.}\ \nu > \frac{3v^2}{\bar{c}\lambda}, \tag{3.90}$$

ignoring the difference between c_P and c_V.

Since v is of the order of $\bar{c} \approx 300\ \mathrm{m\,s^{-1}}$ and λ is about 100 nm (10^{-7} m), this requires a frequency greater than 10^9 Hz. This is extremely high. Further if we do the same substitution for l we find that l must be less than $\bar{c}\lambda/3v$, i.e. it must be less than the mean free path in the gas. Under these conditions a 'normal' sound wave cannot be propagated. We conclude that sound waves are adiabatic because their frequency is not high enough for them to become isothermal. In theory they could become isothermal at frequencies of the order of 10^{10} Hz but such frequencies could not be transmitted through a gas as a normal sound wave.

4

Further theory of perfect gases

In this chapter we shall first describe a rather more sophisticated kinetic theory of perfect gases. Part of the exercise is purely computational and, although it looks more impressive, adds little to our physical understanding. There are, however, a number of points which emerge which are interesting and useful and which shed new light on some of the assumptions made in the simpler forms of the kinetic theory. We shall also discuss the velocity distribution in a gas and the thermal energy of its molecules.

4.1 A better kinetic theory

4.1.1 *Assumptions*

First we recapitulate our basic assumptions. Let us assume that we are dealing with a very large number of molecules uniformly distributed in density; that they have complete randomness of direction and velocity; that the collisions are perfectly elastic; that there are no intermolecular forces; and finally that the molecules have zero volume.

We now consider a way of describing their distribution in space. Thus to each molecule we attach a vector representing its velocity in magnitude and direction (figure 4.1(a)). We then transfer these vectors (*not* the molecules) to a common origin (figure 4.1(b)) and construct a sphere of arbitrary radius r, allowing the vectors to cut the sphere (if necessary by extending their length). Then the velocity vectors intersect the sphere in as many points as there are molecules.

If we postulate randomness of molecular motion all directions are equally probable, so that these points will be uniformly distributed over the surface of the sphere. Suppose we consider ΔN molecules, where N is the total number present. These could be the total number of molecules in the vessel or the number per m^3, or the number per m^3 with a specified velocity range. Then the number of vector points corresponding to an element of area ΔA on the sphere will be

$$\Delta N = \frac{\Delta A}{4\pi r^2} N. \tag{4.1}$$

We can specify the element ΔA in terms of spherical coordinates θ and ϕ as shown in figure 4.2.

$$\Delta A = r\,\mathrm{d}\theta \times r \sin\theta\,\mathrm{d}\phi. \tag{4.2}$$

Thus the number of molecules travelling in a direction between θ and $\theta + \mathrm{d}\theta$ and between ϕ and $\phi + \mathrm{d}\phi$ (relative to some arbitrary axes) is

$$\Delta N_{\theta,\phi} = \frac{\Delta A}{4\pi r^2} N = \frac{N}{4\pi} \times \sin\theta\,\mathrm{d}\theta\,\mathrm{d}\phi. \tag{4.3}$$

We see that r disappears from the result: ΔN depends only on the specified angles.

We could continue to use equation (4.3) in our subsequent calculations, but this would make the arithmetic more complicated than is necessary. In all the cases we are interested in, we only need to consider molecules travelling in a direction between θ and $\theta + \mathrm{d}\theta$ irrespective of their ϕ position. That is to say it is sufficient for our purpose to make ΔA an annulus on the sphere lying between angles θ and $\theta + \mathrm{d}\theta$ (figure 4.2(*b*)).

Then

$$\Delta A = 2\pi r \sin\theta r\,\mathrm{d}\theta, \tag{4.4}$$

so that the total number of molecules travelling in this direction is

$$\Delta N_\theta = \frac{\Delta A}{4\pi r^2} N = \tfrac{1}{2}\sin\theta\,\mathrm{d}\theta\,N. \tag{4.5}$$

Figure 4.1. (*a*) Velocity vectors are attached to the individual molecules. (*b*) The velocity vectors are all transferred to a common origin. A sphere of arbitrary radius *r* is drawn about the origin.

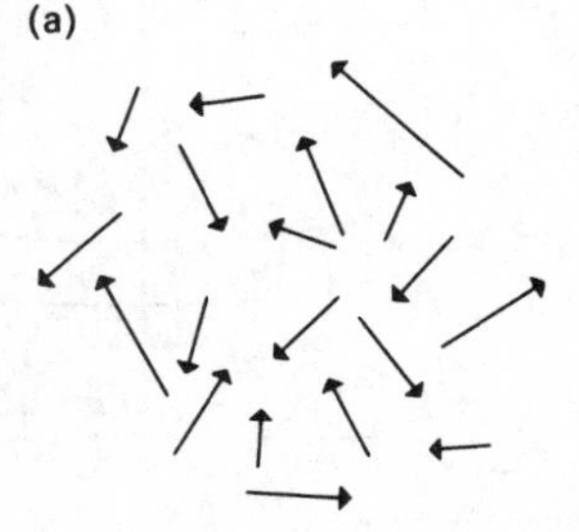

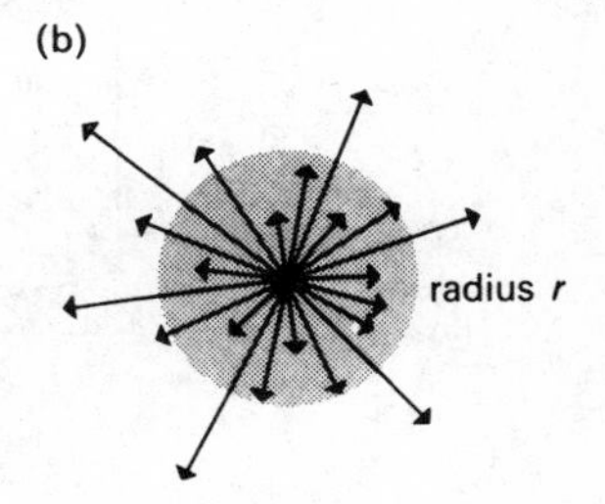

4.1.2 *Number of collisions with a solid wall*

We use the above result to calculate the number of molecules colliding per second with a square metre of a solid surface. Let us first divide the molecules into groups according to their velocities. Suppose there are n_c molecules per m^3 with velocity between c and $c+\mathrm{d}c$. Then the number out of this group that at any instant are travelling towards the surface from a direction θ, $\theta+\mathrm{d}\theta$ is

$$\Delta n_{c,\theta}=\tfrac{1}{2}\sin\theta\,\mathrm{d}\theta\times n_c. \tag{4.6}$$

The only molecules that hit the surface of area α in time $\mathrm{d}t$ are those contained within a prism of basal area α and of sloping height $c\,\mathrm{d}t$ (see figure 4.3). The volume of this is $c\,\mathrm{d}t\,\alpha\cos\theta$. Hence the number of molecules in this group hitting α in time $\mathrm{d}t$ is

$$\Delta n_{c,\theta}\times c\,\mathrm{d}t\,\alpha\cos\theta. \tag{4.7}$$

The number striking each m^2 per second is this number divided by $\alpha\,\mathrm{d}t$,

$$\text{i.e. number}=\tfrac{1}{2}\sin\theta\cos\theta\,\mathrm{d}\theta\,n_c c. \tag{4.8}$$

For all values of θ from 0 to $\pi/2$, i.e. for the whole of the 'half-space' above the surface the total collision rate is

$$\int_0^{\pi/2}\tfrac{1}{2}\sin\theta\cos\theta\,\mathrm{d}\theta\,n_c c=\tfrac{1}{4}n_c c. \tag{4.9}$$

For all molecules of all velocity groups the number striking each m^2 per second is then

$$\tfrac{1}{4}\sum n_c c=\tfrac{1}{4}n\bar{c}, \tag{4.10}$$

Figure 4.2. Specification of an element of area on the velocity sphere (*a*) in terms of angles θ and ϕ. (*b*) in terms only of angle θ.

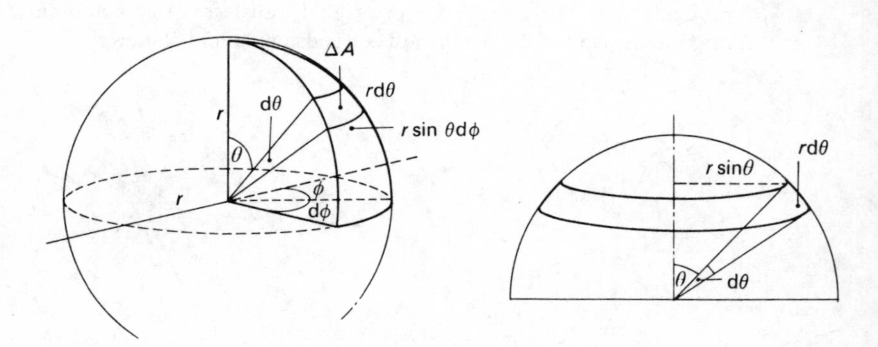

where n is the total number per m^3 and $\bar{c}$ is the mean velocity defined by $\bar{c}=\sum n_c c/n$. This result differs from that given in the previous chapter (equation (3.55)) where the fraction in front of nc was found to be $\frac{1}{6}$ instead of $\frac{1}{4}$.

4.1.3 *Gas equation of state*

We now consider the momentum transfer per second per m^2 of wall, i.e. the pressure exerted by the gas. We first consider the group of molecules n_c per m^3 whose velocity is between c and $c+\mathrm{d}c$. Of these consider the group approaching the surface at a direction θ, $\theta+\mathrm{d}\theta$; each molecule brings with it momentum mc. The vertical component is, therefore, $mc\cos\theta$ and after collision it is reversed so that the net momentum transferred to the wall is $2mc\cos\theta$. The horizontal component $mc\sin\theta$ is unchanged so that no tangential momentum is communicated to the wall. The vertical momentum change per second transferred to each m^2 of wall, i.e. the pressure, is

$$\begin{aligned} &2mc\cos\theta\times(\text{number hitting each m}^2\text{ per second})\\ &\quad=2mc\cos\theta\,\tfrac{1}{2}\sin\theta\cos\theta\,\mathrm{d}\theta\,n_c c\\ &\quad=mn_c c^2\cos^2\theta\sin\theta\,\mathrm{d}\theta. \end{aligned}\tag{4.11}$$

Integrating over the whole half-space we obtain the pressure produced by this group of molecules:

$$\begin{aligned} &mn_c c^2\int_0^{\pi/2}\cos^2\theta\sin\theta\,\mathrm{d}\theta\\ &\quad=mn_c c^2\left[-\frac{\cos^3\theta}{3}\right]_0^{\pi/2}=\tfrac{1}{3}mn_c c^2. \end{aligned}\tag{4.12}$$

Figure 4.3. The number of molecules hitting the area α in time $\mathrm{d}t$ is equal to the number of molecules within the prism of basal area α and sloping height $c\,\mathrm{d}t$.

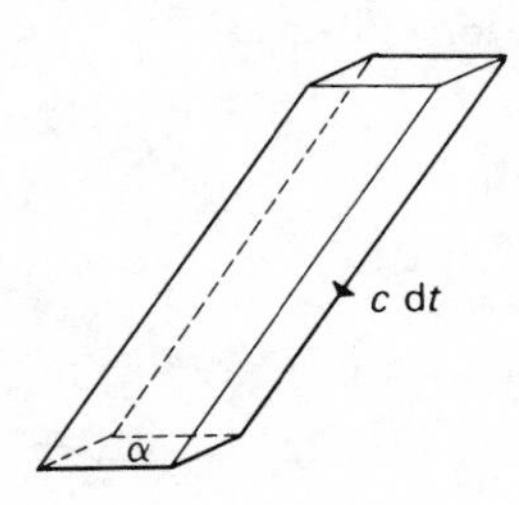

For all groups of molecules we thus obtain

$$p=\tfrac{1}{3}mn\overline{c^2}, \tag{4.13}$$

where the mean square velocity $\overline{c^2}$ is given by $\sum n_c c^2/n$.

4.1.4 *Transport phenomena*

Consider for example, the viscosity of a gas in terms of the momentum transport across a plane as described in the previous chapter. The flow is assumed streamlined parallel to the XX plane, the velocity gradient $\mathrm{d}u/\mathrm{d}y$ in the steady state is constant. We have to calculate the transport of horizontal momentum across each m^2 in the plane XX per second. Consider a small area element $\mathrm{d}\alpha$ in XX. The only molecules we are interested in are those which on an average made their last collision at a distance λ from the element. Draw a hemisphere of radius λ with the element as centre (see figure 4.4). Those coming from the direction θ, $\theta+\mathrm{d}\theta$, are at a distance $\lambda\cos\theta$ from XX so that these molecules bring additional horizontal momentum $m\lambda\cos\theta(\mathrm{d}u/\mathrm{d}y)$. The number coming from this direction and striking each m^2 of the element per second is given by equation (4.8). The transfer of horizontal momentum per m^2 per second is therefore:

$$m\lambda\cos\theta\left(\frac{\mathrm{d}u}{\mathrm{d}y}\right)\times\tfrac{1}{2}\sin\theta\cos\theta\,\mathrm{d}\theta\,n_c c. \tag{4.14}$$

If we integrate this over the whole of the top half-space we obtain

$$\tfrac{1}{6}m\lambda n_c c\times\left(\frac{\mathrm{d}u}{\mathrm{d}y}\right). \tag{4.15}$$

There is a similar contribution from the half-space below so that the

Figure 4.4. Transport of momentum through element of area $\mathrm{d}\alpha$. A hemisphere of radius λ is constructed with $\mathrm{d}\alpha$ as centre. All molecules from this hemisphere have, on average, just made their last collision before proceeding towards $\mathrm{d}\alpha$ from the top half-space.

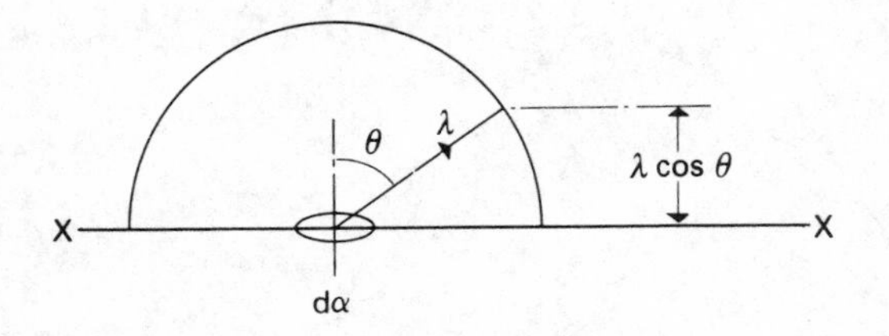

tangential stress in the plane XX is simply

$$\tfrac{1}{3}m\lambda n_c c\left(\frac{\mathrm{d}u}{\mathrm{d}y}\right). \tag{4.16}$$

If we equate this to the viscous stress $\eta(\mathrm{d}u/\mathrm{d}y)$ we obtain

$$\eta=\tfrac{1}{3}mn_c c\lambda, \tag{4.17}$$

for molecules of this group. For all groups of molecules

$$\eta=\tfrac{1}{3}mn\bar{c}\lambda. \tag{4.18}$$

The result is identical with that obtained from the Joule classification.

We may explain this result in a simple way. The number of molecules striking each m^2 in XX per second is $\frac{1}{4}n\bar{c}$ instead of $\frac{1}{6}n\bar{c}$ as in the Joule derivation, but, whereas in the Joule derivation each molecule is assumed to come from a constant distance λ from plane XX, in the present treatment the last collision occurs on the hemisphere of radius λ, so that the last collision distance measured normal to XX ranges from 0 to λ. The average effective distance is $\frac{2}{3}\lambda$. The horizontal momentum transfer is thus the same in both models.

4.1.5 *Mean free path λ and collision frequency ν*

We now extend our treatment to allow for the finite diameter σ of the molecules. As we saw in the previous chapter, a very simple result for the average distance λ between each collision is given by

$$\lambda=\frac{1}{\pi\sigma^2 n}. \tag{4.19}$$

There are many sophisticated ways of improving this relation but they only change λ by a small numerical factor.

We now define the collision frequency ν as the average number of collisions per second made by each molecule.

Then

$$\nu=\frac{\text{average distance travelled per second}}{\text{average distance between collisions}},$$

$$\nu=\frac{\bar{c}}{\lambda}. \tag{4.20}$$

This result does not depend on the way in which λ is derived.

4.1.6 *Distribution of free paths: survival equation*

We wish to follow the fortunes of a group of N molecules. As they travel they collide with themselves and other molecules. Can we estimate the number that, at any specified stage, have not yet made a collision?

Suppose that, at some instant, n have survived without collision. If each is allowed to travel a further distance $\mathrm{d}x$ along its free path further collisions may occur. We *assume* that this number of collisions is proportional to both n and $\mathrm{d}x$. Hence the number removed by these collisions is proportional to $n\,\mathrm{d}x$. To show that n is decreased we write:

$$\mathrm{d}n = -Pn\,\mathrm{d}x, \tag{4.21}$$

where P is the collision probability for the gas. We assume this to be a constant for a given gas under specified conditions. Hence,

$$\frac{\mathrm{d}n}{n} = -P\,\mathrm{d}x,$$

$$n = N\,\mathrm{e}^{-Px} \tag{4.22}$$

and

$$\mathrm{d}n = -PN\,\mathrm{e}^{-Px}\,\mathrm{d}x. \tag{4.23}$$

Since n is the number surviving after travelling a distance x, the number n_c which have suffered collision (see figure 4.5) is simply

$$n_c = N - n.$$

Figure 4.5. Graph showing survival of an initial group of N molecules. After travelling a distance x, n have survived without collision whilst the number n_c suffering collision is given by $n_c = N - n$.

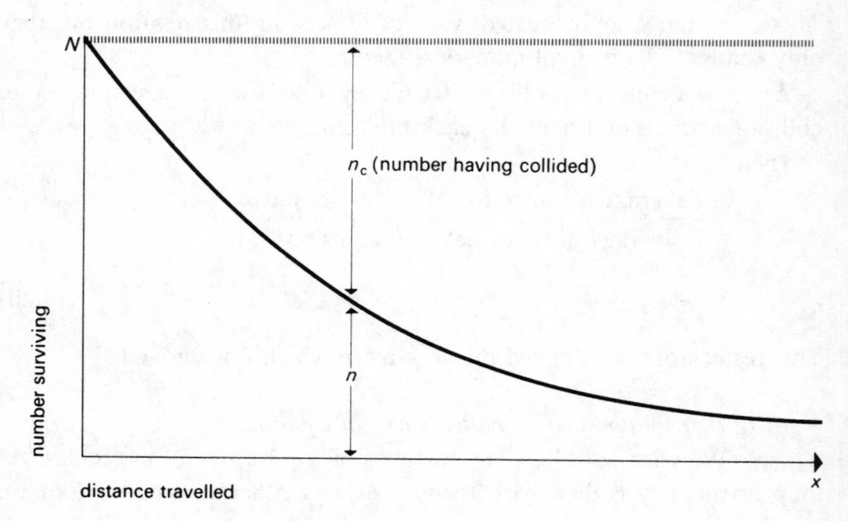

Hence

$$dn_c = d(N-n) = -dn,$$

or

$$dn_c = +PN\,e^{-Px}\,dx.$$

This is the number of molecules with mean free path between x and $x+dx$. The average mean free path x is given by

$$\lambda = \frac{\int_{n=0}^{n=N} x\,dn_c}{\int_0^N dn_c},$$

i.e.

$$\lambda = \frac{\int_{x=0}^{x=\infty} xPN\,e^{-Px}\,dx}{N}$$

$$= -\left[x\,e^{-Px} + \frac{e^{-Px}}{P}\right]_0^{\infty} = \frac{1}{P}.$$

Consequently, equation (4.22) becomes

$$n = N\,e^{-x/\lambda}. \tag{4.24}$$

This is the basic 'survival equation', where n is the number of molecules out of a group N that have not yet made a collision after travelling a distance x.

4.1.7 *Collisions with surfaces: a final treatment*

We may now ask a simple but important question in relation to the kinetic theory as we have so far developed it. We have always assumed that if somewhere there is a group of molecules travelling in some specified direction we can calculate the number striking 1 m^2 of a surface per second by completely ignoring intermolecular collisions. The previous section shows that this is quite unjustified. Molecules a long way from the surface never get there: they are diverted by collisions on the way with themselves and other molecules.

Consider at any instant an element of volume dV in a gas of uniform density containing n molecules per m^3 (figure 4.6(a)). The number in the element is $n\,dV$.

If ν is the collision frequency, the number of collisions occurring within dV in time dt is

$$\tfrac{1}{2}\nu(n\,dV)\,dt. \tag{4.25}$$

The $\frac{1}{2}$ is introduced since otherwise each collision would be counted twice.

Each collision starts off two new free paths so the total number of free paths originating in time $\mathrm{d}t$ from the element $\mathrm{d}V$ is

$$\nu n \,\mathrm{d}V\,\mathrm{d}t.$$

Because all directions of molecular motion are equally probable in any element of the gas, these molecules start off uniformly in all directions. The fraction heading towards the area ΔA is $\Delta\omega/4\pi$, where $\Delta\omega$ is the solid angle subtended by ΔA at the volume element.

Clearly,

$$\Delta\omega = \frac{\Delta A \cos\theta}{r^2}. \tag{4.26}$$

The number heading towards ΔA is thus

$$\Delta n = \frac{\Delta\omega}{4\pi} \times \nu n \,\mathrm{d}V \times \mathrm{d}t. \tag{4.27}$$

However, from the survival equation, the number of molecules reaching ΔA without having made a collision, and so not having been eliminated is

$$\Delta n_1 = \Delta n \,\mathrm{e}^{-r/\lambda}.$$

Consequently, the number of molecules leaving $\mathrm{d}V$ in time $\mathrm{d}t$ and ultimately reaching ΔA without having made an intermediate collision is

$$\Delta n_1 = \frac{1}{4\pi}\,\frac{\Delta A \cos\theta}{r^2}\,\nu n \,\mathrm{e}^{-r/\lambda}\,\mathrm{d}V\,\mathrm{d}t. \tag{4.28}$$

We note that Δn_1 depends on the time element $\mathrm{d}t$, not on the absolute time. This is because on the average any given element of volume appears

Figure 4.6. (*a*) Molecules in element of volume $\mathrm{d}V$, after their last collision start travelling towards ΔA. (*b*) Conical annulus of volume $2\pi r \sin\theta r \,\mathrm{d}r$ defines $\mathrm{d}V$.

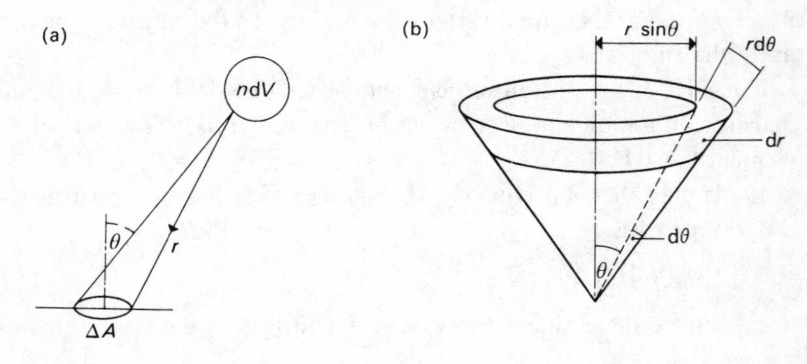

unchanged. (To some extent this observation, in itself implies the conclusion we shall draw in the following paragraphs.) We may thus choose our time element $\mathrm{d}t$ for each volume element $\mathrm{d}V$ at various times, but in such a way that all the undeflected molecules from all the volume elements arrive at ΔA during the same time interval $\mathrm{d}t$. This enables us to integrate for the whole volume of gas above ΔA. Once again we choose as our element of volume a conical annulus of radius $r \sin \theta$, thickness $r\,\mathrm{d}\theta$, height $\mathrm{d}r$ (figure 4.6(*b*)), so that

$$\mathrm{d}V = 2\pi r \sin \theta r\,\mathrm{d}\theta\,\mathrm{d}r. \tag{4.29}$$

We substitute in equation (4.28) and integrate for $r=0$ to $r=\infty$, and for $\theta=0$ to $\theta=\pi/2$. This gives the total number of molecules, from all half-space, from all directions striking ΔA in time $\mathrm{d}t$ as:

$$\nu n\,\Delta A\,\mathrm{d}t \int_0^\infty \int_0^{\pi/2} \frac{\cos\theta \sin\theta}{2}\,\mathrm{d}\theta\,\mathrm{e}^{-r/\lambda}\,\mathrm{d}r$$

$$= \tfrac{1}{4}\nu n\lambda\,\Delta A\,\mathrm{d}t. \tag{4.30}$$

The number striking per m^2 per second is then

$$\tfrac{1}{4}\nu n\lambda. \tag{4.31}$$

But, irrespective of our model for calculating λ,

$$\nu = \frac{\bar{c}}{\lambda}.$$

Hence the number striking each m^2 per second is

$$\tfrac{1}{4}n\bar{c}. \tag{4.32}$$

This is the same as the result we have already obtained. It implies that although molecules are continuously colliding and changing their directions and velocities, other molecules are replacing the ones that are eliminated. For this reason the simple picture which ignores intermolecular collisions in calculating wall collisions, gas pressure and transport phenomena yields the correct answer.

4.2 Sedimentation

4.2.1 *Sedimentation under gravity*

Consider a gas column of unit cross-section and at uniform temperature T (figure 4.7). We shall show that because of gravity there is a density (and pressure) gradient in the gas. Our increments are measured positively in the direction of increasing height h.

At the level A, height h, let the pressure be p and the molecular density n per m^3. At level B, height $h + \mathrm{d}h$, the pressure is $p + \mathrm{d}p$ and the density $n + \mathrm{d}n$.

The resultant force on the layer of gas between A and B is a pressure $\mathrm{d}p$ downwards and a gravitational force $n \times \mathrm{d}h \times mg$ downwards. For equilibrium

$$n\,\mathrm{d}h\,mg + \mathrm{d}p = 0. \tag{4.33}$$

Since $p = \frac{1}{3}nm\overline{c^2}$ and $\overline{c^2}$ is constant at all levels (because of constant temperature)

$$\mathrm{d}p = \tfrac{1}{3}m\overline{c^2}\,\mathrm{d}n.$$

Equation (4.33) becomes

$$\frac{\mathrm{d}n}{n} = \frac{-mg}{\frac{1}{3}m\overline{c^2}}\,\mathrm{d}h.$$

Integrating from $h = 0$ to h and from n_0 to n we obtain

$$\ln\frac{n}{n_0} = \frac{-mg}{\frac{1}{3}m\overline{c^2}}h = \frac{-Nmg}{\frac{1}{3}Nm\overline{c^2}}h,$$

or

$$\ln\frac{n}{n_0} = \ln\frac{p}{p_0} = -\frac{Mg}{RT}h. \tag{4.34}$$

For air, substituting $M \approx 0.030$ kg, $R = 8.3$ J, $T = 300$ K, we find that at a height of 1 km the pressure is reduced by about one-tenth of an atmosphere.

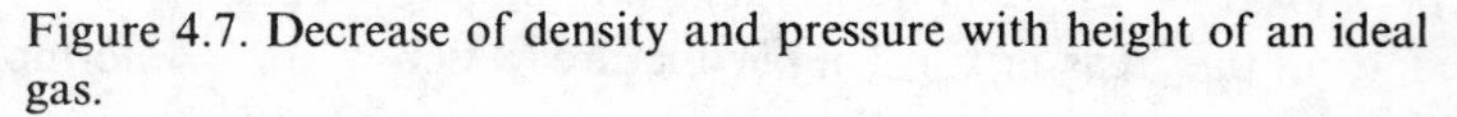

Figure 4.7. Decrease of density and pressure with height of an ideal gas.

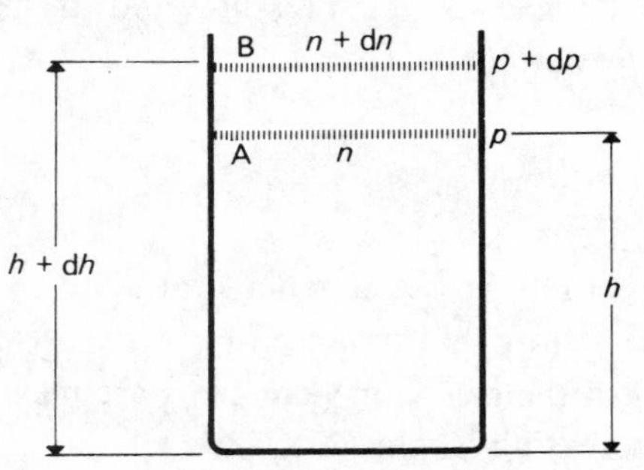

4.2.2 *Sedimentation of particles*

If particles of mass m and density ρ are suspended in a liquid of density ρ_0, then if there is complete thermal equilibrium, the concentration of particles is given by

$$\ln\frac{n}{n_0} = -\frac{Nmgh}{RT}\frac{(\rho-\rho_0)}{\rho}. \tag{4.35}$$

An examination of the variation of n with height for a fine suspension was used by J. Perrin in 1908 to calculate N_A. He obtained a value of 6×10^{23}, which is virtually identical with the value of Avogadro's number obtained by other methods. This shows that fine particles in thermal equilibrium behave like a gas of molecular mass m. It is quite wrong to imagine that this is simply because they are buffeted around by the molecules of the liquid in which they are suspended. They possess this property simply by virtue of being in thermal equilibrium with themselves and their surroundings. Of course the collisions between the particles and the liquid molecules ensure that the whole system is in thermal equilibrium, but the kinetic theory applies to the particles themselves just as it does to the molecules of a conventional gas. Equation (4.35) also shows that by increasing g the major fraction of the particles will be formed in a thin layer near the bottom of the vessel. This is the basis of the action of a centrifuge.

4.2.3 *The Boltzmann distribution*

We may rewrite equation (4.34) in the form

$$\ln\frac{n}{n_0} = -\frac{Mg}{RT}h = -\frac{mg}{kT}h,$$

where $k = R/N =$ the Boltzmann constant. Then

$$n = n_0\exp\left(-\frac{mg}{kT}h\right). \tag{4.36}$$

Since the temperature is constant throughout the gas the kinetic energy is constant at all levels; only the potential energy has been changed, by amount mgh. Then mgh is the amount by which the energy of a molecule at h exceeds that at ground level; call this ε. Then

$$n = n_0\exp\left(-\frac{\varepsilon}{kT}\right). \tag{4.37}$$

Boltzmann showed that for any equilibrium distribution of molecules this relation generally holds. We may write

$$\begin{pmatrix}\text{number of molecules at energy} \\ \text{level } \varepsilon \text{ above ground level}\end{pmatrix} = \begin{pmatrix}\text{number of molecules} \\ \text{at ground level}\end{pmatrix} \times \exp\left(-\frac{\varepsilon}{kT}\right).$$

Again for two energy states ε_1 and ε_2

$$\frac{n_1}{n_2} = \frac{\exp(-\varepsilon_1/kT)}{\exp(-\varepsilon_2/kT)} = \exp\left(-\frac{\Delta\varepsilon}{kT}\right), \tag{4.38}$$

where $\Delta\varepsilon$ is the energy gap between states 1 and 2. The Boltzmann function thus describes the relative population of energy states in an equilibrium system.

As we shall see in section 4.4 (equation (4.48)) the Boltzmann function leads to the result that for particles of mass m, velocity u, kinetic energy $\frac{1}{2}mu^2$, the average thermal energy is $\frac{1}{2}kT$. This is part of a broader conclusion: any degree of freedom which enters quadratically, e.g. $(\text{velocity})^2$, $(\text{displacement})^2$, $(\text{potential})^2$, into the expression for the energy of a system in thermal equilibrium contributes an average of $\frac{1}{2}kT$ to this energy. Several examples will be discussed in section 4.6.

4.3 Temperature variation of reaction rates

A chemical reaction can only occur spontaneously if the final system has a lower free energy state than the initial system. In general, however, before the reaction can take place one or more of the reactants must be excited to a higher energy state (see figure 4.8). The rate of forward reaction is then determined by the number of molecules which have enough energy to get over this barrier. The forward reaction rate is proportional to $\exp(-\varepsilon_f/kT)$, where ε_f is the activation energy.

It is interesting to note that the back reaction rate will be proportional to $\exp(-\varepsilon_b/kT)$, and that the heat of reaction is equal to the difference between the activation energies of forward and backward reaction,

$$q = \varepsilon_b - \varepsilon_f.$$

4.4 Distribution of velocities in a perfect gas

4.4.1 *Velocity distribution in a one-, two- and three-dimensional gas*

We expect to find a wide distribution of molecular velocities in a gas, because even if we could start off with all the molecules travelling in different directions with equal velocities, random collisions are bound to speed some and retard others. Consequently we expect to find the velocities ranging from 0 to ∞ but grouped around a definite average value determined by the temperature.† The velocity distribution is, in fact, of the same type as the Boltzmann function.

If there are no intermolecular forces the only energy the individual molecule has is its kinetic energy $\frac{1}{2}mc^2$. (We ignore here vibrational or rotational energies which are assumed to be unchanged by the gas-collision process.) Then according to the Boltzmann distribution the population density of molecules with velocity c is proportional to

$$\exp\left(-\frac{mc^2}{2kT}\right). \tag{4.39}$$

If we attach velocity vectors to the molecules and transpose them to a single origin (see figure 4.9(a)), the number of velocity vectors ending in the element du, dv, dw gives us the number of molecules with velocities between u and $u+\mathrm{d}u$, v and $v+\mathrm{d}v$, w and $w+\mathrm{d}w$. This number is then

$$\mathrm{d}n = A\exp\left(-\frac{mc^2}{2kT}\right)\mathrm{d}u\,\mathrm{d}v\,\mathrm{d}w, \tag{4.40}$$

where A is a suitable constant.

Figure 4.8. Forward reaction has an activation energy ε_f, the backward reaction an activation energy ε_b. The heat of reaction is $q=\varepsilon_b-\varepsilon_f$.

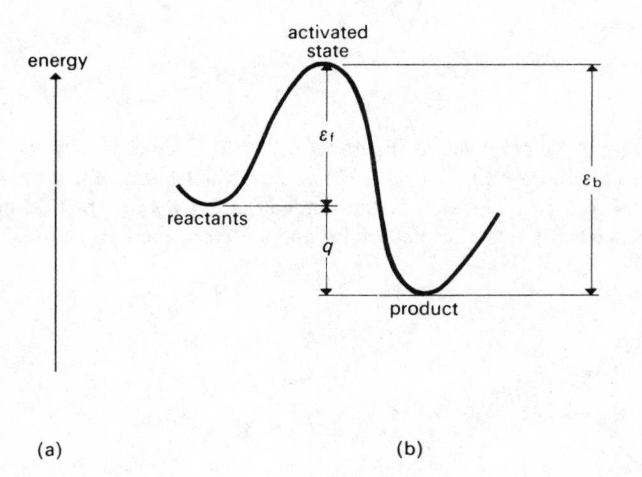

† Because of relativity effects an infinite velocity would imply an infinite effective mass. However, the fraction of molecules with velocities high enough to produce appreciable relativity effects is negligible. In what follows we shall assume that relativity effects may be ignored and that the velocity distribution extends to infinity although, of course, infinite velocities cannot in fact occur.

Before proceeding to use this relation we may split up c^2 into its three components $u^2+v^2+w^2$ so that

$$\mathrm{d}n = A\exp\left(-\frac{mu^2}{2kT}\right)\exp\left(-\frac{mv^2}{2kT}\right)\exp\left(-\frac{mw^2}{2kT}\right)\mathrm{d}u\,\mathrm{d}v\,\mathrm{d}w. \qquad (4.41)$$

Since each of the three components must have the same velocity distributions, the quantity

$$A^{\frac{1}{3}}\exp\left(-\frac{mu^2}{2kT}\right)\mathrm{d}u$$

gives the number of molecules with velocity between u and $u+\mathrm{d}u$ irrespective of the v and w components. The fraction of molecules in this category is

$$\frac{\mathrm{d}n}{n} = B\exp\left(-\frac{mu^2}{2kT}\right)\mathrm{d}u. \qquad (4.42)$$

The constant B may be determined by observing that, if this relation is integrated for all possible velocities from $-\infty$ to $+\infty$, the resulting fraction must be unity. Then

$$\int_{-\infty}^{\infty} B\exp\left(-\frac{mu^2}{2kT}\right)\mathrm{d}u = 1. \qquad (4.43)$$

Figure 4.9. Velocity vector diagrams (*a*) for molecules with velocities between c and $c+\mathrm{d}u+\mathrm{d}v+\mathrm{d}w$, (*b*) velocity distribution for a one-dimensional gas. The area of the shaded band gives the fraction of molecules with velocities between u and $u+\mathrm{d}u$ in the u direction.

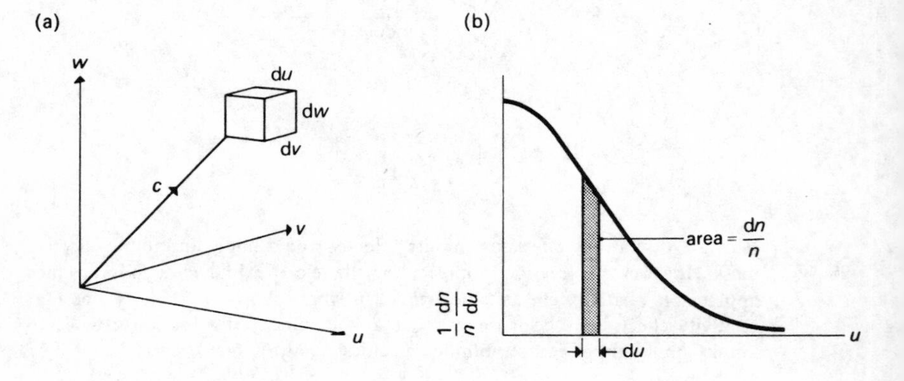

Putting

$$\frac{m}{2kT}=a, \tag{4.44}$$

$$B\int_{-\infty}^{\infty} \mathrm{e}^{-au^2}\,\mathrm{d}u=1.$$

From the table of integrals on p. 121,

$$\int_{-\infty}^{\infty} \mathrm{e}^{-au^2}\,\mathrm{d}u=2\int_{0}^{\infty} \mathrm{e}^{-au^2}\,\mathrm{d}u=\left(\frac{\pi}{a}\right)^{\frac{1}{2}}=\left(\frac{2\pi kT}{m}\right)^{\frac{1}{2}}.$$

Hence

$$B=\left(\frac{m}{2\pi kT}\right)^{\frac{1}{2}}.$$

Consequently

$$\frac{\mathrm{d}n}{n}=\left(\frac{m}{2\pi kT}\right)^{\frac{1}{2}}\exp\left(-\frac{mu^2}{2kT}\right)\mathrm{d}u. \tag{4.45}$$

How can we plot this function? We bring the term du over to the LHS so that we have

$$\frac{\mathrm{d}n}{n}\frac{1}{\mathrm{d}u} \quad \text{or} \quad \frac{1}{n}\frac{\mathrm{d}n}{\mathrm{d}u}=\left(\frac{m}{2\pi kT}\right)^{\frac{1}{2}}\exp\left(-\frac{mu^2}{2kT}\right). \tag{4.46}$$

We may calculate the RHS and plot it as a function of u (see figure 4.9(*b*)). If we take a strip of width du its area is then dn/n which is the fraction of molecules with velocity between u and $u+\mathrm{d}u$. We see that the greatest fraction occurs for zero velocity; this is because of the exponential nature of the Boltzmann factor, and the fact that the 'one-dimensional' gas directly reflects this distribution.

The average energy in the u direction is given by

$$\bar{\varepsilon}=\left[\int_{-\infty}^{\infty}\left(\frac{mu^2}{2}\right)\mathrm{d}n\right]\frac{1}{n} \tag{4.47}$$

$$\bar{\varepsilon}=\frac{m}{2}\left(\frac{m}{2\pi kT}\right)^{\frac{1}{2}}\int_{-\infty}^{\infty} u^2\,\mathrm{e}^{-au^2}\,\mathrm{d}u=\frac{m}{2}\left(\frac{a}{\pi}\right)^{\frac{1}{2}}\int_{-\infty}^{\infty} u^2\,\mathrm{e}^{-au^2}\,\mathrm{d}u.$$

From the table of integrals on p. 121

$$\int_{-\infty}^{\infty} u^2\,\mathrm{e}^{-au^2}\,\mathrm{d}u=\frac{1}{2}\left(\frac{\pi}{a^3}\right)^{\frac{1}{2}}.$$

Hence

$$\bar{\varepsilon}=\frac{m}{2}\left(\frac{a}{\pi}\right)^{\frac{1}{2}}\frac{1}{2}\left(\frac{\pi}{a^3}\right)^{\frac{1}{2}}=\frac{m}{4}\times\frac{1}{a}=\frac{m}{4}\frac{2kT}{m},$$

or

$$\bar{\varepsilon}=\frac{kT}{2}. \tag{4.48}$$

Similarly the fraction of molecules with velocity components between u and $u+\mathrm{d}u$ and between v and $v+\mathrm{d}v$ is simply the product of the two individual probabilities.

$$\frac{\mathrm{d}n}{n}=\left(\frac{m}{2\pi kT}\right)\exp\left[-\frac{m(u^2+v^2)}{2kT}\right]\mathrm{d}u\,\mathrm{d}v. \tag{4.49}$$

Finally the fraction of molecules with velocity components between u and $u+\mathrm{d}u$, v and $v+\mathrm{d}v$, w and $w+\mathrm{d}w$ is

$$\frac{\mathrm{d}n}{n}=\left(\frac{m}{2\pi kT}\right)^{\frac{3}{2}}\exp\left[-\frac{m(u^2+v^2+w^2)}{2kT}\right]\mathrm{d}u\,\mathrm{d}v\,\mathrm{d}w. \tag{4.50}$$

Generally we are not interested in the dependence of n on the individual components of velocity. For practically all purposes we wish rather to know the number of molecules with velocities between c and $c+\mathrm{d}c$, irrespective of direction. We may calculate this easily by drawing a sphere of

Figure 4.10. (*a*) Velocity vector diagram for molecules with velocities between c and $c+\mathrm{d}c$. This is given by the numbers of points ending the velocity vectors which lie within the spherical shell of radii c and $c+\mathrm{d}c$. (*b*) Maxwellian velocity distribution for a three-dimensional gas. The shaded band gives the fraction of molecules with velocities between c and $c+\mathrm{d}c$. The approximate positions of the most probable velocity c_m, the mean velocity $\bar{c}$ and the root mean square velocity $(\overline{c^2})^{\frac{1}{2}}$ are indicated.

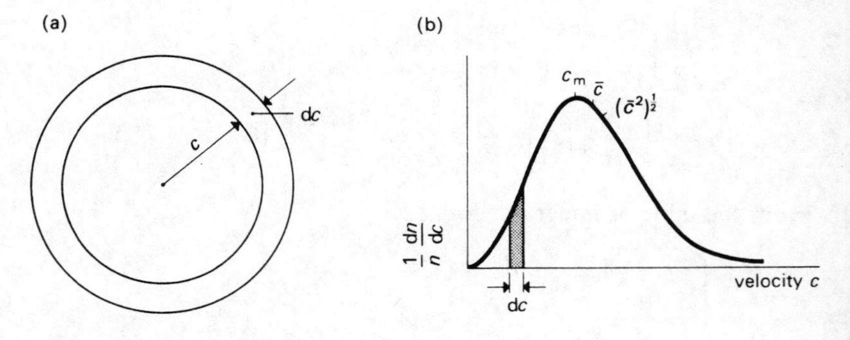

radius c and another of radius $c+\mathrm{d}c$ (see figure 4.10(a)). The number of velocity vectors ending in the spherical shell is then the number required. In velocity space the volume of the shell is $4\pi c^2\,\mathrm{d}c$ and this replaces $\mathrm{d}u\,\mathrm{d}v\,\mathrm{d}w$ in equation (4.50). We then obtain

$$\frac{1}{n}\frac{\mathrm{d}n}{\mathrm{d}c}=4\pi c^2\left(\frac{m}{2\pi kT}\right)^{\frac{3}{2}}\exp\left(-\frac{mc^2}{2kT}\right). \tag{4.51}$$

This is plotted in figure 4.10(b). We note that in any integration to give total quantities, c ranges from zero to ∞, not from $-\infty$ to $+\infty$.

[The reader, if he is interested only in the three-dimensional gas, may pass at once from equation (4.50) to equation (4.51) by writing equation (4.40) in the form

$$\frac{\mathrm{d}n}{n}=C\exp\left(-\frac{mc^2}{2kT}\right)4\pi c^2\,\mathrm{d}c.$$

By specifying that $\int_{c=0}^{c=\infty}(\mathrm{d}n/n)=1$, he will find that C has the value $(m/2\pi kT)^{\frac{3}{2}}$. This results in equation (4.51).]

4.4.2 *Average velocities*

Rewriting equation (4.51) in the form

$$\frac{\mathrm{d}n}{n}=4\pi\left(\frac{a}{\pi}\right)^{\frac{3}{2}}\mathrm{e}^{-ac^2}c^2\,\mathrm{d}c, \tag{4.52}$$

and using, if required, the integrals on p. 121, we may calculate the following three representative velocities: the most probable velocity c_{m}, i.e. the velocity for which equation (4.52) is a maximum, the average velocity $\bar{c}$, and the root mean square velocity $(\overline{c^2})^{\frac{1}{2}}$. We have:

most probable

$$c_{\mathrm{m}}=\sqrt{2}\left(\frac{kT}{m}\right)^{\frac{1}{2}}, \tag{4.53}$$

average

$$\bar{c}=\int_0^\infty c\frac{\mathrm{d}n}{n}=\left(\frac{8}{\pi}\right)^{\frac{1}{2}}\left(\frac{kT}{m}\right)^{\frac{1}{2}}, \tag{4.54}$$

r.m.s.

$$(\overline{c^2})^{\frac{1}{2}}=\left(\int_0^\infty\frac{c^2\,\mathrm{d}n}{n}\right)^{\frac{1}{2}}=\sqrt{3}\left(\frac{kT}{m}\right)^{\frac{1}{2}}. \tag{4.55}$$

Before leaving this we note one of the confusing features of probability distributions. According to equation (4.39) the population density is a maximum when $c=0$. This is also true, as we saw above of the velocity distribution in a 'one-dimensional' gas. As soon, however, as we specify velocity elements in two or three dimensions the most probable velocity (as distinct from the probability density) no longer corresponds to zero velocity. This is specifically brought out in equation (4.51) and figure 4.10 for a three-dimensional gas. A similar point will arise when we discuss the behaviour of rubber molecules in Chapter 7.

4.4.3 *Number of molecules striking a surface per second*

The number of molecules striking a square metre of surface per second may be directly obtained from equation (4.45). If the surface considered is at right angles to the u direction the v and w components are not involved. Consider molecules with velocity u, $u+\mathrm{d}u$; we construct a prism of unit cross-section and length u. Then all the molecules in this volume will reach the surface in one second. This number is $u\,\mathrm{d}n$ where $\mathrm{d}n$ is the number per m^3 with velocity u in the sense *towards* the surface. The total number required is the integral of $u\,\mathrm{d}n$ for all values of u from 0 to infinity: this choice of limits ensures that we are considering only those molecules moving towards the surface. The number is

$$n\int_0^\infty \left(\frac{m}{2\pi kT}\right)^{\frac{1}{2}} \exp\left(-\frac{mu^2}{2kT}\right) u\,\mathrm{d}u \tag{4.56a}$$

$$= n\left(\frac{m}{2\pi kT}\right)^{\frac{1}{2}}\left(-\frac{kT}{m}\right)\exp\left[-\frac{mu^2}{2kT}\right]_0^\infty \tag{4.56b}$$

$$= n\left(\frac{kT}{2\pi m}\right)^{\frac{1}{2}} = \frac{n}{4}\left(\frac{8}{\pi}\times\frac{kT}{m}\right)^{\frac{1}{2}}. \tag{4.56c}$$

Comparison with equation (4.54) shows that this is equal to $\frac{1}{4}n\bar{c}$.

4.4.4 *Maxwell's derivation of the velocity distribution in a gas*

The velocity distribution derived above makes use of the Boltzmann distribution. It is interesting to see how Maxwell derived the velocity distribution before the more general ideas of Boltzmann had been developed. The following is based on Maxwell's paper published in 1860.

Let n be the total number of molecules, and let u, v, w, be the components of the velocity of each molecule in three rectangular directions. Then if the number of molecules for which velocity component u lies between u and $u+\mathrm{d}u$ is $nf(u)\,\mathrm{d}u$, where $f(u)$ is the function to be determined, the number of molecules for which v lies between v and $v+\mathrm{d}v$ will be $nf(v)\,\mathrm{d}v$ and similarly for w, where f always stands for the same function since there is no preferred direction in the gas.

Now the existence of the velocity u does not in any way affect that of the velocities v and w since these are all at right angles to each other and independent. Consequently the number of molecules whose velocity components lie simultaneously between u and $u+\mathrm{d}u$, between v and $v+\mathrm{d}v$ and between w and $w+\mathrm{d}w$ is

$$\mathrm{d}n = nf(u)f(v)f(w)\,\mathrm{d}u\,\mathrm{d}v\,\mathrm{d}w. \tag{4.57}$$

Let this now refer to all those molecules which have a resultant velocity c, where

$$c^2 = u^2 + v^2 + w^2. \tag{4.58}$$

We now consider the condition under which the components u, v, w, can vary whilst c remains constant. This is found by differentiating equation (4.58) and putting $\mathrm{d}c=0$; we obtain

$$u\,\mathrm{d}u + v\,\mathrm{d}v + w\,\mathrm{d}w = 0. \tag{4.59}$$

Since no direction is preferred over any other, it follows that $\mathrm{d}n$ in equation (4.57) must remain constant whatever the individual values of u, v and w, provided these satisfy equation (4.59). This implies that

$$\frac{\mathrm{d}}{\mathrm{d}u}(\mathrm{d}n)\,\mathrm{d}u + \frac{\mathrm{d}}{\mathrm{d}v}(\mathrm{d}n)\,\mathrm{d}v + \frac{\mathrm{d}}{\mathrm{d}w}(\mathrm{d}n)\,\mathrm{d}w = 0. \tag{4.60}$$

From equation (4.57) this gives

$$f'(u)\,\mathrm{d}u\,f(v)f(w) + f'(v)\,\mathrm{d}v\,f(u)f(w) + f'(w)\,\mathrm{d}w\,f(u)f(v) = 0,$$

where $f'(u)$ is the differential of $f(u)$ with respect to u, etc. Dividing throughout by $f(u)f(v)f(w)$ we obtain

$$\frac{f'(u)}{f(u)}\,\mathrm{d}u + \frac{f'(v)}{f(v)}\,\mathrm{d}v + \frac{f'(w)}{f(w)}\,\mathrm{d}w = 0. \tag{4.61}$$

In order to solve equations (4.59) and (4.61), we multiply equation (4.59) by an arbitrary constant λ and add it to equation (4.61). The result is

$$\left(\frac{f'(u)}{f(u)} + \lambda u\right)\mathrm{d}u + \left(\frac{f'(v)}{f(v)} + \lambda v\right)\mathrm{d}v + \left(\frac{f'(w)}{f(w)} + \lambda w\right)\mathrm{d}w = 0.$$

Each term must identically be equal to zero, and since $\mathrm{d}u$, $\mathrm{d}v$, $\mathrm{d}w$, although very small are not themselves zero, the quantities in the brackets must be zero. Hence

$$\frac{f'(u)}{f(u)} + \lambda u = 0 \quad \text{or} \quad \frac{f'(u)}{f(u)} = -\lambda u.$$

Integrating this relation we have

$$\ln f(u) = -\frac{\lambda u^2}{2} + A,$$

where A is an integration constant. This may be written in the form

$$f(u) = B \exp\left(-\frac{\lambda u^2}{2}\right).$$

As Maxwell remarks, if λ were negative, the number of molecules would be infinite, so consequently, λ must be positive. The number of molecules with velocities between u and $u + \mathrm{d}u$ then becomes

$$\mathrm{d}n = nf(u)\,\mathrm{d}u = nB \exp\left(-\frac{\lambda u^2}{2}\right).$$

There is no way of establishing the value of λ except by calculating the mean square velocity $\overline{c^2}$ and making use of the gas equation: $\frac{1}{3}m\overline{c^2} = kT$. It is then found that λ has the value m/kT. The result is

$$\mathrm{d}n = nB \exp\left(-\frac{mu^2}{2kT}\right) \mathrm{d}u,$$

which is identical with equation (4.42).

4.4.5 *Experimental determination of velocity distribution*

In most of the earlier experimental studies of the velocity distribution, a furnace was used to produce a vapour of metal atoms, which behave like a monatomic gas at the temperature of the enclosure. They were allowed to condense on a cold surface and, by the ingenious use of moving shutters, atoms with different velocities were caught at different points on the surface. The intensity of condensed atoms provides then a measure of the relative number of atoms within that velocity range. One method due to I. F. Zartman (1931) and C. C. Ko (1934) is illustrated in figure 4.11(a). Atoms of bismuth are produced by a furnace and the vapour is collimated by a series of slits, S_1, S_2, S_3; the vapour beam reaches a drum rotating at a speed of 6000 rev. per min. and can only enter the drum at the slit S. The atoms then strike the plate P where they condense; the fastest ones reaching A, and slower ones B and the slowest C.

A more elegant and refined experiment by I. Esterman, O. C. Simpson and O. Stern (1947) gave results shown in figure 4.11(b). The agreement with the Boltzmann distribution is surprisingly good.

4.5 Thermal fluctuations in a gas

We discuss here the fluctuations in density of a gas resulting from the wide range of molecular velocities, mean free paths, etc. Instead of tackling this analytically in terms of these parameters we shall proceed in the following way. We consider what would happen if a small change in volume and pressure occurred in a gas: we calculate the stored energy; we find that this is proportional to the square of the volume change. Consequently if such a change is due to thermal fluctuations the Boltzmann principle can be applied and the average value of the stored energy can be put equal to $\frac{1}{2}kT$. We start with N molecules of a gas in a cylinder at volume V, with a piston exerting an equilibrium pressure P. Then

$$PV = NkT. \tag{4.62}$$

If a small pressure p is added externally to the piston the volume of the gas will diminish by v. Under overall isothermal conditions we have

$$(P+p)(V-v) = PV.$$

Hence

$$p = \frac{Pv}{V-v} \approx \frac{Pv}{V}. \tag{4.63}$$

If p is increased to produce a further incremental change dv, the work

Figure 4.11. (*a*) Method due to Zartman and Ko for determining the velocity distribution of bismuth atoms: the atoms are evaporated, collimated by the slits S_1, S_2, S_3 and enter the rotating drum. They fall on the cold plate P where they condense. (*b*) Results obtained in a later experiment by Esterman, Simpson and Stern.

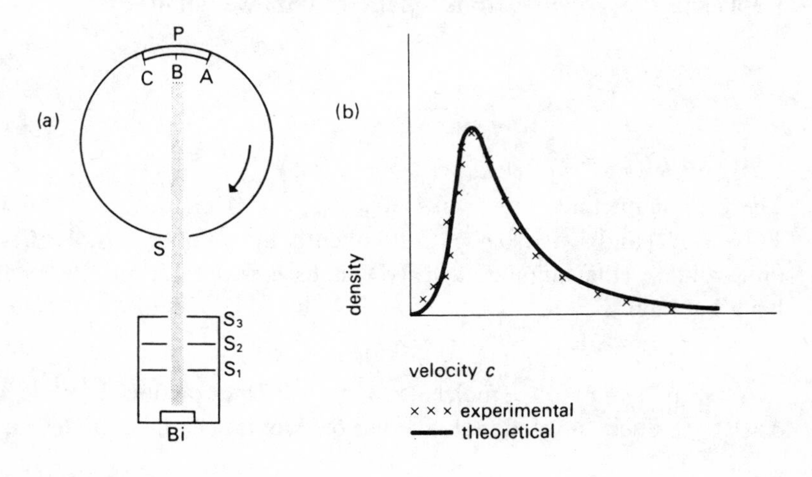

done on the gas is

$$p\,\mathrm{d}v = \frac{P}{V}\,v\,\mathrm{d}v. \tag{4.64}$$

In increasing p from zero to its final value p_0, producing a final volume change v_0, the work done on the gas is

$$w = \int_0^{v_0} p\,\mathrm{d}v = \frac{P}{2V}\,v_0^2. \tag{4.65}$$

It is seen that work is done on the gas whether v_0 is positive or negative. The gas behaves as a 'spring', i.e. as an elastic medium of bulk modulus P.

We may forget about the piston and cylinder and accept equation (4.65) as a description of the extra energy acquired by a volume of gas V, nominal pressure P, if it undergoes a volume change v_0. We now consider the equilibrium between a large number of such volumes, separated from one another by a perfectly flexible, thermally conducting envelope. Then as a result of random molecular motions the volume of one envelope will be somewhat greater, another somewhat less than the equilibrium volume V. The fluctuating energy of each envelope is given by equation (4.65). We can at once apply the Boltzmann concept by noting that this relation gives an energy proportional to the square of a parameter (v_0). The envelope behaves as though its variation in volume corresponds to one degree of freedom. Consequently if $\overline{v_0^2}$ is the mean square fluctuation in volume we have

$$\frac{P}{2V}\,\overline{v_0^2} = \tfrac{1}{2}kT. \tag{4.66}$$

Replacing P by NkT/V from equation (4.62) we obtain

$$\frac{\overline{v_0^2}}{V^2} = \frac{1}{N} \tag{4.67}$$

or

$$(v_0)_{\mathrm{r.m.s.}} = V/N^{\frac{1}{2}}.$$

The volume fluctuates between $V + (v_0)_{\mathrm{r.m.s.}}$ and $V - (v_0)_{\mathrm{r.m.s.}}$, i.e. between $V(1 + 1/N^{\frac{1}{2}})$ and $V(1 - 1/N^{\frac{1}{2}})$. Consequently the number of molecules per unit volume (the 'number density') undergoes a fractional fluctuation lying between

$$(1 - 1/N^{\frac{1}{2}}) \text{ and } (1 + 1/N^{\frac{1}{2}}). \tag{4.68}$$

At s.t.p. $N = 3 \times 10^{28}$ molecules per m^3. The fraction $1/N^{\frac{1}{2}}$ is only 6×10^{-15}. Such a fractional fluctuation would be quite undetectable.

However, if one considers a very small volume containing on average, say, 100 molecules, the number will fluctuate between $100(1-\frac{1}{10})$ and $100(1+\frac{1}{10})$, i.e. between 90 and 110.

4.6 Thermal energy of molecules

4.6.1 *Specific heat capacities, number of degrees of freedom*

We have already seen that for a perfect gas in thermal equilibrium each molecule possesses an average thermal energy of translation of amount $\frac{3}{2}kT$. Since a molecule has 3 independent degrees of translational freedom we may associate energy $\frac{1}{2}kT$ with each degree of freedom. This may be generalized to apply to all particles in thermal equilibrium; however, for macroscopic bodies the resulting velocity is rather small. For example the mean thermal velocity of a cricket ball in equilibrium with its surroundings at room temperature (say 300 K) is about 10^{-10} m s^{-1}.

The result may be generalized even further. If we define the number of degrees of freedom as the number of independent squared terms that enter into the total energy of a particle, then each has associated with it an average thermal energy of amount $\frac{1}{2}kT$. For example the kinetic energy due to translation is determined solely by (linear velocity)^2 in each of these independent coordinates; each degree of freedom has thermal energy $\frac{1}{2}kT$. The energy of rotation of a body about a specified axis is determined by (angular velocity)^2, so that each of these degrees of freedom has energy $\frac{1}{2}kT$. Vibrations present a special problem. If the equation of motion, for a body of mass m, is

$$\ddot{x}=-\omega^2 x, \quad \text{with solution } x=A\sin\omega t, \tag{4.69}$$

the kinetic energy at any instant is $\frac{1}{2}m\dot{x}^2$, whilst the potential energy is easily shown to be equal to $\frac{1}{2}m\omega^2x^2$. By substituting in equation (4.69) we find that the sum of kinetic and potential energy is $\frac{1}{2}mA^2\omega^2$, so it is independent of x and $\dot{x}$. Nevertheless at any instant the kinetic energy is proportional to $\dot{x}^2$ and the potential energy to x^2. Each of these contributes $\frac{1}{2}kT$ to the average thermal energy.

Summarizing we may write: thermal energy per degree of translational freedom $=\frac{1}{2}kT$, per degree of rotational freedom $=\frac{1}{2}kT$ and per vibrational mode (equivalent to two degrees of freedom†) $=kT$.

4.6.2 *Internal energy of a gas*

If we consider the energy which arises solely from thermal motion (not excited electronic states), the internal energy u per molecule of a gas

† Some authorities refer to vibration as possessing a *single* degree of freedom of energy kT.

may be deduced for monatomic, diatomic and polyatomic molecules.

Monatomic gas: only 3 translations $u=\frac{3}{2}kT$ per molecule.

Diatomic gas: 3 translations

2 rotations

$u=\frac{5}{2}kT$ per molecule plus any vibrational energy.

The third rotation about the axis of the molecule does not occur for reasons to be described later.

Polyatomic gas: 3 translations

3 rotations (if non-linear)

$u=\frac{6}{2}kT$ plus any vibrational energy.

These results are summarized in Table 4.1.

For monatomic gases the agreement is excellent. For diatomic gases the agreement is good, but the value given for chlorine seems to indicate that it possesses additional vibrational energy at room temperature. With polyatomic molecules the behaviour is very complicated but it would seem that, in general $C_V > 6 \times R/2\ \text{mol}^{-1}$.

Table 4.1. *Internal energy U and specific heat capacities C of gases*

Gas	Monatomic	Diatomic	Polyatomic
f_{tr}	3	3	3
f_{rot}	0	2	3
f_{vib}	0	assumed 0	assumed 0
Total	3	5	6
u/molecule	$\frac{3}{2}kT$	$\frac{5}{2}kT$	$\frac{6}{2}kT$
$U\ \text{mol}^{-1}$	$\frac{3}{2}RT$	$\frac{5}{2}RT$	$\frac{6}{2}RT$
$C_V=\left(\frac{\partial U}{\partial T}\right)_V$	$\frac{3}{2}R$	$\frac{5}{2}R$	$3R$
$C_P=C_V+R$	$\frac{5}{2}R$	$\frac{7}{2}R$	$4R$
$\gamma=\frac{C_P}{C_V}$	$\frac{5}{3}$	$\frac{7}{5}$	$\frac{4}{3}$
C_V in units of $R/2\ \text{mol}^{-1}$:			
Theoretical	3	5	6
Observed	Ar : 3	H_2 : 4.9	H_2S : 6
	He : 3	O_2 : 5.0	CS_2 : 10
		N_2 : 4.6	
		Cl_2 : 6	

In general, if one knows the number of degrees of freedom f which are 'active' one can write

$$U=f\times\frac{RT}{2},$$

so that

$$C_V=\tfrac{1}{2}Rf$$

and

$$\gamma=\frac{C_P}{C_V}=1+\frac{2}{f}. \tag{4.70}$$

We may express the above discussion in a more general way. If a polyatomic molecule contains N atoms we know that in the free state there are $3N$ degrees of translational freedom. We assume that in the polyatomic molecule these degrees of freedom are retained though they must be shared out amongst translations, rotations and vibrations.

The molecule as a whole will possess three degrees of translational freedom and three degrees of rotational freedom. However if the molecule is linear such as O–O or O–C–O one rotational degree of freedom is inaccessible (see below). We are thus left with

$3N-6$ vibrational degrees of freedom in general
$3N-5$ vibrational degrees of freedom if molecule is linear.

For chlorine $N=2$: the molecule is linear. In this case we have $3N-5=1$ vibrational degree of freedom. The internal energy is thus due to 3 translations$=3(\frac{1}{2}kT)$, 2 rotations$=2(\frac{1}{2}kT)$ and 1 vibration$=2(\frac{1}{2}kT)$, total$\approx 7(\frac{1}{2}kT)$. Hence $C_V=7(R/2)\ \text{mol}^{-1}\ \text{K}^{-1}$. Consequently $C_P=9(R/2)\ \text{mol}^{-1}\ \text{K}^{-1}$, $\gamma=\frac{9}{7}=1.29$. The observed value at 1000 K is 1.30. With oxygen or hydrogen the vibrational mode cannot be excited and the observed value of γ is larger. For O–C–O, $N=3$, again a linear molecule; we have $3N-5=4$ degrees of vibration. These consist of two sets of C–O vibrations and two flexural vibrations about the C atom. Thus there are three translations, two rotations and four vibrations giving a value of $C_V=13(R/2)\ \text{mol}^{-1}\ \text{K}^{-1}$. Then $C_P=15(R/2)\ \text{mol}^{-1}\ \text{K}^{-1}$ and $\gamma=\frac{15}{13}=1.15$. The observed value at room temperature is 1.30; at 800 K it is 1.20 which is near the calculated value.

For CCl_4, $N=5$. Here $3N-6=9$ degrees of vibrational freedom, energy $9kT$; three rotations energy $\frac{3}{2}kT$; three translations, energy $\frac{3}{2}kT$. Total energy$=12kT$. Hence $C_V=12R\ \text{mol}^{-1}\ \text{K}^{-1}$ and $C_P=13R\ \text{mol}^{-1}\ \text{K}^{-1}$. Hence $\gamma=\frac{13}{12}=1.08$. The observed value for γ is 1.13. However, for CH_4 the observed value of γ is 1.31.

The above calculations assume that all available modes are excited. As we shall see in a subsequent section this is not always so.

4.6.3 *Gas kinetic theory of adiabatic expansion*

Before we go on to discuss in greater detail the vibrational and rotational energy of a molecule, we may at once use the ideas expressed in the previous section to deduce the adiabatic gas equation for monatomic and polyatomic gases.

In an adiabatic expansion all the energy is taken from the internal energy of the gas. We first consider a monatomic gas where the internal energy arises solely from the kinetic energy of translation.

From the kinetic theory we have:

$$PV = \tfrac{2}{3}(\text{total kinetic energy } Z)$$

$$Z = \tfrac{3}{2}PV = \text{internal energy } U.$$

The work done by the gas in an adiabatic expansion is

$$P\,\mathrm{d}V = -\mathrm{d}U = -\mathrm{d}Z = -\mathrm{d}(\tfrac{3}{2}PV).$$

Therefore

$$P\,\mathrm{d}V = -\tfrac{3}{2}P\,\mathrm{d}V - \tfrac{3}{2}V\,\mathrm{d}P$$

or

$$\tfrac{5}{2}P\,\mathrm{d}V + \tfrac{3}{2}V\,\mathrm{d}P = 0$$

or

$$5\frac{\mathrm{d}V}{V} + 3\frac{\mathrm{d}P}{P} = 0.$$

Hence

$$PV^{\frac{5}{3}} = \text{constant}. \tag{4.71}$$

For a diatomic gas possessing two rotational degrees of freedom, the internal energy is

$$U = \text{kinetic energy} + \text{rotational energy}$$

$$= \tfrac{3}{2}PV + RT = \tfrac{3}{2}PV + PV$$

$$= \tfrac{5}{2}PV.$$

For an adiabatic expansion $P\,\mathrm{d}V = -\mathrm{d}(\tfrac{5}{2}PV)$. This leads, as above to

$$PV^{\frac{7}{5}} = \text{constant}. \tag{4.72}$$

4.6.4 *Rotational and vibrational degrees of freedom*

At room temperature the specific heat capacity C_V of hydrogen is about $5 \times R/2\ \mathrm{mol}^{-1}$. As the temperature is lowered C_V decreases until at 50 K, C_V has a value of only about $3\times R/2\ \mathrm{mol}^{-1}$. It is apparent that the rotational degrees of freedom have gradually disappeared and that at about 50 K the molecule has only translational energy. The reason for this is that, according to the quantum theory, rotors can possess only discrete energy levels. For a symmetrical diatomic molecule rotating about an axis, for which the moment of inertia is I, the possible energy levels are

$$E_{\mathrm{rot}}=n(n+1)\frac{h^2}{8\pi^2 I}, \qquad (4.73)$$

where h is Planck's constant ($6.6\times 10^{-34}\ \mathrm{J\ s}^{-1}$) and n is an integer. The smallest value of rotational energy permitted occurs when $n=1$ and has the value

$$(E_{\mathrm{rot}})_{\mathrm{min}}=\frac{h^2}{4\pi^2 I}. \qquad (4.74)$$

For a hydrogen molecule, I about an axis normal to the bond is approximately $4.7\times 10^{-48}\ \mathrm{kg\ m}^2$. Therefore

$$(E_{\mathrm{rot}})_{\mathrm{min}}=2\times 10^{-21}\ \mathrm{J}.$$

But the translational energy $E_{\mathrm{tr}}=\frac{3}{2}kT=2\times 10^{-23}T$ J, so that at 50 K

$$E_{\mathrm{tr}}=10^{-21}\ \mathrm{J}. \qquad (4.75)$$

Therefore the translational energy is appreciably less than the lowest rotational energy level. By Boltzmann's principle the fraction of molecules capable of acquiring rotational energy 2×10^{-21} J at 50 K will be

$$\exp\left(-\frac{\Delta\varepsilon}{kT}\right)\approx\exp(-3)\approx\tfrac{1}{20}. \qquad (4.76)$$

Consequently below 50 K very few of the molecules will take up rotational motion. At room temperature (300 K), kT will be large compared with the first rotational energy level, so a large fraction of the molecules will be able to take up the first and higher rotational energy levels.

Equation (4.74) also shows why rotating about the axis itself does not occur. The moment of inertia I is so small that the lowest rotational energy state, $h^2/4\pi^2 I$, is enormous and can never be reached before thermal dissociation occurs.

Similar considerations apply to the vibrational degrees of freedom. If the vibrational frequency is ν, the first energy state (ignoring zero-point energy) demands energy of amount equal to $h\nu$. For the hydrogen molecule $\nu\approx 2.6\times 10^{14}\ \mathrm{s}^{-1}$, so that $h\nu$ is about 2×10^{-19} J. For an appreciable fraction of molecules to be excited to this level kT must have about the same value, so that T must exceed 10 000 K. The molecule dissociates long before this. By contrast the chlorine molecule, which has a much lower natural frequency ν, is able to take up appreciable vibrational energy at temperatures above 600 K so that its specific heat capacity approaches $C_V = 7 \times R/2\ \text{mol}^{-1}$ at elevated temperatures (see figure 4.12).

4.6.5 *Calculation of vibrational energy as a function of temperature*

We reproduce here the standard treatment for a simple oscillator of natural frequency ν. According to quantum theory the possible energy levels are

$$E_{\text{vib}} = (n + \tfrac{1}{2})h\nu, \tag{4.77}$$

where n is an integer.

The zero-point energy $\frac{1}{2}h\nu$ is the vibrational energy the molecule possesses even at absolute zero, but since this is a constant we may ignore it at this stage of the computation and add it at the end. We put

$$E_{\text{vib}} = nh\nu. \tag{4.78}$$

Applying the Boltzmann function, the number of states with quantum number n is

$$N_n = A \exp\left(-\frac{nh\nu}{kT}\right) \tag{4.79}$$

$$= A \exp(-n\beta), \quad \text{where} \quad \beta = \frac{h\nu}{kT}.$$

Figure 4.12. Variation of specific heat capacity with temperature for H_2 and for Cl_2.

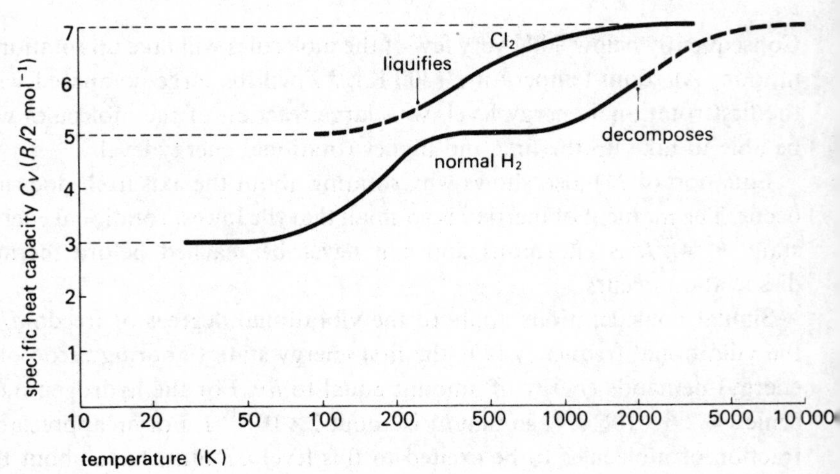

The total number N of atoms is the sum of all N_n from $n=0$ to $n=\infty$,

$$N=A[1+\exp(-\beta)+\exp(-2\beta)+\exp(-3\beta)+\cdots]$$
$$=A[1+\exp(-\beta)+(\exp(-\beta))^2+(\exp(-\beta))^3+\cdots]$$
$$=\frac{A}{1-\exp(-\beta)}. \tag{4.80}$$

This gives A so that N_n becomes

$$N_n=N[1-\exp(-\beta)][\exp(-n\beta)]. \tag{4.81}$$

The total energy is the sum of N_nE_{vib} where $E_{\text{vib}}=nh\nu$.

$$U=N[1-\exp(-\beta)]$$
$$\times[\exp(-n\beta)]nh\nu \text{ summed for all values of } n$$
$$=N[1-\exp(-\beta)]$$
$$\times\{h\nu[\exp(-\beta)]+2h\nu[\exp(-\beta)]^2+3h\nu[\exp(-\beta)]^3+\cdots\}$$
$$=[1-\exp(-\beta)]h\nu[\exp(-\beta)]\{1+2[\exp(-\beta)]$$
$$+3[\exp(-\beta)]^2+\cdots\}$$
$$=N[1-\exp(-\beta)]h\nu[\exp(-\beta)]\frac{1}{[1-\exp(-\beta)]^2}$$
$$=Nh\nu\frac{[\exp(-\beta)]}{[1-\exp(-\beta)]}=\frac{Nh\nu}{\exp(\beta)-1}. \tag{4.82}$$

If we include the zero-point energy $\frac{1}{2}h\nu$ we have

$$U=Nh\nu\{\tfrac{1}{2}+[\exp(\beta)-1]^{-1}\}. \tag{4.83}$$

At high temperatures $\exp\beta=\exp(h\nu/kT)\approx 1+h\nu/kT$ so that

$$U\approx Nh\nu\left(\frac{1}{2}+\frac{kT}{h\nu}\right),$$
$$U=\tfrac{1}{2}Nh\nu+NkT. \tag{4.84}$$

Thus apart from zero-point energy each vibration carries average energy kT, so that the specific heat capacity contribution is $Nk=R$ for each vibration. At lower temperatures $[\exp(h\nu/kT)-1]^{-1}$ and therefore the specific heat capacity falls off and tends to zero when $T=0$ (see figure 4.13). This decrease in vibrational energy with temperature reproduces essentially the right-hand features of the specific heat capacity curves of hydrogen and chlorine in figure 4.12. We may also note in passing that this concept was applied by A. Einstein to explain the specific heat capacity of solids. He suggested that each atom has 3 independent vibrational modes of constant frequency ν. The specific heat capacity (in this case it assumes no volume change and is therefore C_V) is $3R$ at high temperatures and falls to zero at low temperatures. We discuss this in greater detail in a later chapter.

4.6.6 *Rotational energy*

For a symmetrical diatomic molecule rotating about an axis for which the moment of inertia is I, the possible energy levels E_{rot} are as given in equation (4.73). The application of the Boltzmann distribution is complicated because each state of quantum number n consists of g states of practically identical energy. The degeneracy factor g has the value

$$g = 2n + 1. \tag{4.85}$$

One can build up the Boltzmann relation as before where the basic exponential term is

$$\exp\left(-\frac{E_{\text{rot}}}{kT}\right) \text{ or } \exp\left(-n(n+1)\frac{h^2}{8\pi^2 I}\frac{1}{kT}\right).$$

The final integral has to be summed numerically. For high temperatures the energy for N rotors becomes NkT. This is the classical result for a rotor with two degrees of freedom. In general the rotational energy falls

Figure 4.13. Variation of vibrational specific heat capacity with temperature for a simple oscillation of natural frequency ν. Specific heat capacity is in units of $R\,\text{mol}^{-1}$, temperature in units of $k/h\nu$.

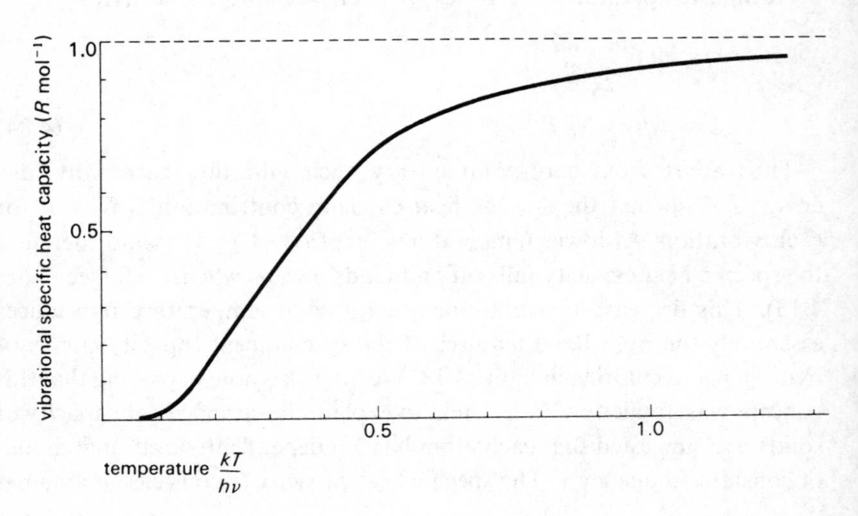

off from NkT when

$$kT < \frac{h^2}{8\pi^2 I}. \tag{4.86}$$

4.6.7 *Translational energy*

For a gas confined within a container, quantization of translational energy does occur, but the number of energy states is enormous and the spacing between energy levels is minute. We may therefore consider all energy states to be available as a continuous distribution, so that the classical Boltzmann distribution applies. Consequently each degree of translational freedom involves energy $\frac{1}{2}kT$ per molecule under all practical conditions.

4.6.8 *Equipartition as a broader principle*

In all that we have so far discussed we have considered systems for which the energy is a function of squared terms: each of these acquires an average energy of $\frac{1}{2}kT$. What of the other energy functions? Dr C. Hall has pointed out the following interesting point.

Suppose the energy is proportional to x^m, say $E = \alpha x^m$, then the average energy is given by

$$\bar{E} = \frac{\int_0^\infty \alpha x^m \exp(-\alpha x^m/kT)\,\mathrm{d}x}{\int_0^\infty \exp(-\alpha x^m/kT)\,\mathrm{d}x}. \tag{4.87}$$

This may be solved as follows. Let us write

$$y^m = \alpha x^m/kT \tag{4.88}$$

so that

$$\mathrm{d}y = (\alpha/kT)^{1/m}\,\mathrm{d}x.$$

Then equation (4.87) becomes

$$\bar{E} = \frac{kT \int_0^\infty y^m \exp(-y^m)\,\mathrm{d}y}{\int_0^\infty \exp(-y^m)\,\mathrm{d}y} = kT\left(\frac{I_\mathrm{N}}{I_\mathrm{D}}\right). \tag{4.89}$$

The denominator I_D may be integrated by parts

$$I_\mathrm{D} = \int_0^\infty \exp(-y^m)\cdot 1 \cdot \mathrm{d}y \tag{4.90}$$

$$= [\exp(-y^m)y]_0^\infty + \int_0^\infty m y^{m-1} \exp(-y^m) y\,\mathrm{d}y.$$

The first term is zero. This leaves only the second term. Hence

$$I_{\mathrm{D}}=m\int_0^\infty y^m \exp(-y^m)\,\mathrm{d}y. \tag{4.91}$$

We note the integral in equation (4.91) corresponds exactly to the integral I_{N} in the numerator of equation (4.89). Thus

$$I_{\mathrm{D}}=mI_{\mathrm{N}}.$$

Consequently

$$E=kT\frac{I_{\mathrm{N}}}{I_{\mathrm{D}}}=\frac{1}{m}kT. \tag{4.92}$$

Typical examples are:

$E=\alpha x$, $\bar{E}=kT$;

this corresponds to a particle in a uniform gravitational field.

$E=\alpha x^2$, $\bar{E}=\frac{1}{2}kT$;

this corresponds to the classical definition of a degree of freedom

$E=\alpha x^3$, $\bar{E}=\frac{1}{3}kT$;

$E=\alpha x^4$, $\bar{E}=\frac{1}{4}kT$;

these do not correspond to common physical phenomena.

4.6.9 *The basic form of the Boltzmann distribution*

This section is included for completeness and may be omitted by the general reader.

In the preceding sections we have provided a rather makeshift version of the Boltzmann distribution. The full form relies on a very subtle theorem due to Liouville. We first consider an assembly of particles, atoms or molecules in equilibrium and specify that a given particle has momentum lying between p_x and $(p_x+\mathrm{d}p_x)$, p_y and $(p_y+\mathrm{d}p_y)$, p_z and $(p_z+\mathrm{d}p_z)$ and has coordinates lying between x and $(x+\mathrm{d}x)$, y and $(y+\mathrm{d}y)$, z and $(z+\mathrm{d}z)$. We then consider how a given particle (or number of particles) will enter and leave this volume element $\mathrm{d}p_x\,\mathrm{d}p_y\,\mathrm{d}p_z\,\mathrm{d}x\,\mathrm{d}y\,\mathrm{d}z$ of 'phase space'. Analysis shows that, for particles obeying the laws of classical mechanics, if a given particle occupies an element of phase space at a given time it will occupy another element of equal volume at a later time. This implies that all phase space elements of the same volume have the same probability of being occupied by a given particle. This is Liouville's theorem.

We may now divide phase space into a number of elementary cells of equal volume, i.e. equal *a priori* probability of occupancy, and consider how we could distribute all the particles of the system in these cells to give a constant (specified) total energy of the system and a maximum total probability, i.e. the one that in fact will occur. Boltzmann showed that the probability of finding a particle with a given momentum at a given place is proportional to $\exp(-E/kT)$ per unit elementary volume of phase space, where E is the *total* energy of the particle. Thus

$$\text{probability} = A \exp(-E/kT)\, \mathrm{d}p_z\, \mathrm{d}p_y\, \mathrm{d}p_z\, \mathrm{d}x\, \mathrm{d}y\, \mathrm{d}z. \tag{4.93}$$

$\mathrm{d}p_z\, \mathrm{d}p_y\, \mathrm{d}p_z$ ↓	$\mathrm{d}x\, \mathrm{d}y\, \mathrm{d}z$ ↓
volume of momentum space	volume of real space

For a particle with kinetic energy K (function of velocity, i.e. of momentum) and potential energy U (function of position) the probability is given by equation (4.93) with $E = K + U$. This can be written as

$$A \exp(-K/kT)\, \mathrm{d}p_x\, \mathrm{d}p_y\, \mathrm{d}p_z \times \exp(-U/kT)\, \mathrm{d}x\, \mathrm{d}y\, \mathrm{d}z. \tag{4.94}$$

↓	↓
Probability of having a given velocity	probability of being at a given place

This is therefore equivalent to the product of two independent probabilities. The splitting-up of the probability into the product of a number of independent terms is one of the characteristic features that follows from the exponential function in the Boltzmann distribution. Many of the calculations in the previous sections are clear examples of this. If no potential energy is involved (as in a perfect gas) we are left with the first term which provides the basis for the Maxwell velocity distribution. Noting that $\mathrm{d}p_x = m\, \mathrm{d}u$, etc., we obtain for the probability

$$A \exp(-E/kT)\, \mathrm{d}u\, \mathrm{d}v\, \mathrm{d}w. \tag{4.95}$$

This is *not* the same as

$$A \exp(-E/kT)\, \mathrm{d}E. \tag{4.96}$$

Indeed, it is generally incorrect to write the probability as

$$\exp(-E/kT)\, \mathrm{d}E.$$

There is only one simple case where it does apply, i.e. to a two-dimensional gas where we can replace $u^2 + v^2$ by, say, c_1^2 and take an annular element of velocity space $2\pi c_1\, \mathrm{d}c_1$. This element then corresponds to $(2\pi/m)\, \mathrm{d}E$.

The Boltzmann distribution has a quantum form which we have already used. Under quantum conditions we do not refer to values of momentum or position but describe the system as being in a quantum state with a probability proportional to $\exp(-E/kT)$.

4.7 Macroscopic examples of equipartition of energy

We conclude this chapter with four examples of thermal motion occurring in systems other than gases.

4.7.1 *Galvanometer mirror*

A delicately suspended galvanometer mirror can be seen to undergo small random oscillations which are due to its thermal energy. The mirror has a single degree of oscillation about its axis so that we attribute to its mean square velocity $(\overline{\omega^2})$, energy $\frac{1}{2}kT$, and to its mean square angular deflection $(\overline{\theta^2})$, energy $\frac{1}{2}kT$. We may write

$$\tfrac{1}{2}kT = \tfrac{1}{2}I(\overline{\omega^2}) = \tfrac{1}{2}\tau(\overline{\theta^2}), \tag{4.97}$$

where I is the moment of inertia, τ is the torsion constant. If we determine the torsional stiffness of the suspension we may then calculate $(\overline{\theta^2})$. This tells us the amount by which any galvanometer reading will fluctuate as a result of thermal motion. These fluctuations do not depend on the presence of air surrounding the mirrors; it is true that gas molecules will buffet the mirror and contribute to the oscillation, but they will also contribute to the damping, so the net effect is unchanged. The mirror behaves like a large molecule with a single mode of torsional oscillation. In a perfect vacuum the oscillations can be considered as arising from the random absorption and radiation of electromagnetic energy associated with the temperature T. The resulting behaviour is the same.

4.7.2 *Sedimentation*

As we saw earlier, particles in thermal equilibrium behave like gas molecules of large molecular weight. If they are in an evacuated container the sedimentation due to gravity is so rapid that they virtually all lie at the bottom of the container. If, however, they are suspended in a liquid of very nearly the same density the effective gravitational field is greatly reduced and there may be an appreciable variation in particle concentration with distance from the bottom of the container, as given by equation (4.35).

4.7.3 *Electrical noise*

The conduction electrons in a metal may be regarded as a gas with random velocities. These give rise to electrical 'noise'. The r.m.s. potential fluctuation across a resistor of resistance Ω is given by

$$\text{r.m.s. potential} = [4\Omega kT(f_2 - f_1)]^{\frac{1}{2}}, \tag{4.98}$$

where $f_2 - f_1$ is the bandwidth frequency over which the measurements are made. This result at once shows that there is a limit to useful amplification. If the original signal is too small (e.g. comparable with the random electrical noise) no amount of amplification will improve reception.

Similarly in the thermionic emission of current I_0 it may be shown that there is a fluctuation i in the emitted current given by

$$|\overline{i^2}|^{\frac{1}{2}} = \left(\frac{I_0 e}{\tau}\right)^{\frac{1}{2}}, \tag{4.99}$$

where e is the electronic charge and τ the time constant of the measuring system. This effect has indeed been used as an independent means of determining e; the result agrees with other methods to within 1 per cent.

4.7.4 *Brownian motion*

The English botanist R. Brown observed in 1827 that finely divided particles in suspension undergo ceaseless movement. This motion was analysed by Einstein in 1906; the following derivation is based on a treatment developed by P. Langevin in 1908. We first assume that the particles possess thermal kinetic energy in one dimension of amount $\frac{1}{2}kT$.

If at any instant the molecules of the liquid collide with an individual particle to produce a net force X in the x direction, we may write the equation of motion of the particle in the form

$$m\ddot{x} + 6\pi a\eta\dot{x} = X, \tag{4.100}$$

where m is the mass, a the radius of the particle (assumed spherical) and η the viscosity of the fluid. In this equation the first term represents the inertial force on the particle and the second the frictional force due to the viscous resistance of the fluid; this is Stokes' well-known relation for viscous forces on a sphere. The R.H. term represents the force due to molecular collisions. Thus from the point of view of collisions we treat the fluid as a collection of individual molecules, but from the point of view of the resistive forces we treat the fluid as a continuum. Finally, this equation refers only to motions of the particle produced by collisions; the intrinsic motion of the particle by virtue of its own kinetic energy ($\frac{1}{2}kT$) emerges from the analysis.

We multiply both sides of equation (4.100) by x and then average over a large number of collisions. Then for any fixed specified value of x, for every value of X there is an equal chance of $-X$ occurring so that $\sum xX=0$.

$$\sum m\ddot{x}x+6\pi a\eta\sum \dot{x}x=\sum xX=0 \tag{4.101}$$

or

$$m\overline{\ddot{x}x}+6\pi a\eta\overline{\dot{x}x}=0.$$

We note that $\mathrm{d}(\dot{x}x)/\mathrm{d}t=\ddot{x}x+\dot{x}^2=\ddot{x}x+u^2$, where u is the velocity in the x direction. Then

$$m\frac{\mathrm{d}}{\mathrm{d}t}\overline{(\dot{x}x)}-m\overline{u^2}+6\pi a\eta\overline{\dot{x}x}=0. \tag{4.102}$$

It turns out that the first term implies a transient at the beginning of the collision which has a negligible effect on the subsequent displacement.† Hence

$$6\pi a\eta\overline{\dot{x}x}=m\overline{u^2}=kT, \tag{4.103}$$

or

$$3\pi a\eta\frac{\mathrm{d}}{\mathrm{d}t}(\overline{x^2})=kT. \tag{4.104}$$

In a time interval τ the mean square displacement $\overline{\Delta x^2}$ is then

$$\overline{\Delta x^2}=\frac{kT\tau}{3\pi a\eta}=\frac{RT\tau}{3\pi a\eta N_{\mathrm{A}}}, \tag{4.105}$$

where N_{A} is Avogadro's number. A given particle is observed say every 30 seconds and the displacement Δx in the x direction is observed for each interval of time. These values are squared and the mean value is calculated; this gives $\overline{\Delta x^2}$. This is an average path for the time interval $\tau=30$ seconds. The value of $\overline{\Delta x^2}$ so found may be compared with the

† More precisely we may write $\dot{x}x$ as $\frac{1}{2}\,\mathrm{d}x^2/\mathrm{d}t$ which we may call z. Equation (4.102) then becomes

$$m\dot{z}+6\pi a\eta z=m\overline{u^2}=kT.$$

The solution is

$$z=\frac{kT}{6\pi a\eta}+A\exp\left(\frac{-6\pi a\eta t}{m}\right).$$

where A is a constant of integration. The quantity $6\pi an/m$ is extremely large so that even for minute values of t the exponential term becomes negligible. During the total time of the collision, z is thus given essentially by the first term on the RHS of the equation. This is indeed the same equation as (4.104).

right-hand term of equation (4.105). Experiments show that the results agree very well with calculation.

There are two points of interest. First, the displacement does not depend on the mass of the particle (only on its radius). Secondly, although the individual particle has intrinsic motion, which has nothing to do with the surrounding liquid, the distance it moves about an equilibrium position *is* determined by the viscous damping of the liquid. In a descriptive way we may say that a viscous force $6\pi a\eta\dot{x}$ acting over a distance x consumes the thermal energy $\frac{1}{2}mu^2$. The particle then reacquires its thermal energy and the process is repeated. Apart from a numerical factor this is essentially the result implied by equation (4.103).

If then equation (4.103) is a description of the way in which viscous damping consumes thermal energy, is it not possible to apply it to the behaviour of a gas molecule within its own gas? We have already derived a relation for the viscosity of a gas, namely $\eta=\frac{1}{3}mn\bar{c}\lambda$. However, this relation can only be used when the distance between the moving bodies is larger than the mean free path λ. On the other hand, in the derivation of Stokes' law for the viscous resistance experienced by a particle moving through a gas, the particle must be large compared with the mean free path of the gas so that it experiences the buffeting by the gas molecules as though the gas were a continuum. Thus we cannot really use Stokes' law if the sphere is the gas molecule itself. However, as an extreme case let us see what happens if we take a heavily condensed gas where $\lambda=a$. We have

$$\eta=\tfrac{1}{3}mn\bar{c}a; \quad \dot{x}\approx u; \quad x\approx a.$$

Then the LHS of equation (4.103) becomes

$$6\pi a\eta\overline{\dot{x}x}=6\pi a\tfrac{1}{3}mn\bar{c}aua=2\pi a^3nm\bar{c}u.$$

But

$$\bar{c}\approx\sqrt{3}(u).$$

Hence

$$6\pi an\overline{\dot{x}x}\approx 10a^3nmu^2. \tag{4.106}$$

But each sphere of radius a occupies a cube of volume $8a^3$, so that $8a^3n=1$. The value of equation (4.106) is thus approximately mu^2, which agrees with the RHS of equation (4.103). This is, of course, pushing the model to an extreme, but it shows again that the particle considered in the analysis of Brownian motion is essentially no different from a molecule, except that it is very much larger.

This has been a long chapter, and we now conclude it with a relevant quotation from Lucretius:

This process, as I might point out, is illustrated by an image of it that is continually taking place before our very eyes. Observe what happens when sunbeams are admitted into a building and shed light on its shadowy places. You will see a multitude of tiny particles mingling in a multitude of ways in the empty space within the light of the beam, as though contending in everlasting conflict, rushing into battle rank upon rank with never a moment's pause in a rapid sequence of unions and dis-unions. From this you may picture what it is for the atoms to be perpetually tossed about in the illimitable void. To some extent a small thing may afford an illustration and an imperfect image of great things. Besides, there is a further reason why you should give your mind to these particles that are seen dancing in a sunbeam: their dancing is an actual imitation of underlying movements of matter that are hidden from our sight. There you will see many particles under the impact of invisible blows changing their course and driven back upon their tracks, this way and that, in all directions. You must understand that they all derive this restlessness from the atom. It originates with the atoms, which move themselves. Then those small compound bodies that are least removed from the impetus of the atoms are set in motion by the impact of their invisible blows and in turn cannon against slightly larger bodies. So the movement mounts up from the atoms and gradually emerges to the level of our senses, so that those bodies are in motion that we see in sunbeams, moved by blows that remain invisible.

(It is interesting to note that almost two thousand years later Perrin used the identical argument in his studies of Brownian motion of gamboge particles in a liquid: he suggested that this could be projected optically before an audience to demonstrate the molecular movement within the liquid.)

Professor S. Sambursky, in his perceptive study *The Physical World of the Greeks*,† adds the comment that this remarkable description, 'perfectly describes and explains the Brownian movement by a wrong example. The movement of dust particles as seen by the naked eye in sunlight are caused by air-currents: the real phenomenon postulated by Lucretius on the basis of abstract reasoning can only be seen in a microscope.' Further, as we have pointed out previously, the random motion of fine particles is an intrinsic property of any assembly of free entities in thermal equilibrium, whether they be gas molecules or solid particles. Nothing, however, can detract from the beauty and elegance of this inferential type of argument.

† S. Sambursky, *The Physical World of the Greeks* (Routledge, 1956).

Integrals

Values of

$$\int_0^\infty x^n \, e^{-ax^2} \, dx:$$

$$\int_0^\infty e^{-ax^2} \, dx = \frac{1}{2}\left(\frac{\pi}{a}\right)^{\frac{1}{2}}$$

$$\int_0^\infty x \, e^{-ax^2} \, dx = \frac{1}{2a}$$

$$\int_0^\infty x^2 \, e^{-ax^2} \, dx = \frac{1}{4}\left(\frac{\pi}{a^3}\right)^{\frac{1}{2}}$$

$$\int_0^\infty x^3 \, e^{-ax^2} \, dx = \frac{1}{2a^2}$$

$$\int_0^\infty x^4 \, e^{-ax^2} \, dx = \frac{3}{8}\left(\frac{\pi}{a^5}\right)^{\frac{1}{2}}.$$

In general

$$\int_{-\infty}^\infty x^n \, e^{-ax^2} \, dx = 2\int_0^\infty x^n \, e^{-ax^2} \, dx \quad \text{for } n \text{ even}$$

$$= 0 \quad \text{for } n \text{ odd.}$$

5

Imperfect gases

Many gases even at room temperature deviate markedly from the 'ideal' gas relation $PV=RT$. Similar deviations are observed with noble gases at higher pressures and lower temperatures. In this chapter we shall discuss the behaviour of such 'imperfect' gases and show that their properties can be explained in terms of the 'ideal' gas theory, modified to allow for the finite volume of the gas molecules and for the existence of intermolecular forces.

5.1 Deviations from perfect gas behaviour

5.1.1 *The virial equation*

In general, instead of the relation $PV=$constant at a fixed temperature, imperfect gases obey an empirical power equation of the form

$$PV=A+BP+CP^2+DP^3. \tag{5.1}$$

This is known as a virial equation and A, B, C and D as virial coefficients. The most important coefficient, apart from A, is B, since C and D are usually very small. At low temperatures B is negative, at higher temperatures positive and at some intermediate temperature T_B it has the value zero. At this temperature, if the pressures are not too high we see at once that equation (5.1) reduces to

$$PV=A=\text{constant}, \tag{5.2}$$

i.e. Boyle's law becomes approximately true again. For this reason T_B is called the Boyle temperature.

Plots of PV against P are shown schematically in figure 5.1. If we measure the slope of the curves near the PV axis, i.e. near $P=0$, we have

$$\begin{aligned}\frac{\partial(PV)}{\partial P}&=B+2CP+\cdots\\ &=B \text{ for } P\to 0.\end{aligned} \tag{5.3}$$

The slope near $P=0$ will, therefore, be zero at the Boyle temperature. It is also easy to show that the curve connecting the minima is a parabola.

5.1.2 *Andrews' experiments on carbon dioxide*

T. Andrews (1813–85) carried out a series of important and classical experiments on the compressibility of gases and the conditions under which they can be liquefied. His first paper appeared in 1861 and one of his major surveys was given in his Bakerian lecture of 1869. We summarize his main conclusions as exemplified by the behaviour of CO_2 (see figure 5.2).

(*a*) Above a temperature of about 50 °C, the compressibility of CO_2 resembles that of a perfect gas.

(*b*) At, say, 21 °C compression produces liquefaction.
From B to C the material behaves as a gas, from C to D it condenses at a constant pressure, and at D it is completely liquid. From D to K the slope is very steep because a liquid is relatively incompressible.

(*c*) At 31.1 °C there is no liquefaction although the isotherm is very distorted from the ideal gas isotherm.

(*d*) At 30.9 °C liquefaction just occurs under compression.
Above this temperature, liquefaction cannot be produced. This is known as the critical temperature T_c, the temperature above which liquefaction cannot be achieved, however high the pressure. The pressure at which

Figure 5.1. Variation of PV with P for a real gas. At one special temperature, where the initial tangent is horizontal, PV is constant in the lower pressure range; this is the Boyle temperature T_B.

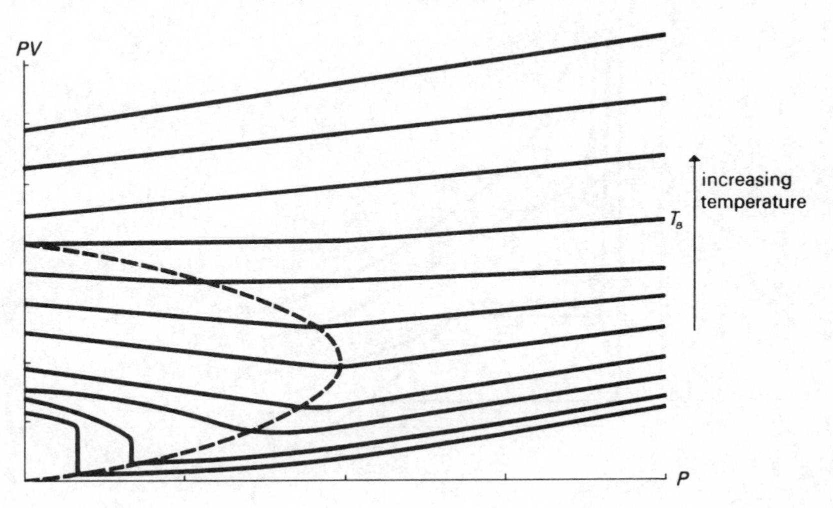

liquefaction just occurs is called the critical pressure P_c, and the corresponding volume (for one mole of the substance) is called the critical volume V_c.

5.1.3 *Continuity of liquid and gaseous state*

Let us consider two isothermals, one above T_c and one below. Consider the points J and K on these isothermals at some pressure level above P_c. At J the substance is purely a gas, at K purely a liquid; this means that if we keep the pressure constant at J but steadily reduce the temperature we may pass to the liquid phase at K. This, in turn, implies that above P_c it is possible, simply by cooling, to pass from the gaseous to the liquid state without any mixture of phases.

5.1.4 *Significance of T_c*

The critical temperature may be easily understood in terms of intermolecular forces. Suppose figure 5.3 represents the potential energy between one molecule and a single neighbour, as a function of separation. It is clear that if the thermal energy of the molecule is greater than $\Delta\varepsilon$, one molecule can always escape from its neighbours. However closely they are pressed together (within limits) the thermal energy will always be sufficient to overcome the attraction between the molecules.

Now the thermal energy of each molecule is of order kT, where k is Boltzmann's constant and equals 1.4×10^{-23} J K^{-1} and T is the absolute

Figure 5.2. Variation of pressure P with volume V for CO_2 (based on Andrews' results).

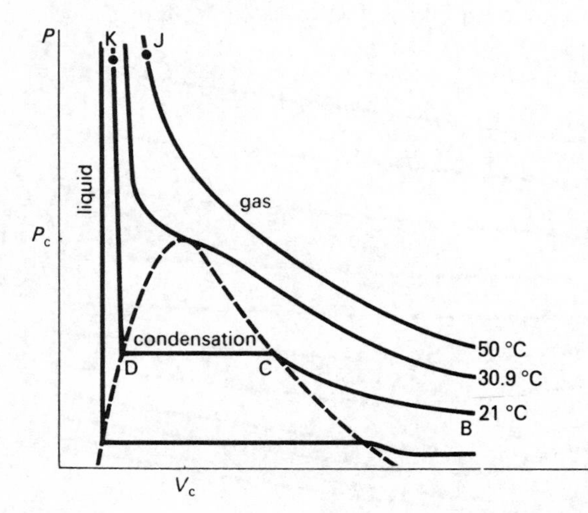

temperature. Thus a molecule can always escape from its neighbour if $kT > \Delta\varepsilon$ or

$$T > \frac{\Delta\varepsilon}{k}. \tag{5.4}$$

The substance is therefore unliquefiable for temperatures above

$$T_c = \frac{\Delta\varepsilon}{k}.$$

Some typical results are given in Table 5.1. It is seen that the agreement is good, indeed surprisingly good, in view of the extremely simple assumptions made. This model deals only with the problem of a molecule escaping from a single neighbour. In section 5.2.9.2 we shall consider this behaviour in somewhat greater detail and, using a different model, establish essentially the same results.

Table 5.1. *Critical temperatures*

Gas	$\Delta\varepsilon$ (J) (theoretical)	$\frac{\Delta\varepsilon}{k}$ (K)	T_c (K) (observed)
He	0.8×10^{-22}	6	5.2
H_2	4×10^{-22}	30	33.1
N_2	13×10^{-22}	100	121

Figure 5.3. Potential energy between one molecule and a single neighbour as a function of separation.

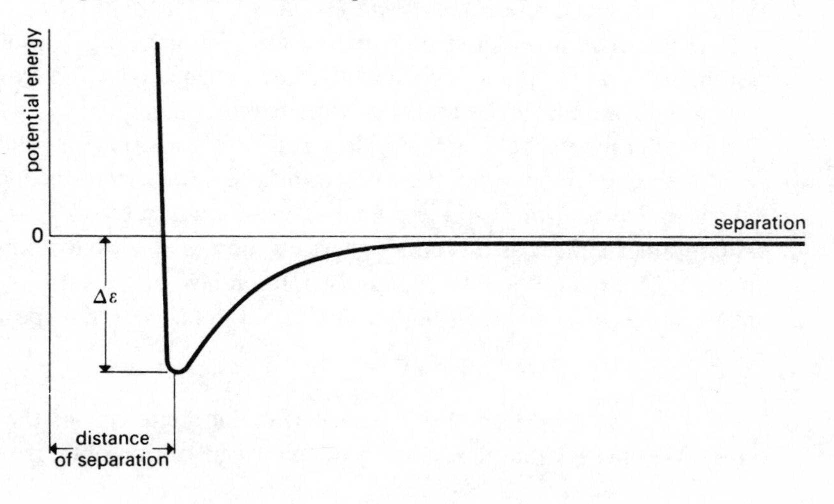

5.1.5 *The miscibility of liquids and gases*

The continuity of the liquid and gaseous states has an interesting bearing on the solubility or miscibility of liquids. Our starting point is that, unlike solids and liquids, all gases are completely soluble in one another (see p. 59). Let us now consider the behaviour of, say, water and oil: these are mutually insoluble in the liquid state and if mixed together will separate into two distinct phases. However, water vapour and hydrocarbon vapour form a homogeneous mixture in the gaseous phase, in which there is a perfect random mixing of molecules of the two materials. If now the vapour is very highly compressed it can be brought to a density of the ordinary liquid; but provided the temperature is high enough – above the 'effective critical temperature' of the system – it will behave as a gas and still remain as a single-phase material. We shall thus have achieved complete miscibility of oil in water. Of course, this can be seen from a completely different angle. If water and oil are enclosed in a container at constant volume and heated, a temperature will be reached at which the thermal energy is sufficient to overcome the intermolecular attraction of like and unlike molecules – the material will then cease to consist of two separate phases and complete miscibility will have been achieved. Alternatively we may say that by increasing the temperature the entropy of the system is increased and the randomization which this implies may be sufficient to mix the materials on a molecular scale, i.e. to achieve miscibility. Is the single-phase material now a high-density gas or a high-temperature liquid? To some extent this is a matter of semantics.

5.2 **Kinetic theory of an imperfect gas: van der Waals equation**

5.2.1 *Derivation of van der Waals equation*

We shall now derive an equation for an imperfect gas, using the assumption that the main deviations from ideal behaviour arise from the finite size of the molecules and the forces between them.

We first assume that if real and ideal gases are in thermal equilibrium, i.e. at the same temperature, the mean translational energy of the individual gas molecules ($\frac{1}{2}m\overline{c^2}$) still implies a value of $\overline{c^2}$ given by $3RT/M$. This assumption, in effect, is a broader generalization of the discussion given on p. 56. (Note that the Boltzmann distribution law, $n_\varepsilon = N\exp(-\varepsilon/kT)$, applies only if ε includes both kinetic and intermolecular potential energy.)

Size of molecule. We first deal with the finite size of the molecules. We may say that the collisions with the wall of the container which determine the observed gas pressure are exactly the same as for zero-size molecules, i.e. an ideal gas, but that the available volume is reduced by some amount b depending on the number and size of the molecules. We may, therefore, write for one mole

$$P(V-b)=BT. \tag{5.5}$$

There are several ways of estimating b.

(*a*) Molecular collisions between the walls. If σ is the effective diameter of a molecule (volume $v_{\mathrm{m}}=\frac{1}{6}\pi\sigma^3$) the mean free path between collisions is reduced from λ to $\lambda-\sigma$. Thus as the molecules travel between the walls of the vessel they will travel $(\lambda-\sigma)$ between collisions where, on the ideal gas model, they would have to travel a distance λ. Thus the number of collisions on the wall per second is increased in the ratio $\lambda/(\lambda-\sigma)$. This means that the pressure is increased by the factor $1/(1-\sigma/\lambda)$. Instead of the ideal gas equation

$$PV=\tfrac{1}{3}Nm\overline{c^2}=RT,$$

we therefore have

$$PV=\frac{RT}{\left(1-\dfrac{\sigma}{\lambda}\right)} \tag{5.6}$$

or

$$P\left(V-V\frac{\sigma}{\lambda}\right)=RT. \tag{5.7}$$

The correction term b in equation (5.5) therefore has the value

$$\begin{aligned} b &= V\frac{\sigma}{\lambda}=V\sigma(\pi\sigma^2 n) \\ &= Vn\pi\sigma^3 \\ &= N_{\mathrm{A}}6v_{\mathrm{m}}, \end{aligned} \tag{5.8}$$

where we have used the simple value of $\lambda=1/\pi\sigma^2 n$ derived in Chapter 3. On this model, therefore, b is about 6 times the total volume occupied by the molecules.

(*b*) Excluded volume. If a molecule approaches another within a distance σ between centres contact occurs. Thus each molecule appears to carry a sheath of radius σ, i.e. of volume $8v_{\mathrm{m}}$ which excludes other molecules. If, therefore, we start with a container of volume V, the first molecule has a free volume V; the second a free volume of $V-8v_{\mathrm{m}}$; the third a free volume of $V-2\times 8v_{\mathrm{m}}$; the fourth a free volume of $V-3\times 8v_{\mathrm{m}}$. We then have to find an effective average value for all the N_{a} molecules in the gas. The standard procedure is to take the geometric mean, i.e. the N_Ath root of the product

$$V(V-8v_m)(V-2\times 8v_m)(V-3\times 8v_m)\ldots,$$

i.e.

$$N_A\text{th root of}$$

$$V^{N_A}\left\{1-\frac{8v_m}{V}[1+2+3+\cdots+(N_A-1)]\right\}$$

$$\approx V^{N_A}\left(1-\frac{8v_m}{V}\frac{N_A^2}{2}\right). \tag{5.9}$$

One thus obtains as the average available volume a value very close to

$$V-V\left(\frac{8v_m}{V}\frac{N_A}{2}\right)=V-N_A 4v_m,$$

so that

$$b=4N_A v_m. \tag{5.10}$$

(*c*) Excluded volume simplified. A much simpler approach (no less valid) is to say that each molecule carries an excluding volume of $8v_m$. But if we are concerned only with double collisions (ignoring triple and higher collisions) this means that on the average $8v_m$ is the excluded volume for a pair of molecules. The average is $4v_m$ so that once again we have

$$b=4N_A v_m. \tag{5.11}$$

(*d*) Collisions with the wall. Jeans gives a subtle, rather more basic way of calculating b. He suggests that all molecules are surrounded by a sheath of radius σ which excludes the other molecules. The excluded volume is therefore 8 times the volume of the molecules themselves. For one mole containing N_A molecules this is therefore $8N_A v_m$. There is thus a zero chance of finding a molecule within the excluded volume and a chance

$$\frac{dV}{V-8N_A v_m}, \tag{5.12}$$

of finding it in an element of volume dV outside the excluded volume. He then considers the space available for molecules about to collide with the wall and argues that, since only the farther halves of the spherical sheaths are excluded when collisions are imminent, the chance of dV being free in space is

$$\frac{V-4N_A v_m}{V}. \tag{5.13}$$

The chance of a molecular collision with the wall is found by multiplying the two probabilities together. The result is

$$\frac{\mathrm{d}V}{V}\times\frac{V-4N_\mathrm{A}v_\mathrm{m}}{V-8N_\mathrm{A}v_\mathrm{m}}\approx\frac{\mathrm{d}V}{V}\,\frac{V}{V-4N_\mathrm{A}v_\mathrm{m}}.$$

For a gas of negligible size the probability is $\mathrm{d}V/V$ so that for the real gas the collision probability, and hence the pressure, are increased by the ratio $V/(V-4N_\mathrm{A}v_\mathrm{m})$. This is equivalent to replacing V by $V-4N_\mathrm{A}v_\mathrm{m}$ in the ideal gas equation so that

$$b=4N_\mathrm{A}v_\mathrm{m}.$$

We have given these four different approaches to show that there is a simple numerical factor relating the excluded volume to the volume of the molecule but that its value depends on the model used to derive it. Experiments agree well with a value of about 4 but readers should realize that this is a 'good' value, not a 'correct' value.

5.2.2 *Intermolecular forces – the van der Waals approach*

We now turn to the effect of intermolecular forces using the approach proposed by van der Waals in 1873. Within the bulk of the gas the molecular forces, on an average, act symmetrically on one another so that the net effect is zero. Consequently within the bulk of the gas the molecules behave as though they were in a gas without attraction, so that the effective pressure is the same as for an ideal gas P_ideal. Near the walls of the vessel, however, the molecules have to escape from their neighbours before they collide with the wall. In overcoming this molecular attraction some kinetic energy is lost, the molecular velocity is reduced and the momentum imparted to the wall on impact and rebound is less than it would be for an ideal gas. Suppose the pressure defect is ΔP. Then the observed pressure P at the walls is less than P_ideal by an amount ΔP.

If we write the ideal gas equation, modified to allow for this, we have

$$P_\mathrm{ideal}V_\mathrm{ideal}=RT,$$

or

$$(P+\Delta P)(V-b)=RT. \qquad (5.14)$$

The pressure defect is proportional to the number of molecules striking the surface per unit area per second, i.e. to the density ρ. It is also proportional to the number of molecules attracting the molecules and reducing their impact. This is also proportional to ρ. As a result ΔP is proportional to ρ^2. It could be written as aP^2 since P is nearly proportional to R, but it turns out that it is better to write this as a/V^2; this correction term appearing to operate over a larger range than a term aP^2. We may thus regard the correction term a/V^2 as the internal pressure in the gas, i.e. the pressure that has to be exerted to pull the molecules away from one another in overcoming intermolecular forces. Our final relation is

$$\left(P+\frac{a}{V^2}\right)(V-b)=RT. \tag{5.15}$$

This celebrated relation is known as the van der Waals equation and is a very convenient way of describing the behaviour of a 'real' gas. The model on which it is derived is admittedly a very crude one but it is generally felt that the equation holds better than it ought. Its main merit is that it is simple and that it allows for molecular volume and molecular attraction with the use of only two parameters a and b.

If a and b are determined for one mole of gas, the van der Waals equation of state for n moles becomes

$$\left(P+n^2\frac{a}{V^2}\right)(V-nb)=nRT.$$

5.2.3 *Intermolecular forces – another approach*

The van der Waals approach assumes that the gas pressure has its 'ideal' value in the bulk of the gas and that the molecules suffer a loss of energy (and momentum) as they escape from the bulk and collide with the walls. However, this treatment (also given by Jeans as late as 1940) is somewhat misleading. We fix our attention on some plane in the gas normal to the x direction and consider the transfer of momentum across it. If the plane is parallel to the wall of the vessel, whatever momentum is transferred across the plane will ultimately be communicated to the wall. This momentum is made up of two parts. The first part is that which is due to the kinetic energy of the molecules ($\frac{3}{2}kT$). The momentum flux per unit area in, say, the x direction is then $2mu$ as for an ideal gas. The second part arises from the loss of energy and momentum as a molecule on the near side of the plane pulls away from all the molecules on the far side of the plane. This will produce a net reduction of momentum transfer across the plane proportional to (density)^2, i.e. to V^{-2}, exactly as in the van der Waals argument. Thus the pressure which is transmitted to the walls is again given by

$$P=P_{\text{ideal}}-a/V^2.$$

At first sight it would seem that the molecule which loses momentum by pulling itself away from molecules outside the plane will gain momentum because of the attraction of the molecules on its other side. This may be so but the overall transmission of momentum across the plane is determined by what happens at the plane itself. This may be seen in a different way if we consider the energy density of the gas as a whole. There will be kinetic energy ($\frac{3}{2}kT$) per molecule and also a potential energy term arising from the attraction between the molecules. This is negative (as for all systems involving attractive forces) so that the total energy is of the form

$$E_{\text{total}} = \text{kinetic energy} + \text{potential energy},$$

where the potential energy is a negative quantity. If we now consider a small isothermal compression (volume change ΔV) we increase the energy density and this provides a measure of the work done $P\Delta V$ by the external pressure P. Clearly P will be less than P_{ideal} since the energy density is less.

The pressure defect (a/V^2) may be regarded as the cohesive energy of the gas. If the gas were so heavily compressed that the molecules were in close contact there would also be a repulsive term which would act in a sense opposite to that of the attractive forces, but the attractive forces would dominate and produce an extremely large pressure defect. Since there is nothing in this model which refers to the state of the medium it will apply to liquids as well as gases. Thus for water under standard atmospheric conditions the kinetic pressure term is of order 1312 atmospheres;† this large value is due to the large molecular density as compared with say water vapour at s.t.p. But the internal pressure is −1311 atmospheres. The net pressure exerted on the walls is simply 1 atmosphere.

Although this treatment emphasizes the conditions in the bulk of the gas, we can, of course, draw our imaginary plane wherever we wish. If we draw it at a distance from the wall equal to the mean free path the momentum transfer process resembles the model used in the van der Waals treatment. We shall use this idea in the following section and later, where we calculate the pressure defect in terms of intermolecular forces.

5.2.4 *Pressure defect: the Dieterici equation*

In the van der Waals treatment it is assumed that every molecule striking the wall has less energy than it would have had if there had been no intermolecular attraction. (Some molecules, in fact might never reach the wall.) If E is the average kinetic energy of the equivalent ideal gas and ΔE is the energy loss, the pressure defect ΔP, expressed as a ratio of

† See E. H. Kennard, *Kinetic Theory of Gases* (McGraw-Hill, 1938).

the ideal pressure P, is given by

$$\frac{\Delta P}{P}=\frac{\Delta E}{E},$$

since we are assuming the same molecular density at the wall as in the bulk.

By contrast, Dieterici stresses that the temperature of the gas is everywhere constant, even near the walls, so that the Boltzmann distribution for the total energy applies to the molecules striking the wall as well as to those in the bulk. Since the wall-striking molecules have reduced total energy, the mean density n is lower than the ideal number n_0 by the factor

$$n=n_0 \exp(-\Delta E/kT)\approx n_0(1-\Delta E/kT),$$

so that

$$\frac{\Delta n}{n_0}\approx\frac{\Delta E}{kT}.$$

Since the gas pressure for a gas at a specified temperature is proportional to the density this again gives for the pressure defect

$$\frac{\Delta P}{P}=\frac{\Delta E}{kT}\approx\frac{\Delta E}{E},$$

since kT (or $\frac{3}{2}kT$) is the average thermal energy of the molecules. This is not really a different model but a different way of calculating the effect. It implies that at a fixed temperature the same Boltzmann distribution applies to the wall molecules as to the bulk – but the density at the wall is less. This is analogous to the decrease in density of a gas in a gravitational field.

The energy needed for a single molecule to escape is proportional to the mean molecular density n, i.e. to $1/V$ so that ΔE is of the form α/V. Hence for a mole of gas

$$P=P_{\text{ideal}} \exp\left(-\frac{\Delta E}{RT}\right)=P_{\text{ideal}} \exp\left(\frac{-\alpha}{VRT}\right).$$

Since

$$P_{\text{ideal}}V_{\text{true}}=RT,$$

we have, assuming that the excluded volume is b,

$$P(V-b)=RT\exp\left(\frac{-\alpha}{VRT}\right). \tag{5.16}$$

This is the Dieterici equation of state. A little manipulation shows that for small values of a this is identical with the van der Waals equation of state.

5.2.5 *Clausius' and Berthelot's equations of state*

There are two other equations of state which at various times have been proposed. R. Clausius (1822–88) attempted to allow for the effect of temperature on the cohesive forces and suggested in 1880

$$\left[P+\frac{a}{T(V+c)^2}\right](V-b)=RT. \tag{5.17}$$

This involves three arbitrary constants. D. Berthelot in 1897 found that if he used a different value for a he could get about as good a fit with experiment by putting $c=0$. He therefore used

$$\left(P+\frac{a}{TV^2}\right)(V-b)=RT. \tag{5.18}$$

These equations are mainly of historical interest. We shall not consider them further.

5.2.6 *Attraction of the walls*

We have spent some space in discussing intermolecular attraction as the cause of a pressure 'defect' in real gases. What of the attraction between the molecules and the walls of the container? Does the pressure exerted by a gas depend on the nature of the container? Will the gas pressure be higher for a polar container than for an inert one? The answer is simple. The wall can have no net effect. Attraction may increase the impulse during the collision but the impulse will be reduced by an exactly equal amount due to the extraction of momentum from the wall. The net effect will be zero.

This is shown schematically in figure 5.4. We draw our arbitrary plane (see section 5.2.3 above) at AA′ at a distance from the wall equal to the mean free path λ. The velocity of a molecule in the bulk of the gas is, say, c. When it escapes from the attraction of its neighbours its velocity has fallen to c' (position B). At B it comes under the attractive forces of the wall. Consider the interaction with a wall molecule X of mass M. Suppose X is initially at rest. The molecule is accelerated towards X but also draws X out of the wall to X′. When collision occurs the velocity of the molecule (mass m) has been increased to c'' whilst the velocity of the wall molecule is, say, v. The gain in momentum of the molecule $m\,(c''-c')$ is exactly equal to the gain in momentum in the opposite direction Mv

of the wall molecule. Thus the wall molecule is given an increased blow because of the attractive forces, but the increased momentum is exactly nullified by the momentum withdrawn from the wall. The net momentum transfer is the same as if no attraction had existed. On rebound the molecule leaves with increased velocity so imparting a larger momentum to the wall molecule, but as the gas molecule returns to position B′ the attractive forces decrease its velocity to its original value c': at the same time they decrease the effective impulse on M, and it returns to the wall with the same momentum as though no attractive force had existed. The total momentum change is thus the same as if the wall exerted no force on the gas molecule. (This is simply a description of Newton's laws of momentum in terms of molecular interactions.) By the time the gas molecule reaches A′ its velocity is back to c and the bulk behaviour of the gas now resembles that of a perfect gas.

The problem in this section is often avoided in elementary texts. Usually there is a bland statement that, by applying Newton's laws of momentum, the molecular momentum is reversed after collision with the wall. This is, of course, perfectly correct but it is equivalent to assuming that the wall exists at ZZ. It ignores the detailed interaction which takes place in the region between ZZ and the wall molecules themselves.

5.2.7 *Van der Waals equation and the Boyle temperature*

If we plot isothermals of PV against P we note that the Boyle temperature occurs when the slope of the curve is zero, at $P \to 0$. We

Figure 5.4. Behaviour of a gas molecule as it escapes at A from the attraction of its neighbours and then approaches at B to within the range of action of the surface forces of the walls of the container.

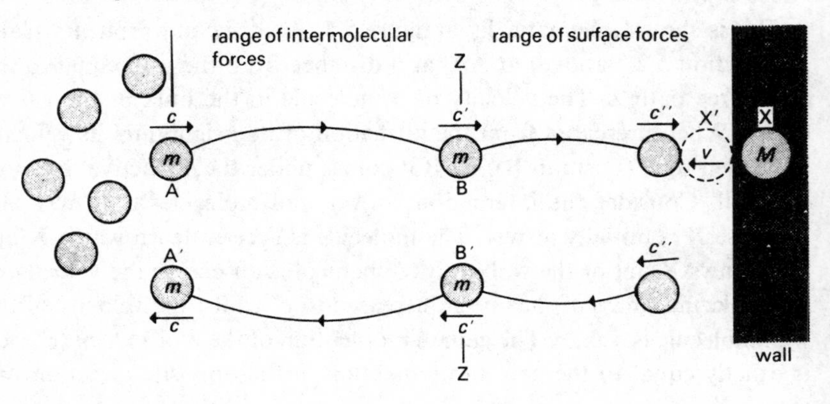

therefore use the van der Waals equation, determine $\partial(PV)/\partial P$ and put it equal to zero at $P=0$.

From the original equation (5.15) we have

$$P=\frac{RT}{V-b}-\frac{a}{V^2},$$

so that

$$PV=\frac{RTV}{V-b}-\frac{a}{V}. \tag{5.19}$$

Then

$$\frac{\partial}{\partial P}(PV)=\left[RT\frac{1}{V-b}-RT\frac{V}{(V-b)^2}+\frac{a}{V^2}\right]\left(\frac{\partial V}{\partial P}\right)_T$$

$$=\left[-\frac{RTb}{(V-b)^2}+\frac{a}{V^2}\right]\left(\frac{\partial V}{\partial P}\right)_T. \tag{5.20}$$

If we specify that, at $T=T_B$, this is zero when $P=0$, this is equivalent to considering V as approaching infinity so that $(V-b)$ is indistinguishable from V. From equation (5.20) we therefore have

$$-RT_Bb+a=0,$$

or

$$T_B=\frac{a}{Rb}. \tag{5.21}$$

At all temperatures above T_B, PV will increase with increasing P, so that there will be no inflexion in the PV curves. At T_B itself we may substitute $T=T_B=a/Rb$ in the van der Waals equation. Multiplying out equation (5.15) we have

$$PV+\frac{a}{V}-bP-\frac{ab}{V^2}=RT.$$

On the LHS we may, if the pressures are small, replace a/V by aP/RT, and neglect the fourth term. We are left with

$$PV+\frac{aP}{RT}-bP=RT.$$

From our value of the Boyle temperature ($T_B=a/Rb$) we see that the second and third terms on the LHS are equal and opposite. Hence at T_B

$$PV=RT_B$$

for a real van der Waals gas.

5.2.8 *Van der Waals equation and the critical points*

The van der Waals equation is a cubic in V. If we plot P against V we obtain a family of curves shown schematically in figures 5.5 and 5.6. Let us consider a typical isotherm XDBACEY in figure 5.5 below the critical temperature. Join D and E by an isobar which cuts the curve at A.

Consider now the behaviour at A. Suppose we started with a homogeneous phase. Along BC the slope $\partial P/\partial V$ is positive. This means that if a small increase in volume occurs this is accompanied by an *increase* in gas pressure. There is thus a spontaneous expansion towards C. Similarly a small decrease in volume would lead to a spontaneous contraction towards B. These processes constitute the initial separation into two phases, gaseous at C liquid at B. The part BD is unstable; the liquid can remain in this region (region of superheating), but generally the phase spreads to D. Similarly CE is unstable; the gas can remain in this region of supercooling but generally the phase spreads to E.† In passing we may note that some isotherms can give negative pressures, i.e. B may be below the V axis. This only occurs in the liquid phase and as we shall see this provides a means of estimating the tensile strength of a liquid.

Figure 5.5. Stylized behaviour of a van der Waals gas.

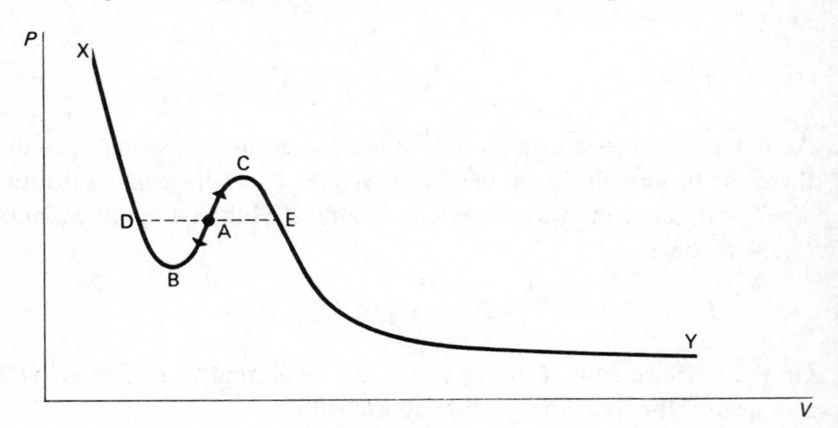

† A thermodynamic argument quoted in J. K. Roberts, *Heat and Thermodynamics*, 5th edn (Wiley, 1960) shows the line EAD should cut the curve at such a position that the areas DBA and ACE are equal. The reasoning is as follows. In carrying the gas through a reversible cycle DBACED (i.e. from D back to D) the gas is restored to its initial state so that there is no change in entropy. Since the whole process is at a constant temperature this implies that the net work done is zero. The work done is given by the $P\,dV$ areas in each part so that the area DBA must be equal to the area ACE.

The two turning points B and C coalesce into a single region at Z on the critical isotherm (figure 5.6). We thus have two conditions determining the critical isotherm:

$$\left(\frac{\partial P}{\partial V}\right)_T = 0 \text{ for turning points such as B and C,}$$

$$\left(\frac{\partial^2 P}{\partial V^2}\right)_T = 0 \text{ for the coalescence of these at the point of inflexion Z.} \tag{5.22}$$

Multiplying out the van der Waals equation we have

$$PV + \frac{a}{V} - Pb - \frac{ab}{V^2} = RT.$$

Differentiating with respect to V at constant temperature we have

$$P + V\left(\frac{\partial P}{\partial V}\right)_T - \frac{a}{V^2} - b\left(\frac{\partial P}{\partial V}\right)_T + \frac{2ab}{V^3} = 0. \tag{5.23}$$

Inserting the condition that for all turning points $(\partial P/\partial V)_T = 0$ we obtain

$$P = \frac{a}{V^2} - \frac{2ab}{V^3}. \tag{5.24}$$

This is the equation of the curve BKZLC (figure 5.6). The tip of this curve corresponds to Z. This is where $(\partial P/\partial V)_T$ of equation (5.24) is zero. We have

$$\left(\frac{\partial P}{\partial V}\right)_T = -\frac{2a}{V^3} + \frac{6ab}{V^4} = 0. \tag{5.25}$$

Figure 5.6. The van der Waals isotherms at various temperatures.

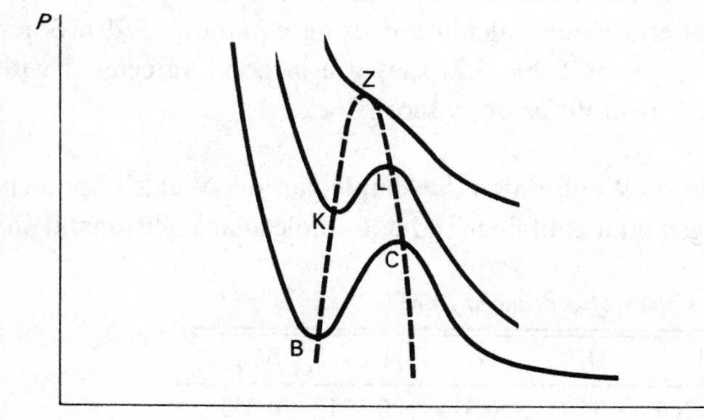

We call the volume at which this occurs the critical volume V_c. From equation (5.25) we obtain

$$V_c = 3b. \tag{5.26}$$

Substituting in equation (5.24) we obtain for the critical pressure

$$P_c = \frac{a}{27b^2}. \tag{5.27}$$

Substituting V_c and P_c in the van der Waals equation we find the critical temperature T_c from the relation

$$\left(P_c + \frac{a}{V_c^2}\right)(V_c - b) = RT_c.$$

This gives

$$T_c = \frac{8a}{27Rb}. \tag{5.28}$$

The nature of the van der Waals equation is such that all isotherms below T_c have two turning-points; no isothermals above T_c have even a single point of inflexion. We may note that empirically the boiling point (K) of a liquid at atmospheric pressure is about $\frac{2}{3}T_c$. There is no simple theory for this.

5.2.9 *Magnitudes of b and a in the van der Waals equation*

5.2.9.1 *Meaning of b*. We have already seen that for a van der Waals gas b is four times the volume of the molecules, treating them as spheres of molecular diameter σ. In a mole therefore

$$b = 4N_A \tfrac{4}{3}\pi \left(\frac{\sigma}{2}\right)^3 = \frac{2\pi}{3} N_A \sigma^3. \tag{5.29}$$

From a study of the compressibility of a gas we can find the best values of b to fit the data and hence calculate σ using equation (5.29). Some typical results are given in Table 5.2; they are in good agreement with values of σ deduced from diffusion, viscosity, etc.

5.2.9.2 *Meaning of a*. We use here the simple van der Waals approach. The pressure of a gas on a container is due to molecular collisions. If the

Table 5.2. *Values of σ deduced from b*

Gas	He	H_2	N_2	CO_2	SO_2
σ (nm)	0.266	0.276	0.314	0.324	0.355

velocity component normal to the container wall is u, then as we saw in Chapter 3 we may say:

$$\text{Number of collisions m}^{-2}\,\text{s}^{-1} = \frac{nu}{2}$$

$$\text{Momentum change per collision} = 2mu. \tag{5.30}$$

The rate of change of momentum $\text{m}^{-2}\,\text{s}^{-1}$ which is the product of these is the ideal gas pressure

$$P_{\text{ideal}} = nmu^2 = \tfrac{1}{3}nmc^2. \tag{5.31}$$

In escaping from the attractive force of the last few neighbouring molecules the collision velocity is reduced from u to $u - \Delta u$, so that the momentum transfer is reduced to $2m(u - \Delta u)$. The number of collisions $\text{m}^{-2}\,\text{s}^{-1}$ which we may call the 'flux' is still $nu/2$. At first sight it might seem that this should also be diminished to $2(u - \Delta u)/n$. This, however, is not so. Although the molecules are retarded they are not, *in this model*, prevented from reaching the wall, so continuity in flux from the bulk of the gas to the wall is maintained. The flux does, of course, depend on the molecular velocity but near the wall where retardation occurs there is a slight 'crowding-up' of molecules, i.e. a compensating increase in density of the gas. In this way flux-continuity is maintained. This model should be contrasted with the Dieterici treatment described above. The observed pressure P therefore becomes

$$P = \frac{nu}{2}\,2m(u - \Delta u) = nmu^2\left(1 - \frac{\Delta u}{u}\right)$$

or

$$P = P_{\text{ideal}}\left(1 - \frac{\Delta u}{u}\right). \tag{5.32}$$

Clearly $P_{\text{ideal}}\Delta u/u$ is the van der Waals term a/V^2. We may write this as

$$P_{\text{ideal}}\frac{\Delta u}{u} = nmu^2\frac{\Delta u}{u}$$

$$= \Delta(\tfrac{1}{2}mu^2)n.$$

Now $\Delta(\frac{1}{2}mu^2)$ is the loss in kinetic energy per molecule in the u direction and $n = N/V$. Consequently the van der Waals term

$$\frac{a}{V^2} = \frac{N_{\text{A}}}{V}\ \text{(loss of kinetic energy per molecule)} \tag{5.33}$$

or

$$a = N_{\text{A}}V\ \text{(loss of kinetic energy per molecule)}, \tag{5.34}$$

assuming no changes occur on the average in the v or w directions.

We could now assume some law of force between the molecules and integrate for all molecules and all distances, extending from the position of the last collision near the wall to all the molecules throughout the bulk of the gas. Indeed the van der Waals treatment is basically concerned with relatively long-range forces. However, as we are interested only in an order of magnitude calculation we can arrive at a reasonable value of a by considering only the nearest neighbours. Because of the simple way this is done it actually overestimates the effect of the attractive forces. The method is illustrated in figure 5.7. Every molecule within the hemi-spherical shell bounded by $r_1=\sigma/2$ and $r_2=3\sigma/2$ is a neighbour which actually touches the molecule considered. The work done in dragging the molecule away from these will be this number of neighbours multiplied by $\Delta\varepsilon$. The volume of the shell is $\frac{1}{2}\frac{4}{3}\pi(r_2^3-r_1^3)=\frac{1}{6}\pi 13\sigma^3=13v_\mathrm{m}$, where v_m is the theoretical volume of a single molecule. The number of neighbours is then $n\times 13v_\mathrm{m}$ so that the energy lost in pulling away the molecule, i.e. its loss of kinetic energy in the u direction is

$$\Delta(\tfrac{1}{2}mu^2)=n\times 13\times v_\mathrm{m}\times\Delta\varepsilon.$$

The van der Waals term is then

$$a=N_\mathrm{A}Vn13v_\mathrm{m}(\Delta\varepsilon). \tag{5.35}$$

Since

$$b=4N_\mathrm{a}v_\mathrm{m} \text{ and } N=Vn,$$

we have

$$a=\tfrac{13}{4}N_\mathrm{A}b(\Delta\varepsilon). \tag{5.36}$$

Inserting this in our value for T_c,

$$T_\mathrm{c}=\frac{8a}{27Rb}=\frac{8}{27}\frac{13}{4}N_\mathrm{A}b(\Delta\varepsilon)\frac{1}{Rb}=\frac{26}{27}N_\mathrm{A}\frac{\Delta\varepsilon}{R}=\frac{26}{27}\frac{\Delta\varepsilon}{k},$$

Figure 5.7. The range of attraction of surrounding gas molecules as a single gas molecule collides with the wall.

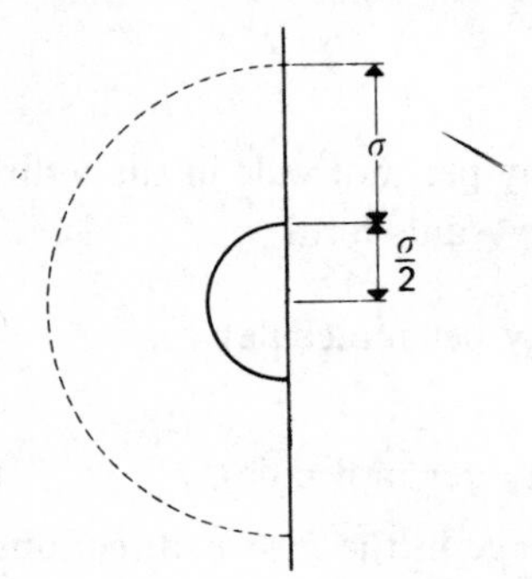

i.e.

$$kT_c \approx \Delta\varepsilon. \tag{5.37}$$

This extremely crude model gives a reasonable value for a and explains why the critical temperature is nearly equal to $\Delta\varepsilon/k$. It must be emphasized that this is an 'order of magnitude' calculation and that the final constant in front of $\Delta\varepsilon$ ($\frac{26}{27}\approx 1$) is largely fortuitous. The model itself, however, is basically valid.

5.3 Some properties of the critical point

5.3.1 *Critical volume*

$V_c = 3b = 12 \times$ volume of molecules themselves. Hence the mean separation between the centres of the molecules at the critical point is

$$l = (12)^{\frac{1}{3}}\sigma \approx 2.4\sigma. \tag{5.38}$$

There is just about enough room between molecules to squeeze another.

5.3.2 *Mean free path*

Using our simple derivation,

$$\lambda = \frac{1}{\pi n \sigma^2},$$

$$\frac{\lambda}{\sigma} = \frac{1}{\pi n \sigma^3} = \frac{1}{6 n v_m}.$$

But nv_m is the volume occupied by the molecules in 1 m^3 of space. At the critical point this is $\frac{1}{12}$ m^3. Hence

$$\frac{\lambda}{\sigma} = \frac{1}{6 \times \frac{1}{12}}$$

$$= 2. \tag{5.39}$$

Thus σ, λ and the mean molecular separation are all roughly equal at the critical point. For ideal gases at ordinary temperatures and pressures the relevant quantities are roughly 0.3, 100 and 3 nm, respectively.

5.3.3 *Critical coefficient*

The ratio RT_c/P_cV_c is known as the critical coefficient. For all equations of state which involve only two arbitrary constants the critical coefficient is an explicit dimensionless number. The values obtained for the Berthelot, Dieterici and van der Waals equations are summarized in Table 5.3. We see that, in general, for a simple gas RT_c/P_cV_c lies between 3 and 3.5 if the attractive forces are of a van der Waals nature. For this

Table 5.3. *Critical points and critical coefficients*

Equation of state	V_c	P_c	T_c	$\frac{RT_c}{P_cV_c}$
Berthelot see equation (5.18)	$3b$	$\frac{RT_c}{2b}-\frac{a}{9T_cb^2}$	$\frac{8a}{27Rb}$	$\frac{8}{3}\approx 2.7$
Dieterici see equation (5.16)	$2b$	$\frac{a}{4e^2b^2}$	$\frac{a}{4Rb}$	$\frac{e^2}{2}\approx 3.7$
van der Waals see equation (5.15)	$3b$	$\frac{a}{27b^2}$	$\frac{8a}{27Rb}$	$\frac{8}{3}\approx 2.7$

Experimental values for the critical coefficients are as follows:

Gas	He	H_2	N_2	CO_2	Water	Acetic acid
$\frac{RT_c}{P_cV_c}$	3.1	3.0	3.4	3.5	4.5	5.0

particular parameter the Dieterici equation gives better agreement than the van der Waals equation. For larger and more polar molecules the critical coefficient is generally greater than 4.5.

5.4 Law of corresponding states

We first consider how, on dimensional grounds, we could formulate the equation of state of a real gas if we knew the law of force between the molecules as a function of separation. For an intermolecular law of force of a given shape we need two parameters to define the scale of the separation axis and the scale of the potential axis. Any two parameters will do. For convenience we may choose the molecular diameter σ and the depth of the potential well at equilibrium $\Delta\varepsilon$ (see figure 5.3). The other parameters likely to be involved are the number density n, the molecular mass m and the thermal energy kT. Dimensional analysis at once shows how we can derive the gas pressure P. The quantities $kT/\Delta\varepsilon$ and $V/n\sigma^3$ have zero dimensions, and $\Delta\varepsilon/\sigma^3$ has the dimensions of pressure. There is no way of including m. Thus

$$P=\frac{\Delta\varepsilon}{\sigma^3}f\left(\frac{kT}{\Delta\varepsilon},\frac{V}{n\sigma^3}\right)$$

or

$$\frac{P}{P^*}=f\left(\frac{T}{T^*},\frac{V}{V^*}\right),$$

where

$$P^* = \Delta\varepsilon/\sigma^3,$$
$$kT^* = \Delta\varepsilon,$$
$$V^* = n\sigma^3.$$

Thus the reduced equation of state is the same for all gases having the same type of force-law; the characteristic pressure P^*, temperature T^* and volume V^* are related to the quantities σ and $\Delta\varepsilon$ which fix the shape and scale of the potential curve.† The choice of the parameters σ and $\Delta\varepsilon$ will determine the values of P^*, T^* and V^* for any particular gas under consideration. This result is very useful because it allows measurements on one gas to be applied to others of the same type since the mass of the molecule is not involved.

More specifically we may express the behaviour of a gas in terms of their critical values. We put

$$\theta = \frac{T}{T_c}, \qquad \pi = \frac{P}{P_c}, \qquad \phi = \frac{V}{V_c}, \tag{5.40}$$

then the van der Waals equation becomes

$$\left(\pi + \frac{3}{\phi^2}\right)(3\phi - 1) = 8\theta, \tag{5.41}$$

thus

$$\pi = f(\theta, \phi).$$

Similarly the Dieterici equation becomes

$$\pi(2\phi - 1) = \theta \exp\left[2\left(1 - \frac{1}{\theta\phi}\right)\right]. \tag{5.42}$$

The assumption that these reduced equations of state are equally applicable to all gases is known as the 'law of corresponding states'. As the dimensional analysis shows, this is only valid if the intermolecular law of force has the same shape for all gases being considered. It implies that, by a simple extension or contraction of the scales on which P and V are plotted, all gases will give identical isotherms at the appropriate corresponding temperatures.

† A good reason for choosing these two quantities is that even if the laws of force vary somewhat from one gas to another, σ and $\Delta\varepsilon$ are the most important quantities determining molecular interactions, whatever the precise laws of force may be.

There is one beautiful extension of the dimensional argument due to de Boer. With very light atoms or molecules one might expect quantum effects to become important. In that case m comes into the picture in the form of the dimensionless parameter $h/(\sigma^2 m\Delta\varepsilon)^{\frac{1}{2}}$, where h is Planck's constant. In this way he was able to predict (correctly) the critical temperature of He^3 before the measurements had been made.†

5.5 Internal energy and specific heat capacities of a van der Waals gas

Consider a gas the molecules of which have f degrees of freedom. When the molecules are far apart we may write the internal (kinetic) energy per mole as

$$U=(f/2)RT. \tag{5.43}$$

As the molecules come closer together the attractive forces between them lower their potential energy (see 5.6.2 below) by the amount

$$\int_{\infty}^{V}(a/V^2)\,\mathrm{d}V=a/V. \tag{5.44}$$

The total internal energy is the sum of the kinetic and potential energies so that

$$U=(f/2)RT-a/V, \tag{5.45}$$

where the first term on the RHS is the kinetic and the second the potential energy.

We may now calculate the specific heat capacity (per mole) when the volume is kept constant:

$$C_V=\left(\frac{\partial U}{\partial T}\right)_V=(f/2)R. \tag{5.46}$$

To calculate the specific heat capacity when the pressure is kept constant, i.e. v is allowed to change, we note that one part comes from the change in U, the other from external work done, $\mathrm{d}V$.

$$C_P\,\mathrm{d}T=(\partial U)_P+P\,\mathrm{d}V.$$

From (5.45) we see that

$$(\mathrm{d}U)_P=C_V\,\mathrm{d}T+(a/V^2)\,\mathrm{d}V,$$

therefore

$$C_P\,\mathrm{d}T=C_V\,\mathrm{d}T+(P+a/V^2)\,\mathrm{d}V. \tag{5.47}$$

We may derive a relation for $\mathrm{d}V$ at constant pressure from the van der

† See A. B. Pippard, *Forces and Particles* (Macmillan, 1977).

Waals equation of state:

$$(P+a/V^2)(V-b)=RT.$$

Neglecting the term ab/V^2 we have

$$PV-Pb+a/V=RT. \tag{5.48}$$

For a change in which P is constant we have

$$P\,\mathrm{d}V-(a/V^2)\,\mathrm{d}V=R\,\mathrm{d}T, \tag{5.49}$$

$$\mathrm{d}V=R\,\mathrm{d}T/(P-a/V^2). \tag{5.50}$$

Substituting into equation (5.47) we have

$$C_P\,\mathrm{d}T=C_V\,\mathrm{d}T+\left(\frac{P+a/V^2}{P-a/V^2}\right)R\,\mathrm{d}T,$$

or

$$C_P-C_V=\left(\frac{P+a/V^2}{P-a/V^2}\right)R\approx R\left(1+\frac{2a}{PV^2}\right)\approx R\left(1+\frac{2a}{RTV}\right). \tag{5.51}$$

5.6 Expansion of gases

5.6.1 *Reversible adiabatic expansion of a real gas*

For an ideal gas adiabatic expansion gives a temperature drop given by

$$\frac{T_2}{T_1}=\left(\frac{V_1}{V_2}\right)^{\gamma-1}. \tag{5.52}$$

For a real gas the temperature drop is greater because of the energy expended in overcoming intermolecular attraction.

5.6.2 *Joule expansion into a vacuum*

For an ideal gas the internal energy is independent of the volume so that there is no net temperature change. For a real gas work is done solely against the intermolecular attractions. We may consider this as work done against the internal pressure a/V^2, and write it as

$$W=\int_{V_2}^{V_1}\frac{a}{V^2}\,\mathrm{d}V=a\left(\frac{1}{V_1}-\frac{1}{V_2}\right), \tag{5.53}$$

where V_1 is the initial and V_2 the final volume.

5.6.3 *Joule–Kelvin expansion*

This is a type of expansion suggested by Lord Kelvin and carried out by him in collaboration with Joule in 1853. The purpose was to study (under more carefully controlled conditions than Joule's earlier experiments) the heat changes occurring when a gas undergoes free expansion. The general principle is illustrated in figure 5.8(*a*). Gas in the left-hand part of the tube is driven at a constant pressure P_1 through an orifice or porous plug and withdrawn on the right-hand part at a constant, lower, pressure P_2. Joule and Kelvin studied the temperature changes occurring after a steady state had been reached for a system that was thermally insulated. For the purpose of our treatment, however, we shall assume that heat sources or sinks can be applied so that the temperature T on both sides is maintained constant.

Let V_1 be the volume per mole on the LHS, V_2 the volume on the RHS. During the passage of one mole of gas

$$\text{external work done on the gas on LHS} = P_1V_1,$$

$$\text{external work done by the gas on RHS} = P_2V_2,$$

$$\text{hence net work done by the gas} = P_2V_2 - P_1V_1. \tag{5.54}$$

There is also work done by the gas against the internal forces. We have already calculated this in equation (5.53). The amount is

$$\frac{a}{V_1} - \frac{a}{V_2}. \tag{5.55}$$

Figure 5.8. The Joule–Kelvin expansion. (*a*) The gas is driven at constant pressure P_1 through a porous plug and withdrawn at a lower pressure P_2. (*b*) Isotherms showing that for an expansion from a higher to a lower pressure, the change in PV is positive at low temperature and negative at high temperature.

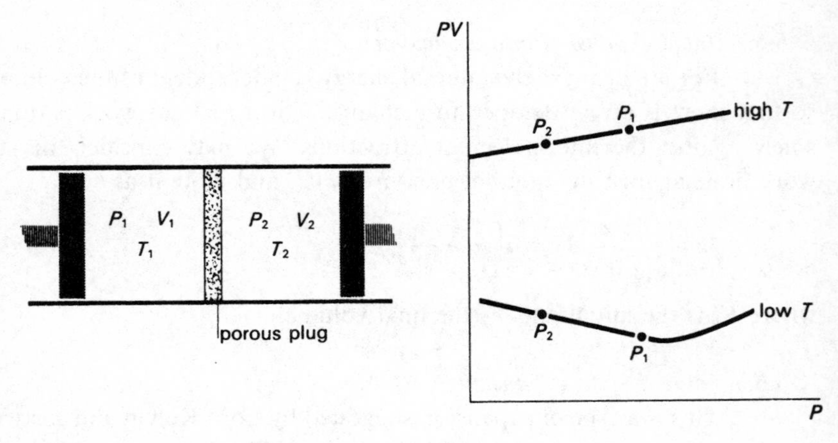

Hence the total work done by the gas when one mole flows is

$$P_2V_2 - P_1V_1 + \frac{a}{V_1} - \frac{a}{V_2} = \left(P_2V_2 - \frac{a}{V_2}\right) - \left(P_1V_1 - \frac{a}{V_1}\right). \tag{5.56}$$

If there is to be no cooling of the gas, this amount of heat has to be supplied. We may calculate the amount of heat involved assuming the gas obeys the van der Waals equation. We start with

$$\left(P + \frac{a}{V^2}\right)(V - b) = RT,$$

and multiply out. Neglecting the term in ab/V^2 we have

$$PV = RT - \frac{a}{V} + bP, \tag{5.57}$$

or

$$PV - \frac{a}{V} = RT - \frac{2a}{V} + bP. \tag{5.58}$$

In the second term on the RHS we replace V by RT/P, so that

$$PV - \frac{a}{V} = RT - P\frac{2a}{RT} + bP. \tag{5.59}$$

If T is to be maintained constant the amount of heat to be added is

$$\left(PV - \frac{a}{V}\right)_2 - \left(PV - \frac{a}{V}\right)_1 = (P_1 - P_2)\left(\frac{2a}{RT} - b\right). \tag{5.60}$$

Since $P_1 > P_2$ the heat supplied to maintain T constant must be positive if $2a/RT - b$ is positive, i.e. if $2a/RT > b$. This means that if heat is *not* supplied there will be *cooling*.

Experiments show that with nearly all gases there is a cooling at room temperature after a Joule–Kelvin expansion. With hydrogen, however, there is a rise in temperature unless the gas is first cooled to about −80 °C.

If the system is thermally isolated we may estimate the temperature difference produced during a Joule–Kelvin expansion in the following way. We write equation (5.59) more fully as

$$\left(PV - \frac{a}{V}\right)_2 - \left(PV - \frac{a}{V}\right)_1 = RT_2 - RT_1 + (P_1 - P_2)\left(\frac{2a}{RT} - b\right). \tag{5.61}$$

We cannot complete this rigorously without the use of the second law of thermodynamics. It turns out that the LHS, which is the net work done by the gas, is equal to $C_V\Delta T$, where ΔT is the temperature drop $T_1 - T_2$.

Equation (5.61) becomes

$$C_V \Delta T = -R\Delta T + (P_1 - P_2)\left(\frac{2a}{RT} - b\right).$$

Since $C_V \approx C_P - R$ we have finally for the cooling

$$\Delta T = \frac{1}{C_P}(P_1 - P_2)\left(\frac{2a}{RT} - b\right). \tag{5.62}$$

We see that there is a critical temperature below which there is a cooling, above which a heating of the gas. This inversion temperature occurs when

$$\frac{2a}{RT_i} = b$$

or

$$T_i = \frac{2a}{Rb} = 2T_B. \tag{5.63}$$

5.6.4 *The reason for an inversion temperature*

It is instructive to consider separately the two main terms involved in the cooling of a gas subjected to a Joule–Kelvin expansion. If the gas expands from P_1 to P_2, i.e. if P_1 is greater than P_2, the work done by the gas which may cause cooling is

$$P_2V_2 - P_1V_1 + \left(\frac{a}{V_1} - \frac{a}{V_2}\right)$$

$$= \Delta[PV]_1^2 + \text{work against internal attraction.} \tag{5.64}$$

For a real gas in which there is molecular attraction the second term on the right must be positive and therefore always lead to cooling. This is not necessarily true of the first term. At low temperatures, in moving from P_1 to P_2, the quantity $\Delta[PV]_1^2$, is positive (see figure 5.8(*b*)). Consequently additional work must be done by the gas in expanding and this causes increased cooling. At high temperatures, however, well above the Boyle isotherm, $\Delta[PV]_1^2$ is negative, i.e. work is done on the gas as it expands and this opposes the cooling effect of the second term in equation (5.64). At some intermediate temperature the two effects just balance and there is neither heating nor cooling. This is the inversion temperature T_i.

Although this is helpful in showing how the inversion temperature arises it is, in some ways, artificial since it attempts to separate the 'imperfect' properties of the gas into two independent characteristics. As the van der Waals equation shows these two characteristics are, in fact, linked.

5.6.5 *An isoenthalpic expansion in the Joule–Kelvin experiment*

When a gas is expanded reversibly through nozzles or a porous plug under conditions of thermal insulation we have by the first law of thermodynamics

$$\Delta Q = 0 = U_2 - U_1 + P_2V_2 - P_1V_1,$$

so that

$$U_2 + P_2V_2 = U_1 + P_1V_1$$

$$H_2 = H_1.$$

The porous-plug expansion is thus an isoenthalpic process. The temperature change associated with the pressure change must therefore be expressed in the form $(\partial T/\partial P)_H$. For a small change in H we have

$$\mathrm{d}H = \left(\frac{\partial H}{\partial T}\right)_P \mathrm{d}T + \left(\frac{\partial H}{\partial P}\right)_T \mathrm{d}P. \tag{5.65}$$

If we are specifying that H is constant, $\mathrm{d}H = 0$ and we are left with

$$\left(\frac{\partial H}{\partial T}\right)_P + \left(\frac{\partial H}{\partial P}\right)_T \left(\frac{\partial P}{\partial T}\right)_H = 0.$$

The first term on the left is C_P. Consequently

$$\left(\frac{\partial T}{\partial P}\right)_H = -\frac{1}{C_P}\left(\frac{\partial H}{\partial P}\right)_T. \tag{5.66}$$

In the next step we have to make use of thermodynamic functions based on the second law, viz:

$$\left(\frac{\partial H}{\partial P}\right)_T - V = -T\left(\frac{\partial V}{\partial T}\right)_P. \tag{5.67}$$

The Joule–Kelvin temperature change is therefore given by

$$\left(\frac{\partial T}{\partial P}\right)_H = -\frac{1}{C_P}\left(\frac{\partial H}{\partial P}\right)_T = \frac{T\left(\frac{\partial V}{\partial T}\right)_P - V}{C_P},$$

or

$$\Delta T = \frac{1}{C_P}\Delta P\left[T\left(\frac{\partial V}{\partial T}\right)_P - V\right]. \tag{5.68}$$

Using equation (5.57) a little manipulation shows that the term

$$T(\partial V/\partial T)_P - V$$

in equation (5.68) is equal to $2a/RT - b$. Thus equation (5.68) becomes identical with the relation given in equation (5.62) for the Joule–Kelvin cooling. Consequently the inversion temperature is again given by

$$T_i = \frac{2a}{Rb}.$$

The above treatment suggests that there is a single inversion temperature which is independent of pressure. This results from assuming that the pressures are low, so that ab/V^2 can be neglected. If the higher terms are included it may be shown that for the van der Waals equation of state, T_i is a parabolic function of the pressure – indeed at pressures above $a/3b^2$ the gas expansion shows inversion at all temperatures, i.e. at high pressures it will heat up on expanding.

This is illustrated schematically in figure 5.9, where curves of constant enthalpy have been plotted as a function of temperature and pressure. In expanding from say point D to point E the temperature of the gas will

Table 5.4. *Inversion temperatures T_i of some gases compared with $2T_B$ (K)*

Gas	T_B	$2T_B$	T_i
He	16	32	25
H_2	114	228	190
N_2	420	840	650

Figure 5.9. Curves of constant enthalpy as a function of temperature and pressure (from F. W. Sears, *Thermodynamics, the Kinetic Theory of Gases and Statistical Mechanics*, 2nd edn (Addison-Wesley, 1953)).

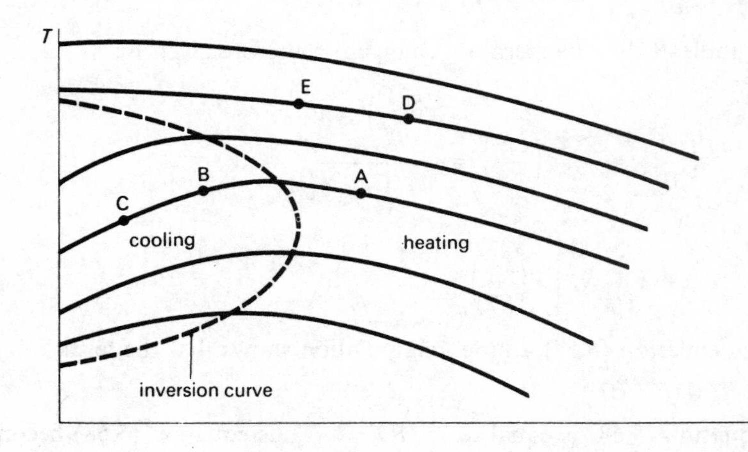

rise – whereas an expansion from point A or point B to point C will produce a drop in temperature.

In effect the temperature T_i given in equation (5.63) for a low pressure expansion is the highest value of the inversion temperature. Some approximate values of T_i at low pressures are given in Table 5.4 and compared with $2T_B$.

6

The solid state

We pass at once from gases to solids. In gases the atoms and/or molecules are almost completely free; in solids they are almost completely lacking in mobility. Indeed in a solid the thermal motion is so greatly reduced that individual atoms can move from their fixed positions only with the greatest difficulty. It is this which imparts to solids their most characteristic macroscopic property – they maintain whatever shape they are given, they have appreciable stiffness. Although the atoms are in fixed locations they possess some thermal energy – in some cases individual atoms can diffuse through the solid but in general the process is extremely slow. Their main thermal exercise consists in vibrating about an equilibrium position.

6.1 **Types of solids**

There are three main types of solids: crystalline, amorphous and polymeric. The first part of this chapter will deal with crystalline solids; amorphous solids will be described briefly at the end while a separate chapter (Chapter 13) will be devoted to polymers.

6.1.1 *Crystalline solids*

The main feature here is long-range order. The molecules or atoms are in regular array over extended regions within each individual grain or crystal. With large single crystals the regular array may extend over an enormous number of atoms or molecules. For example with a single crystal of copper of side 1 cm, along any one direction there will be 30 000 000 copper atoms in regular array, the arrangement of the last few being (in the ideal case) in perfect step with the first.

In spite of this extraordinary degree of regularity the individual atoms can vibrate over relatively large amplitudes about their equilibrium positions. Thermal vibrations can easily exceed one-tenth of the mean spacing; yet the mean spacing is precise and determinate to an extremely high degree of reproducibility.

Some common crystalline structures are shown in figure 6.1. Parts (*a*), (*b*), (*c*) are the structures most commonly observed with metals. Figure 6.1(*a*) is the face-centred cubic (f.c.c.) structure; this is the structure of metals such as aluminium, copper, gold, silver, lead and most of the inert gases. It is a close-packed structure in which each atom has 12 nearest neighbours. Figure 6.1(*b*) is the hexagonal close-packed (h.c.p.) structure

Figure 6.1. (*a*), (*b*) and (*c*): the common structures in metals. (*a*) f.c.c. with 12 nearest neighbours; (*b*) h.c.p. with 12 nearest neighbours; (*c*) b.c.c. with 8 nearest neighbours. (*d*), (*e*) and (*f*): typical structures in ionic crystals of the type X^+Y^- or $X^{2+}Y^{2-}$. (*d*) CsCl structure in which each ion of one type is surrounded by 8 ions of the opposite sign; (*e*) NaCl structure in which each ion is surrounded by 6 of the opposite sign; (*f*) ZnS (zinc blend) structure in which each ion is surrounded by 4 ions of the opposite sign at the corners of a tetrahedron. If all the atoms in (*f*) are of the same type we have the diamond structure, for example silicon, germanium, diamond.

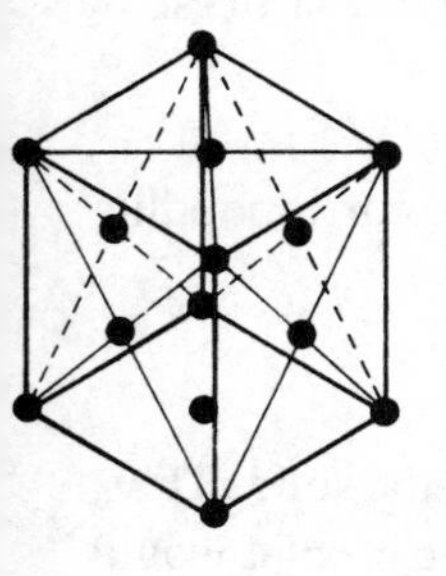

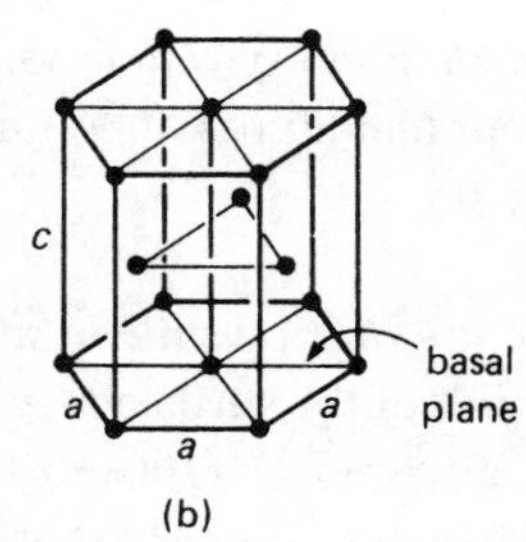

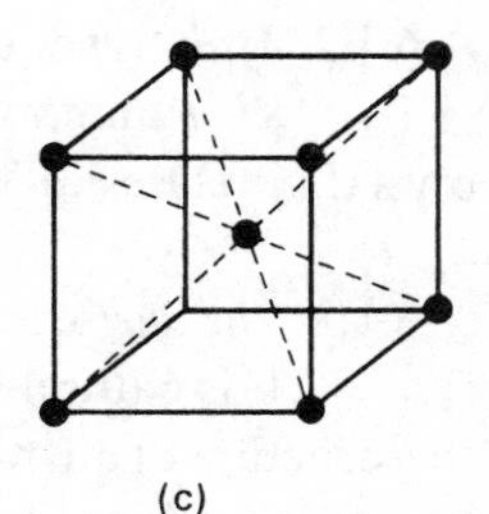

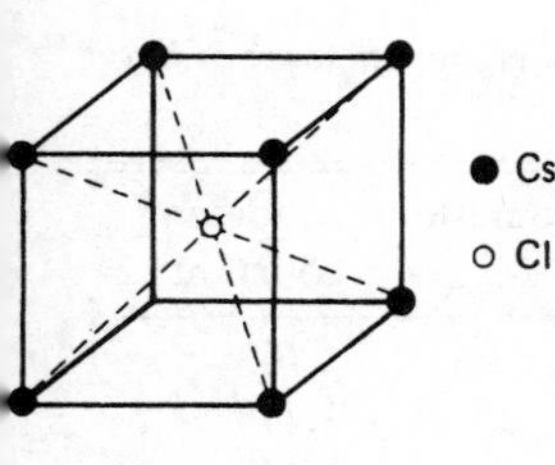

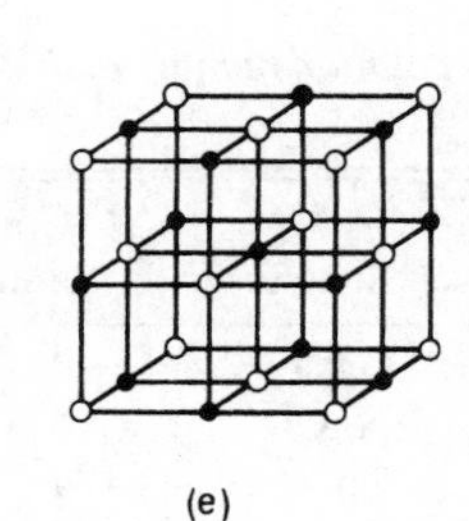

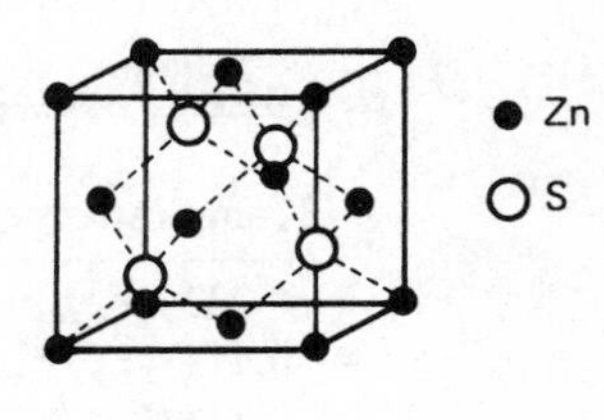

and again each atom has 12 nearest neighbours. Amongst the more common metals it is represented by cadmium and zinc. Figure 6.1(*c*) is the body-centred cubic (b.c.c.) structure. This is less closely packed than f.c.c. and h.c.p., each atom having only 8 nearest neighbours. The commonest examples of these are the alkali metals, α-iron, and a number of the transition metals such as molybdenum and niobium. Figures 6.1(*d*), (*e*) and (*f*) show the three main crystalline arrangements of equi-valent ionic crystals of the type X^+Y^- or $X^{2+}Y^{2-}$. The structures depend on the ratios of the effective radii of the positive and negative ions and represent a subtle balance between the energy of interaction of the ions and their packing. Figure 6.1(*d*) is the CsCl structure which occurs when r_+/r_- is greater than 0.732 and less than 1. The coordination number is 8, i.e. each ion is surrounded by eight ions of opposite sign. Figure 6.1(*e*) is the NaCl structure which occurs when r_+/r_- is greater than 0.414 and less than 0.732: the coordination number is 6. Finally figure 6.1(*f*) is the ZnS (zinc blend) structure which occurs when r_+/r_- is greater than 0.225 and less than 0.414. The coordination number is 4. These details are summarized in Table 6.1. We may also note that if the two atoms in the zinc-blend structure are identical we have the diamond structure. This is represented by silicon, germanium and (naturally) diamond. Each atom has four nearest neighbours in tetrahedral array. The structure may also be regarded as being equivalent to two interpenetrating f.c.c. lattices.

6.1.2 *Main types of bonding in crystalline solids*

We summarize four main types of bonding and the corresponding crystalline states in Table 6.2.

6.1.3 *The surface structure of a crystalline solid*

It is natural to think of the surface as a section of the solid. Thus, the structure of an atomically smooth (100) surface on a f.c.c. solid would be a series of face centred squares. The (111) surface would be a hexagonal

Table 6.1. *Structure, coordination number and Madelung constant of ionic crystals*

Ionic ratio r_+/r_-	Structure	Coordination number	Madelung constant
$0.732 < r_+/r_- < 1$	CsCl	8	1.763
$0.414 < r_+/r_- < 0.732$	NaCl	6	1.748
$0.225 < r_+/r_- < 0.414$	ZnS	4	1.638

Table 6.2. *Main types of bonding*

Bonding	Bond strength	Examples	Crystalline state
Van der Waals	Weak	Solid H_2, Kr paraffins	Close packing of weakly attracting units
Ionic	Strong	NaCl crystal	Giant aggregates of positive and negative ions closely packed in a way consistent with neutrality of charge
Covalent	Strong	Diamond, Si, Ge	Giant molecules with directed bonds, packing determined by valency number and valency directions
Metallic	Strong	Metals	Metal atoms give up their valency electrons leaving metal ions in a sea of electrons. Forces between ions and electrons are central giving close packing. Strong attraction gives strength. Mobile electrons give conductivity.

array. The structural properties of surfaces cannot easily be studied with X-rays since they are too penetrating. A far more effective technique is the diffraction of low velocity electrons. An electron accelerated through a potential of V volts has associated with it a wavelength λ given by

$$\lambda = \sqrt{\frac{150}{V}}\,\text{Å} \quad \text{or} \quad \sqrt{\frac{1.5}{V}}\,\text{nm} \tag{6.1}$$

so that for potentials of the order of 10–100 V, the wavelength is of the order of atomic spacing and therefore well suited for diffraction studies. In this range of electron energies, the penetration depth is less than 2 nm, so that of the beam will sample perhaps the first couple of atomic layers. An even more discriminating tool is the diffraction of neutral atoms (usually helium). These bounce off the surface and rarely penetrate farther. Such studies with simple metals show that the ordering of the surface atoms often agrees with the expected structure but the spacing between the two outermost layers is often less than between layers in the bulk: this is because of asymmetric forces at the surface arising from the absence of atoms above the surface.

However, the surface structure of some solids shows remarkable distortions that are not at first sight to be expected. These distortions are marked with the transition metals where there is an appreciable amount of d-orbital bonding. But the most marked effects are observed with covalent solids such as silicon and germanium. Here, the interatomic bonding is covalent so that in the free surface electrons are left unpaired. These electrons are redistributed amongst the surface and subsurface atoms (back-bonding) and a new lower-energy surface structure emerges. This is shown in figure 6.2 for the (111) surface of atomically clean Si, where the disarrangement and discontinuities in the first and second layers are striking. The third layer is also out of register with the original lattice. By the fourth layer, the bulk structure is regained. If the clean surface is exposed to molecular hydrogen the molecule dissociates and the individual hydrogen atoms mop up the unpaired surface electrons of the silicon. As a result the surface structure at once reverts to that of the original undistorted lattice. The silicon remains covered with a tightly bonded surface layer of hydrogen.

In the remarkable phase changes illustrated in figure 6.2 the energy change is relatively large. Theoretical analysis suggests that the energy of the distorted lattice is about 25 per cent less than that of the undistorted surface if it could retain its dangling bond and its undistorted structure. This is because of the large energy associated with the back-bonded covalent bond. In other systems where phase changes occur without a change in the nature of the bond, e.g. when a metal changes from f.c.c. to h.c.p., the energy change is usually extremely small. An example of this is the structure of solid inert gases (see Chapter 7, p. 186).

Some transition metal surfaces such as tungsten and molybdenum also show surface reconstruction, although the displacements of the surface atoms are less extreme than in the case of semiconductor surfaces. The transition metals play a very important part in the field of catalysis particularly in the petrochemical industry. The surface tends to cause dissociation of adsorbed gases and to activate reaction between them: it is indeed probable that these catalytic properties are directly related to details of the surface structure.

6.2 Solid–liquid transitions: surface melting and bulk melting

6.2.1 *Surface melting*

If we heat a solid, the molecules or atoms acquire additional thermal energy: the atoms vibrate more vigorously, the amplitude depending on the temperature and the strength of the bonding between

Figure 6.2. The 7×7 structure of the (111) surface of silicon. (*a*) The atomically clean surface: extensive reconstruction occurs due to the back-bonding of dangling surface bonds. (From A. Zangwill, *Physics at Surfaces*, C.U.P., 1988.) (*b*) The result of exposing the clean surface to atomic hydrogen. The unpaired surface electrons are mopped up and the structure reverts to that of the undistorted 7×7 lattice.

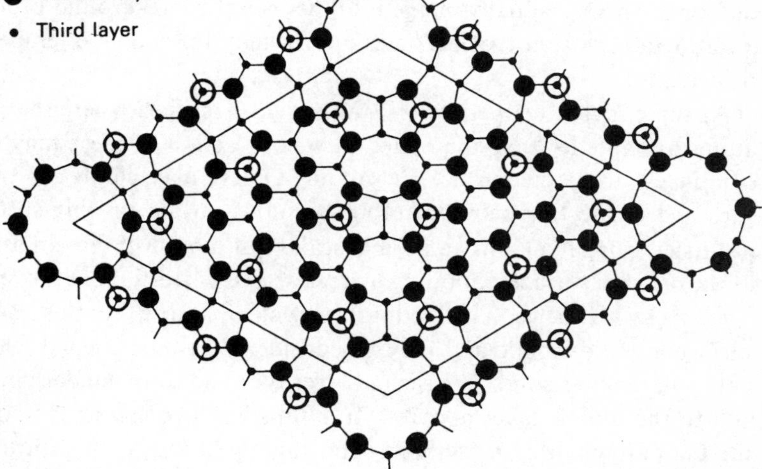

(*a*)

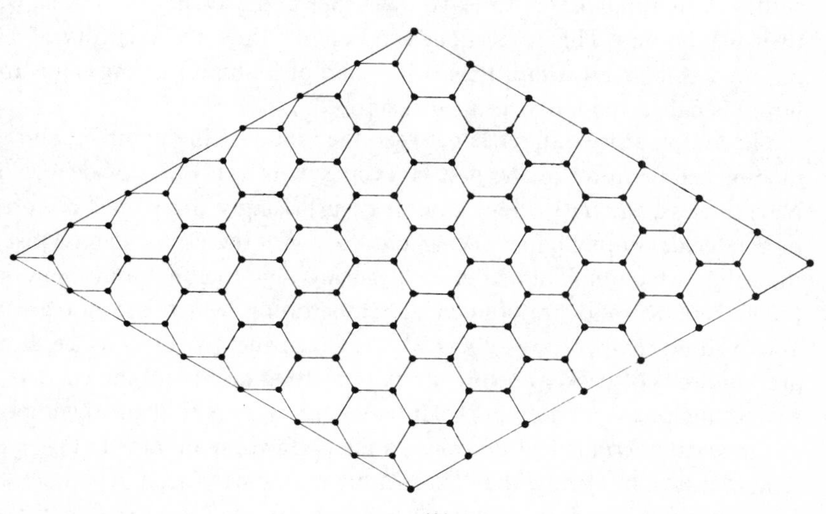

(*b*)

the atoms. The surface atoms having fewer neighbours are more loosely bound than in the bulk so that their vibrational amplitude is greater. This may be observed directly using low energy electron diffraction or proton scattering. The diffraction spots produced by surface atoms become more diffuse. This is because their vibrational velocity produces an increase in the amount of inelastic scattering while the intensity of the elastically scattered (i.e. the diffracted) beams decreases as a result of the displacement of the atoms from their equilibrium position. Consequently the diffracted spots gradually merge into the inelastic background producing a diffraction pattern that becomes increasingly fuzzy as the temperature is increased.

At some higher temperature a stage may be reached where the surface atoms are able to leave their sites. A small fraction of these may escape completely from the surface as vapour. Others may climb out of their sites and locate themselves at other positions on the existing surface to produce 'roughening' on an atomic scale. But most will remain attached to the original surface exhibiting a great degree of freedom to move about over it (two dimensional freedom). Structure is lost. At this stage the surface layer is mobile and may be considered to have melted while the bulk still remains solid. The second layer is bound to the underlying solid and to the mobile layer above it. It is thus less strongly anchored to its site than atoms deep in the bulk. At a slightly higher temperature, it too may melt. The third layer will melt at yet a higher temperature. Ultimately, at yet a higher temperature, the atoms in the bulk will have sufficient thermal energy to leave their sites and exchange positions with their neighbours. This corresponds to bulk melting. The quantity of heat required to convert a unit mass of a solid at its bulk melting point to a liquid is called the latent heat of fusion (L_F).

The temperature difference between the bulk melting point T_m and the melting temperature of the first layer may be several tens of degrees K. Nevertheless, the bulk melting point is surprisingly sharp as is discussed in greater detail in Chaper 10, which deals with the liquid state. Most of the solids which are studied in catalysis are high melting point materials (1000–2000 K) and are studied at temperatures which are not greatly above room temperature (300–600 K). Consequently, the surface atoms are hundreds of degrees below the bulk melting point and the concept of surface melting does not apply. However, surface melting may be important in another context. If two clean solid surfaces, maintained at elevated temperatures, but below the bulk melting point, are placed in contact, the liquid surface films may readily coalesce: as soon as they do so, they cease to be surface films and become solid. They will thus facilitate solid–solid bonding without the application of pressure. This is why, as Faraday pointed out in 1842, ice cubes well below 0 °C stick to one another so readily and so strongly.†

Another point may be raised concerning the naturally occurring water film on ice. Some workers have suggested that this film accounts for the low friction of ice. However, detailed studies show that although a surface liquid film is necessary for low friction its formation is mainly due to frictional heating, generated by the sliding process itself.‡

6.2.2 *Bulk melting*

The bulk melting point of a solid depends on the pressure to which it is subjected. For a solid which is more dense than the liquid (this is the usual case) the melting point increases with pressure: for a solid which is less dense (e.g. ice) the melting point decreases with pressure (see figure 6.3(*a*) and (*b*)). If we combine this with the liquid–vapour transition

Figure 6.3. Melting curves (solid–liquid equilibrium) as a function of pressure for (*a*) a solid more dense than the liquid, (*b*) a solid less dense than the liquid.

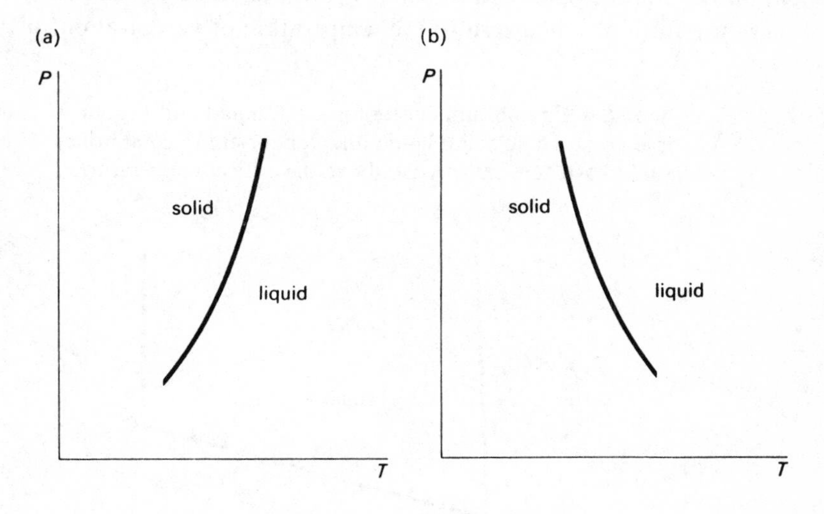

† J. G. Dash, *Contemp. Phys.* **30**, 89–100, 1989: Dash treats surface melting in thermodynamic rather than atomic terms.

‡ F. P. Bowden and D. Tabor, *Friction – an introduction to tribology*, Heinemann, 1974. For fuller accounts by the same authors *see Friction and Lubrication Part I.* 1954, Reprint 1986 and Part II, Chapter VIII, 1964, Oxford University Press.

we obtain a phase diagram as shown in figure 6.4. We note the following characteristics:

(*a*) for the liquid–vapour phase equilibrium the curve ends at C. Above this point, which corresponds to the critical temperature T_c, the liquid ceases to exist;

(*b*) by contrast, for the solid–liquid phases there is no evidence that there is an end-point to OY above which only one phase can exist.

A simple theory of bulk melting due to Lindemann is described in Chapter 10, section 10.8.

6.3 Consequences of interatomic forces in solids

We describe here a few of the properties of solids which are directly explicable in terms of interatomic forces.

6.3.1 *Heat of sublimation*

If we could take a van der Waals solid such as solid argon at absolute zero, and heat it until all the constituent atoms evaporate we should have pumped into the solid sufficient energy to break the bonds between all the atoms. This is the theoretical heat of sublimation L_S. It is equal to the latent heat of fusion (L_F) plus the latent heat of vaporization (L_V) plus a specific heat capacity term involved in raising the temperature from absolute zero to the temperature of vaporization. If there

Figure 6.4. Equilibrium curves for solid, liquid and vapour. O is the triple point where solid, liquid and vapour are in equilibrium. The liquid phase does not persist above the critical temperature.

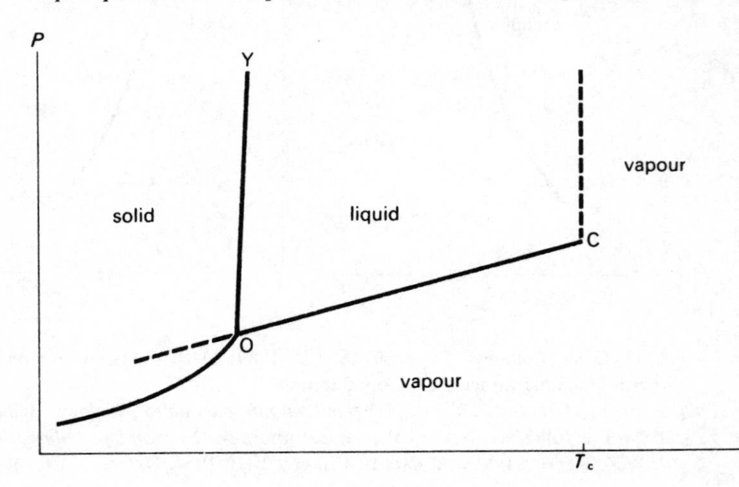

were an exact energy balance the released atoms would just leave the solid with a minute escape velocity; in practice they will have considerable thermal energy. Thus precise comparisons between bond strengths and energy of sublimation is difficult. However, as an order of magnitude calculation such correlation is easy, as we shall now show.

If the potential energy between each atom (or each molecule) and its neighbour is $\Delta\varepsilon$, and if each atom (or molecule) has n nearest neighbours, the energy required to break the bonds in one mole will be

$$L_S = \tfrac{1}{2} N_A n \Delta\varepsilon. \tag{6.2a}$$

This is allowing only for nearest-neighbour interactions, where N_A is Avogadro's number and the factor $\frac{1}{2}$ is introduced to avoid counting each bond twice. For close packing, such as in an f.c.c. structure or an h.c.p. structure, n is 12.

Hence

$$L_S = 6\Delta\varepsilon N_A. \tag{6.2b}$$

If $\Delta\varepsilon$ is in Joules and we wish to express L_S in kJ per mole we obtain

$$L_S \approx 36 \times 10^{20}\ \Delta\varepsilon. \tag{6.3}$$

Some typical results are given in Table 6.3, the 'observed' values of L_S being simply the sum of L_F and L_V. The values of $\Delta\varepsilon$ are obtained either from theoretical calculations of the van der Waals forces, or from the behaviour of the substance in the gaseous state (e.g. the deviation from ideal gas behaviour at high pressures).

Apart from helium, where the effects of 'zero-point energy' are important, the agreement is very satisfactory. A more detailed consideration of lattice energy, latent heat of sublimation and intermolecular forces will be given in the next chapter. At this stage we merely draw attention to

Table 6.3. *Heats of sublimation*

		L_S (kJ mol^{-1})	
Substance	$\Delta\varepsilon$ (J)	Calculated	Observed
He	0.9×10^{-22}	0.33	0.08
Ne	4.8×10^{-22}	1.7	1.3
Ar	16.5×10^{-22}	5.9	7.3
Kr	25×10^{-22}	9.0	10.0
H_2*	4×10^{-22}	1.4	1.0
N_2*	13×10^{-22}	4.7	5.4

* Not f.c.c. structure.

the direct agreement with the simplest types of solids. We may also note that the latent heat of fusion is generally much lower than the latent heat of vaporization. From the point of view of *thermal energy* the liquid state, near the melting point, is much nearer to the solid state than it is to the gaseous state.

6.3.2 *Surface energy*

In spite of the discussions in sections 6.1.3 and 6.2.2 we shall greatly simplify our approach by assuming that the surface is a simple section of the crystalline solid, that no change in spacing of the surface layers occurs and that the temperature is far below that at which surface melting occurs. We now consider how the atoms in the surface differ from those in the bulk.

The atoms in the bulk of a solid are subjected to the attraction of atoms all round them, but at the free surface they are subjected only to the inward pull of the atoms within the solid. Consequently the surface layers have a higher energy than those in the bulk and this excess is called the free surface energy γ.

We may calculate this in terms of the bond energy for, say, a typical f.c.c. solid where all atoms except those in the surface have 12 nearest neighbours. An atom X in the surface has only 8 nearest neighbours. This is shown in figure 6.5(a). If the crystal contains a total of N atoms and the surface contains N_s atoms the total bond energy of the crystal, coun-

Figure 6.5. f.c.c. crystal of side $2a$. (a) The surface atom X has 8 nearest neighbours at B, B, B, B and C, C, C, C; whereas in the bulk each atom has 12 nearest neighbours. (b) If the crystal is pulled apart in the plane PQRS each atom such as X breaks 4 bonds.

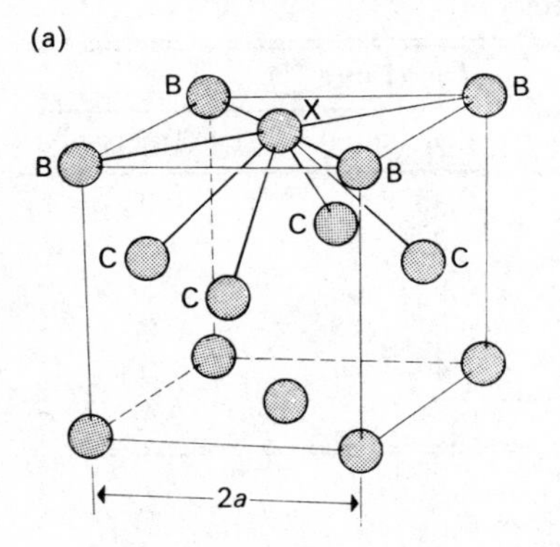

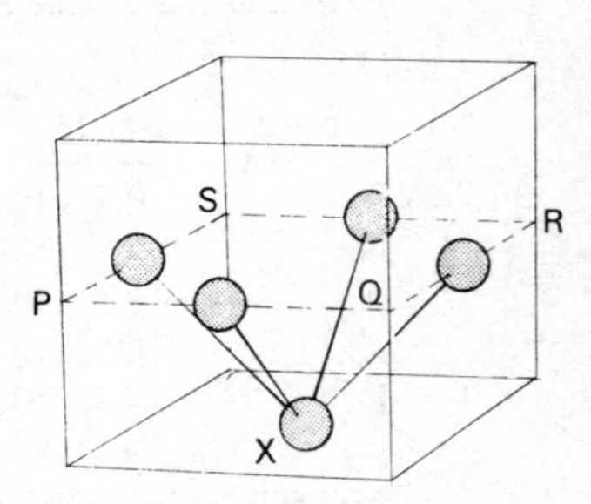

ting only nearest-neighbour interactions, is then

$$\tfrac{1}{2}\Delta\varepsilon[12(N-N_s)+8N_s]=\tfrac{1}{2}\Delta\varepsilon[12N-4N_s]. \tag{6.4}$$

(This assumes that there is no change in lattice spacing in the surface layers; in fact a small change in spacing does occur.)

If there were no surface atoms the energy would be $\tfrac{1}{2}\Delta\varepsilon\times 12N$. Thus the energy is reduced by $\tfrac{1}{2}\Delta\varepsilon\times 4N_s$. Since $\Delta\varepsilon$ is negative this is a positive increase in energy which is indeed the surface energy. If the surface area of the crystal is A,

$$\gamma A=\tfrac{1}{2}\Delta\varepsilon\times 4N_s. \tag{6.5}$$

We may obtain this result more simply by considering a rectangular bar of crystal of cross-sectional area A. Any section across the crystal contains N_s atoms, so if we pull the crystal in two each surface atom breaks four bonds and the work done per atom is $4\Delta\varepsilon$ (see figure 6.5(*b*)). Hence the total work done is $4\Delta\varepsilon\times N_s$. But the area exposed is $2A$. Hence

$$2\gamma A=4\Delta\varepsilon\times N_s. \tag{6.5a}$$

This is identical with equation (6.5). If in the crystal orientation we have considered there are z atoms per m^2 we see from equations (6.5) and (6.5*a*) that

$$\gamma=2\times\Delta\varepsilon\times z. \tag{6.6}$$

But the energy necessary to sublime z atoms from the bulk of the crystal (see equation (6.2)) is

$$L_S=6\Delta\varepsilon z. \tag{6.7}$$

Hence the surface energy is equal to one-third of the heat of sublimation of a single atomic layer of atoms. For different orientations the fraction will be a little different but generally it will be of the order of $\tfrac{1}{3}$ to $\tfrac{1}{2}$. For an f.c.c. lattice of side $2a$ as shown, the volume occupied by each atom is $2a^3$. If M is the molecular weight and ρ the density

$$\frac{M}{N_A\rho}=2a^3 \text{ or } a^3=\frac{M}{2N_A\rho}.$$

For the (100) face that we have considered the area occupied by each atom is $2a^2$. The number of atoms z per m^2 is $1/2a^2$. Hence

$$z=\frac{1}{2a^2}=\frac{1}{2}\left(\frac{2N_A\rho}{M}\right)^{\frac{2}{3}} \tag{6.8}$$

$$=0.8\left(\frac{N_A\rho}{M}\right)^{\frac{2}{3}}.$$

Hence

$$\gamma = \frac{0.8}{3}\left(\frac{L_S}{N_A}\right)\left(\frac{N_A\rho}{M}\right)^{\frac{2}{3}}. \tag{6.9}$$

This of course refers to a specific face and to a specific structure. For other faces and structures we should merely change the numerical factor by a small amount. A reasonable average value for the factor in front of equation (6.9) is 0.3. This relation should apply best to van der Waals solids such as solid neon, argon and krypton. However, the surface energy of such materials is not known. We have therefore applied the model to metals, the results being given in Table 6.4. The quantitative agreement between the calculated and observed values is poor, but the agreement in relative values is surprisingly good. If all the calculated values are multiplied by a correction factor of 0.4 the calculated values all agree with observation to within about 10 per cent. The basis for this correction factor is discussed in the next chapter.

There is another way of estimating the surface energy of van der Waals solids. We make use of the van der Waals forces given in equation (1.22*a*). We take two parallel blocks of unit area and allow them to come into atomic contact. The work done to separate them to infinity is 2γ. Ignoring retardation effects the attractive force per unit area is $A/6\pi x^3$. If we allow for repulsive forces we expect a term of the type B/x^n where n is a large number of the order 9–12 (a power law is only a crude approximation).

Table 6.4. *Surface energies of some metals in solid state*

Metal	MP (°C)	L_S (kJ mol^{-1}) at 25 °C	M (kg)	ρ (kg m^{-3})	γ (mJ m^{-2}) Calc.	Obs.*
Indium	156	240	0.115	7.3×10^3	1300	630
Lead	327	200	0.207	11.3×10^3	1200	560
Gold	1063	380	0.197	19.3×10^3	3300	1400
Iron	1537	400	0.056	7.9×10^3	4000	2100
Platinum	1769	570	0.195	21.5×10^3	5000	2500
Tungsten	3380	850	0.184	19.3×10^3	6700	2900

* Observed measurements are generally made at elevated temperatures just below the MP. Values of γ increase as the temperature is reduced reaching a maximum at 0 K where γ is 5 to 10 per cent greater than the values quoted in the table. See W. R. Tyson, 'Surface energies of solid metals', *Canadian Metallurgical Quarterly*, vol. 14 (1975), pp. 307–14.
Note: mJ m^{-2} = erg cm^{-2}.

Then for the net force per unit area

$$\sigma=\frac{A}{6\pi x^3}-\frac{B}{x^9}=\frac{A}{6\pi}\left[\frac{1}{x^3}-\frac{B^1}{x^9}\right] \tag{6.10}$$

where B^1 is an appropriate constant. At equilibrium separation there is no force between the surfaces, $\sigma=0$, so that equation (6.10) becomes

$$\sigma=\frac{A}{6\pi}\left[\frac{1}{x^3}-\frac{x_0^6}{x^9}\right]. \tag{6.11}$$

Integrating $\sigma\,\mathrm{d}x$ for x ranging from x_0 to infinity we have

$$\int_{x_0}^{\infty}\sigma\,\mathrm{d}x=\frac{A}{12\pi x_0^2}\left[1-\tfrac{1}{4}\right]=2\gamma. \tag{6.12}$$

We see that the value obtained is only 25 per cent less than we would have obtained if we had ignored repulsion (see figure 6.6). A more realistic relation for repulsion would give an even smaller error. For this reason, in order-of-magnitude calculations of this sort the repulsion is often neglected. Using a typical value of $A=2\times10^{-19}$ J and $x_0=3$ Å (3×10^{-10} m) we obtain $\gamma=29$ mJ m^{-1}. This is a good value for van der Waals solids

Figure 6.6. The force per unit area between two parallel surfaces as a function of separation x. The work done in separating the surfaces from atomic contact ($x=x_0$) to infinity is 2γ. Continuous curve – allowing for repulsive forces. Broken curve – assuming no repulsive forces until the surfaces come into atomic contact (hard-wall model).

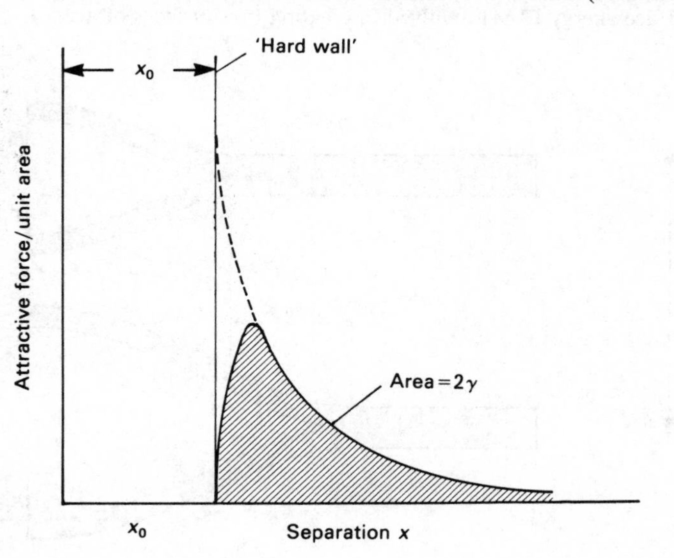

such as hydrocarbons and non-polar polymers. In fact for most polymer surfaces γ lies between 20 and 60 mJ m^{-2} being highest for those containing polar groups. For this reason most polymers are not wetted by water for which $\gamma = 72$ mJ m^{-1}.

With liquids, the surface energy is relatively easy to determine since it is the same as the surface tension. With solids for reasons to be described this is not so. With metals the surface energy may be determined by the zero-creep method (see figure 6.7(*a*)). Here a series of wires of fixed radius r are allowed to deform at high temperatures under different tensile loads W (more accurately, forces F). Heavily loaded wires extend; lightly loaded wires contract due to the surface energy. A load W_0 (force F_0) is found at which zero creep occurs. At first sight one would equate F_0 to $2\pi r\gamma$ but this is not correct: we have to consider a small notional extension of the surface and then specify that at F_0 the surface is unchanged. If A is the surface area of the wire, $A = 2\pi rl$, the surface energy is γA. Then for a notional extension dl we have

$$F\,\mathrm{d}l = \gamma\,\mathrm{d}A + A\,\mathrm{d}\gamma. \tag{6.13}$$

Figure 6.7. Determination of surface energy of solids (*a*) by finding the force F_0 at which there is zero creep: this occurs when $F_0 = \pi r\gamma$, (*b*) by bifurcation of a mica strip. The work done in opening up the strip is equal to the change in elastic strain energy plus the increase in surface energy ($2\gamma A$) involved in creating two surfaces of area A.

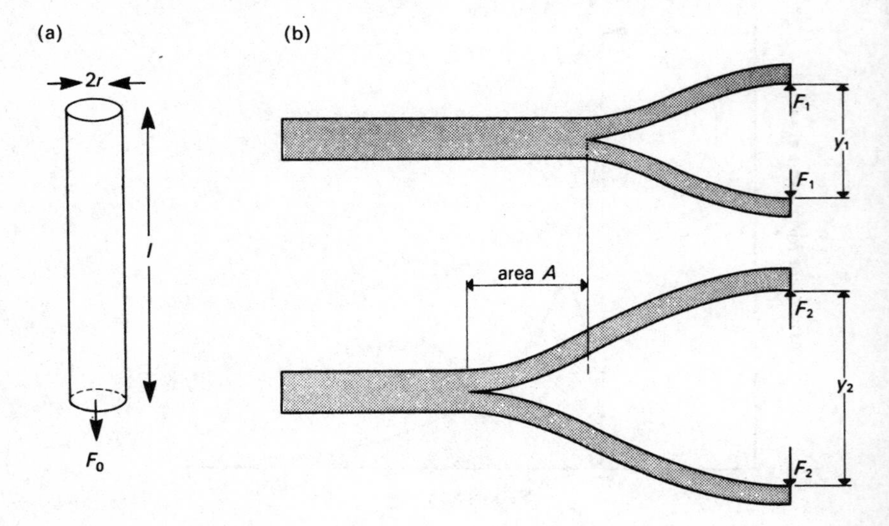

Zero creep occurs at F_0 when $d\gamma = 0$. Hence

$$F_0 \, dl = \gamma \, dA = \gamma 2\pi(r \, dl + l \, dr). \tag{6.14}$$

Assuming constant volume, $r^2 l$ = constant, so that

$$2r \, dl + r^2 \, dl = 0 \quad \text{or} \quad r \, dl = -(r/2) \, dr. \tag{6.15}$$

Substituting in equation (6.14) we have

$$F_0 = \pi r \gamma. \tag{6.16}$$

For most metals γ is of order 500–2000 mJ m^{-2}.

With a 'perfectly brittle' solid like mica the surface energy can be determined by finding the force required to open a crack. If we consider a cleaved specimen as shown in figure 6.7(*b*) we may allow the force F to move and to open up the crack as shown. The work done $= \int_{y_1}^{y_2} F \, dy$; this is expended in changing the elastic strain energy in the two mica leaves and also in creating two new surfaces each of area A. The work done can be measured, the elastic energy calculated and the difference may then be put equal to $2\gamma A$. For mica in a good vacuum γ has a value of about 5000 mJ m^{-2}; in air about 300 mJ m^{-2}. In general such experiments cannot be carried out satisfactorily with ductile solids such as metals or polymers, since the plastic work absorbed by deformation at the tip of the crack may be several hundred times greater than the surface energy. For example values of γ often come out to be of the order of 10^5 mJ m^{-2}. Therefore the experiments can be carried out only on brittle solids. Reasonable values have been obtained on glass (A. A. Griffith 1920) and as mentioned above on mica (J. W. Obreimov 1931, A. I. Bailey 1955). Recently Gilman has shown that the technique can be applied successfully to many crystalline solids if the experiments are carried out at liquid air temperature – he showed that at this temperature the ductility is 'frozen out'. Surface energies of the order of a few hundred mJ m^{-2} are obtained for materials such as NaCl, LiF.

We may mention here a simple feature involved in the surface energy of solids which differentiates it markedly from that of liquids. Suppose we have a surface of area A possessing surface energy γ. If we increase the area by a small increment the work done per unit increase may be written

$$\frac{d(A\gamma)}{dA} = \gamma + A \frac{\partial \gamma}{\partial A}. \tag{6.17}$$

For liquids the term $\partial\gamma/\partial A$ is zero, since, on account of atomic mobility, the structure of a liquid surface is unchanged when the surface area is increased. This is not so with solids; when the surface is stretched (even only notionally) the atoms are pulled apart and γ diminishes so that $\partial\gamma/\partial A$ is negative. This implies that with solids the surface energy is not the same as the surface force (more strictly the line tension in the surface; see Chapter 10 on the surface tension of liquids); with liquids they are identical.

Although one cannot equate surface energy with surface force there is generally some tension in the free surface. Even if it is as large as an effective line-tension of 1 N m^{-1}, this is minute compared with the elastic modulus of the solid, usually about 10^{11} N m^{-2}. Clearly forces of this magnitude can have very little visual effect on metals; so they do not influence the shape of a solid in the same way as they do liquids. A centimetre cube of copper is not dragged into spherical shape at room temperature because of surface tension. However, with very minute particles some such effects may be observed.

Surface energy of solids plays a very important part in sintering; and it is a crucial factor in the wetting of solids by liquids. From the point of view of atomic mechanisms it is worth emphasizing that the same forces which produce the pressure defect in a real gas, are responsible for the surface energy of a solid and the surface tension of a liquid.

6.3.3 *Elastic stiffness*

If atoms or molecules in a solid are subjected to a stress and thereby displaced from their equilibrium positions the interatomic forces will tend to restore them to their original positions. If the distortions are not too large, when the stress is removed the atoms will (under ideal conditions) return to exactly the same situation as before; the elastic modulus is defined as the ratio of the applied stress to the deformation produced. This will be discussed in detail in a later chapter. Here we merely quote the result for Young's modulus which is defined as

$$E = \frac{\text{tensile stress increment } \Delta\sigma}{(\text{increase in length } \Delta x)/(\text{original length } x)} = x\frac{\Delta\sigma}{\Delta x}. \tag{6.18}$$

The elastic modulus of a solid may be easily demonstrated if one considers the force–displacement curve. Figure 6.8 shows the force F between two neighbouring atoms as a function of separation. At low temperatures, where thermal energy can be neglected, the equilibrium separation occurs where the attractive force equals the repulsive force. Thus OA is the value of the equilibrium separation x. The slope of the line ZAY through A is dF/dx. Hence the quantity $x(dF/dx)$ is the exact *one-dimensional* analogue of Young's modulus as defined above.

If the solid is heated, thermal expansion (see below) shifts the equilibrium positions from A to A′. The slope dF/dx is seen to be smaller and this decrease is appreciably greater than the corresponding increase in the new equilibrium separation O′A′. Thus the effective modulus is reduced. If we tried to construct a simple 3-D cubic structure out of these pairs of atoms the number in any square metre of an atomic plane would be $1/x^2$ so that the stress (as distinct from the force) is F/x^2 and the strain is dx/x. Then Young's modulus becomes $(1/x)\, dF/dx$. In this case we see that the slight increase in x, with rise in temperature, augments the decrease in dF/dx so that both terms of the modulus are reduced by a rise in temperature. However, as already mentioned, the change in x is in general small compared with the change in dF/dx since F is usually a 'strong' power function of x.

We conclude that if the behaviour of the solid as a whole resembles that of the diatomic model discussed above the elastic modulus decreases as the temperature is increased. This is true of practically all solids except rubber-like solids.

Figure 6.8. (*a*) Force-separation diagram illustrating the decrease in elastic modulus with expansion. (*b*) Potential energy–separation diagram illustrating the increase in separation which occurs when the temperature is increased (thermal expansion).

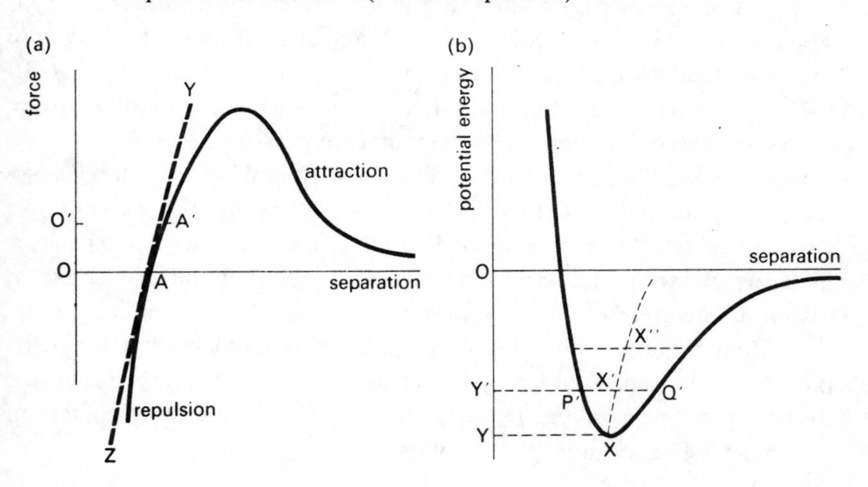

6.3.4 *Thermal expansion*

If the length of a solid specimen at 0 °C is l_0 and at t °C is l_t one can usually write

$$l_t = l_0(1 + \alpha t),$$

where α is the coefficient of linear expansion and for most solids is of order 10^{-5} °C^{-1}.

This behaviour may again be explained in terms of the forces between the individual atoms. It is most convenient to plot the potential energy curve as a function of separation. If we do this for two atoms we obtain the typical curve shown in figure 6.8(*b*). At absolute zero the equilibrium position (ignoring zero-point energy) is at X, so the equilibrium separation is YX. As the temperature is raised the energy rises to, say, Y′ and one of the atoms will oscillate relative to the other between positions P′ and Q′ about a mean position X′. Because of asymmetry in the shape of the potential energy curve the length Y′X′ is greater than YX. Thus the mean separation increases with temperature along the curve XX′X″ . . . .

If an assembly of atoms in a solid behaved similarly this would explain thermal expansion. Of course there are added complications. In a solid one ought to treat this problem in terms of the free energy. This involves concepts beyond the scope of this book; but we shall discuss briefly the main ideas in Chapter 9, pp. 237–41. Again in the diatomic model the atoms dissociate when the energy reaches 0. This is equivalent to sublimation of the solid and tells us nothing of the melting process that occurs somewhere *en route*. This is because melting is a cooperative process involving many atoms and a satisfactory model must be multi-dimensional to allow for their behaviour (see Chapter 10). Finally we may note that increased temperature may give rise to transverse oscillations. In some types of crystals this may actually lead to a contraction in one crystallographic direction as the temperature is raised.

Nevertheless the simple model given above provides a reasonably satisfactory explanation of thermal expansion in solids. In a later chapter we shall indeed use it to derive an explicit relation for the thermal expansion in terms of intermolecular forces. We shall see that the results are in quantitative agreement with experiment. In general the potential energy-separation curve provides a very useful way of approaching many of the physical and mechanical properties of solids. It is particularly applicable to the crystalline state but may also be used, perhaps with less precision to the behaviour of amorphous solids.

6.4 Amorphous solids (glasses)

6.4.1 *The amorphous state*

In the amorphous state the individual atoms or molecules are in ordered array over a short range but there is no long-range order. They are indeed, in structure, very much like an instantaneous snapshot of a liquid and are sometimes called supercooled liquids. However, molecular movement, a characteristic feature of liquids, is minute and the attempt to describe their flow properties in terms of an equivalent viscosity leads to astronomical values of the viscosity. For example, from the viscosity of certain silicate glasses at elevated temperature it is possible to extrapolate and deduce a room temperature value of the order of 10^{69} Pa s^{-1}.

It has been pointed out by F. W. Preston† that this is equivalent to stating that if the glass were in the form of a rod of length equal to the circumference of the Universe and was subjected to half its breaking tension for a period of time equal to the age of the Universe, its total extension could amount to less than one thousandth of the diameter of an electron. For such materials, room temperature viscosity is clearly not a very useful physical concept.

6.4.2 *Time–temperature transformation*

In principle, all crystalline solids can also exist in the amorphous or glassy state if they can be cooled fast enough from the liquid state. The general idea is that as the liquid cools, crystal nuclei are formed; as the temperature drops, the nucleation rate and the size of the nuclei increase. However, the liquid becomes increasingly viscous and this slows up the growth rate until further crystallization is virtually impossible. Typical changes in viscosity with temperature are shown in figures 6.9(*a*) and (*b*). Figure 6.9(*a*) is for slow cooling where, below the melting point T_m, the material becomes totally crystalline: the viscosity, if such a term can be used for a crystal, rises abruptly. Figure 6.9(*b*) is for rapid cooling where below T_m the viscosity of the supercooled liquid continuously increases. It is customary to regard a supercooled liquid for which the viscosity is greater than 10^{15} poise as a glassy or amorphous *solid.*

The cooling rate necessary to achieve the glassy state depends on the ease of crystal formation and on the viscosity of the cooling liquid as its temperature is reduced. If crystal growth is slow and the liquid has a high viscosity over an appreciable temperature range, it is relatively easy to form a glass. This applies to inorganic glasses and to polymers (to be

† F. W. Preston, *Journal of Applied Physics*, vol. 13 (1942), p. 623.

discussed in Chapter 13). With simpler solids, where crystallization is easy and the liquid has a narrow viscosity-temperature range, high cooling rates are necessary of the order of 10^3–10^5 K s^{-1}. This applies to pure metals, metallic alloys and solids such as silicon and germanium. The highest cooling rate (10^7 K s^{-1}) may be obtained by evaporating or sputtering the substance onto a cold surface. This virtually short-circuits the problems of nucleation and growth as indicated by the dotted line *XY* in figure 6.10 which is a typical time–temperature transformation curve.

The structure of amorphous silicon and germanium has been widely studied because of their extensive use in electronic devices. An attempt to indicate the three-dimensional arrangement of atoms in amorphous silicon is shown in figure 6.11(*a*). The distances between first and second nearest neighbours are similar to those in the crystalline form but there is distortion in the bond angles. It is not surprising that the radial distribution function (see Chapter 10 on the liquid state) resembles that of a liquid.

Figure 6.9. The effect of temperature on the viscosity of crystalline and amorphous materials. (*a*) With crystalline materials the transition from liquid to solid is sharp at the melting point, unless exceptionally high cooling rates are employed. (*b*) With amorphous materials the viscosity increases rapidly as the temperature is reduced. When the viscosity exceeds 10^{15} poise the material is considered to be a glassy solid.

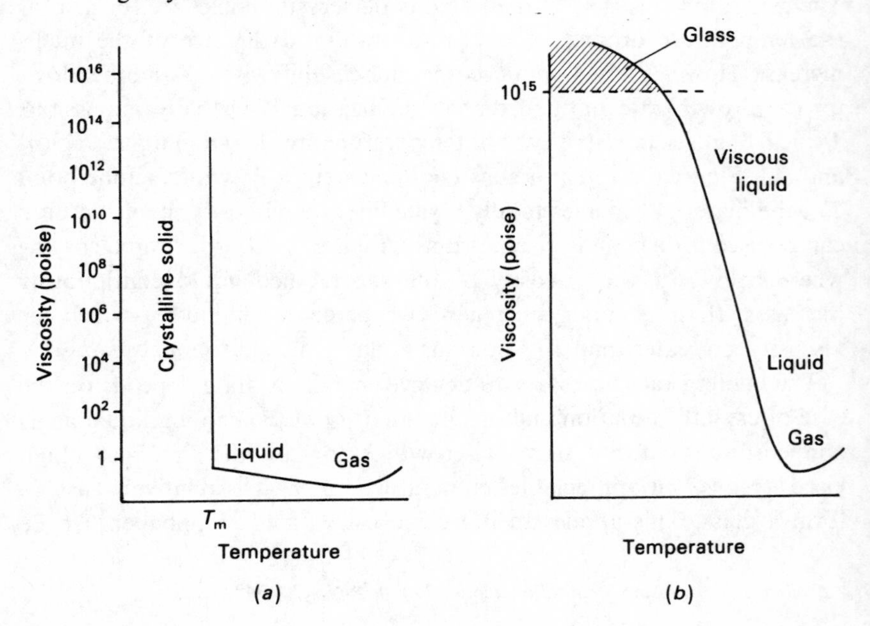

Figure 6.10. Schematic time–temperature–transformation (T–T–T) curve. The time, expressed logarithmically, is the time taken for a certain fraction of the material, say 10^{-6}, to become crystalline. Above T_m the material is liquid. At A crystallization commences, nucleation is extremely slow – large times are involved. As the temperature decreases nucleation is more rapid and the time is shorter (points B and C). Viscosity increases as temperature falls. At D the material is crystallizing at its maximum rate (temperature T_d). At lower temperatures the increase in viscosity makes nucleation more difficult until at T_g (point F) crystallization is virtually impossible and the material is glassy. The dotted line XY shows the effect of extremely rapid cooling.

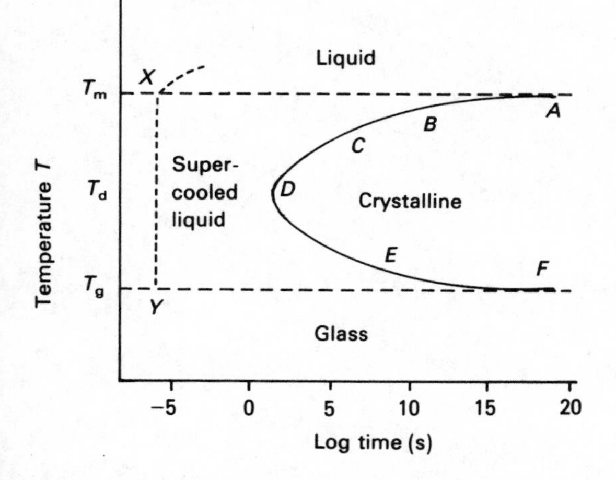

Figure 6.11. Two-dimensional projection of (*a*) amorphous silicon (*b*) amorphous silica; ● Silicon, ○ Oxygen. From Z. H. Zachariasen, *J. Am. Chem. Soc.* **54**, 3841 (1932). Later work shows that the silicon is four-fold coordinated.

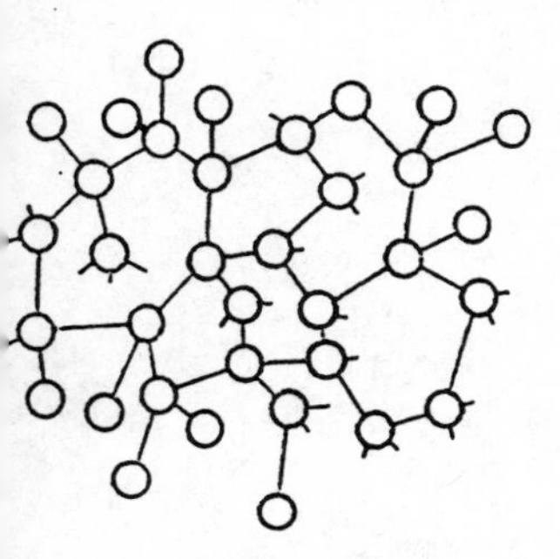

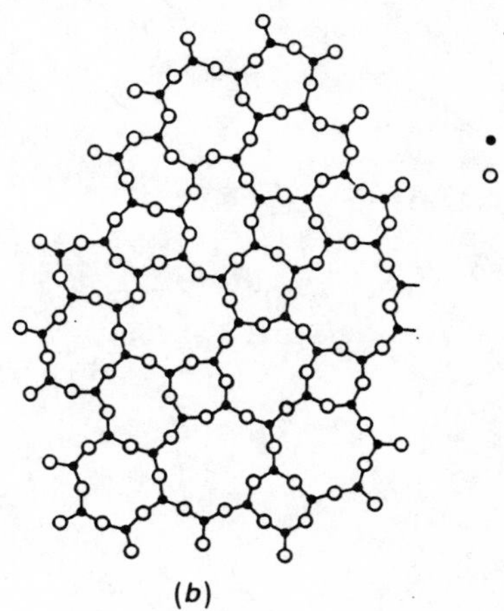

Indeed, as the previous discussion has indicated, the amorphous state is like a liquid in which molecular motion has been frozen out.

The crystalline state has, thermodynamically, a lower energy than the amorphous state. However, this does not mean that amorphous materials are unstable since the activation barrier (see pp. 44–5) to achieve structural changes in the solid state may be very high. This is one reason why glass utensils do not often spontaneously devitrify, i.e. recrystallize. In practice, appropriate constituents are added to the glass to make recrystallization more difficult.

7

The elastic properties of solids

If we subject a solid rod to tension we pull the atoms further apart; if we compress it we push them closer together. The interatomic forces resist these changes and, if the distortions are not too large, the body will return to its original dimensions when the external forces are removed; the deformations are then called elastic.

7.1 Some basic elastic properties

7.1.1 *Elastic moduli*

If a solid rod of uniform cross-sectional area A is pulled with a tensile force F the tensile stress is defined as

$$\sigma = \frac{F}{A}.$$

If, as a result, the rod increases its length from l to $l + \Delta l$ the fractional increase $\Delta l/l$ is called the linear strain ε_l (see figure 7.1(*a*)). Within the elastic range Hooke showed that the stress is proportional to the strain.

Figure 7.1. (*a*) Deformation in tension, (*b*) deformation in shear, (*c*) deformation under hydrostatic pressure.

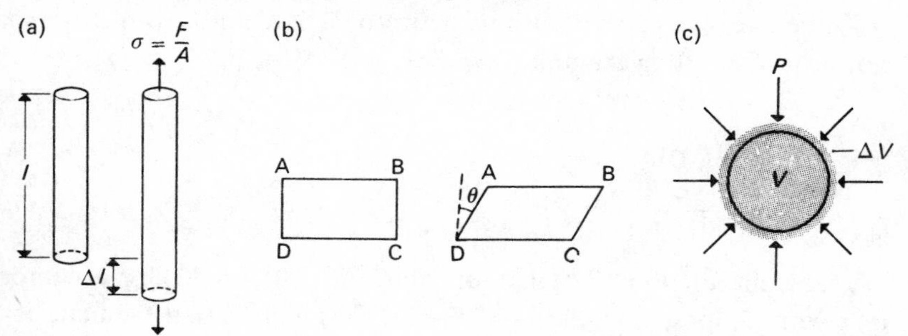

The ratio is defined as Young's modulus E:

$$E=\frac{\text{tensile stress}}{\text{linear strain}}=\frac{F/A}{\Delta l/l}. \tag{7.1}$$

A solid may also be deformed in shear. For example if we consider a rectangular element as in figure 7.1(*b*), we may apply a shear stress τ (force ÷ area) across the faces AB and CD. This must be accompanied by equal shear stresses τ along CB and AD so that the net couple is zero (see Chapter 8, section 8.1.2). The element is then distorted through a small angle θ which is termed the shear strain. The strain is again found to be proportional to the stress and the ratio is termed the rigidity modulus n:

$$n=\frac{\tau}{\theta}. \tag{7.2}$$

A solid may also be deformed by a hydrostatic pressure P. If the solid has volume V, and the change in volume is $-\Delta V$, then the volume strain is $-\Delta V/V$ and this is found to be proportional to P. The ratio is the bulk modulus K.

$$K=\frac{P}{-\Delta V/V}. \tag{7.3}$$

The reciprocal of this modulus is known as the compressibility, β.

Finally, when a body is extended in tension to give a linear strain ε_l, it contracts in directions at right angles to the direction of tension. The fractional contraction ε_t in the transverse direction is found to be proportional to ε_l. The ratio

$$\frac{\text{contractile strain}}{\text{linear strain}}=\frac{\varepsilon_t}{\varepsilon_l}=\text{Poisson's ratio } \nu. \tag{7.4}$$

For most solids ν has a value between $\frac{1}{4}$ and $\frac{1}{3}$ and this implies a decrease in volume under tension. If $\nu=\frac{1}{2}$ (as is nearly the case with rubber) there is no volume change in tension or compression.

There are several relationships between E, n, K and ν which we shall not prove here. For example

$$n=\frac{E}{2(1+\nu)}, \tag{7.5}$$

$$K=\frac{E}{3(1-2\nu)}. \tag{7.6}$$

We see from equation (7.6) that for values of $\nu=0.25$ to 0.33, K is comparable with E; for a value of $\nu=\frac{1}{2}$, $K=\infty$. This implies that the material is incompressible, which is what would be expected since tension and compression produce no volume change in such a material.

7.1.2 *Torsional rigidity*

If a circular tube of internal radius r_1, external radius r_2, length l is held at one end and subjected to couple Γ at the other end, every part of the tube is subjected to shear. Suppose the free end is twisted through an angle ϕ. The ratio Γ/ϕ is known as the torsional rigidity. We shall not derive it but simply quote the relationship for Γ/ϕ to be:

$$\frac{\Gamma}{\phi}=\frac{n\pi}{2l}(r_2^4-r_1^4), \tag{7.7}$$

where n is the rigidity modulus.

7.1.3 *The bending of a cantilever*

If a bar is encastered at one end and deflected sideways by a force F at the other end, the deflection produced depends on F, on the section of the bar, its length and its Young's modulus. Although we shall not derive the equation here, we shall indicate the principle of the analysis involved. Suppose figure 7.2(*a*) represents, in exaggerated form, the bending of a beam. The sections p, q, r, s, t indicate how the beam is deformed. We make a notional cut through the bar at section r; then the material to the left of r is subjected to a bending moment Fx. If the beam is to be in equilibrium this must be counterbalanced by a reverse moment of the same magnitude. The material above AB is stretched and so exerts a tensile force to the left. The material below AB is compressed and so exerts a resisting force to the right. The beam bends to just that point where these two forces produce a couple exactly equal to Fx. This implies that between the stretched and compressed portions there is a line AB which is unchanged in length. AB is, in fact, a part of the line XX′ which represents the original length of the unstressed bar; it is known as the neutral axis. We may note in passing that the section between s and t has to carry the smallest bending moment and is therefore distorted least: the end part of the beam is indeed straight. On the other hand, the bending moment is a maximum at CD; here the curvature is a maximum. This raises an interesting point which is usually ignored by physicists and engineers. The encastered portion CDEF is considered completely rigid and subjected to zero bending moment. This implies that at the boundary CD the bending moment suddenly drops from a maximum value on the right to zero on the left. This is impossible, for there must be a transitional region HK within the encastered zone (figure 7.2(*b*)) where the bending moment gradually falls to zero. In practice this implies a certain amount of slip between the bar and the clamp at ED and FC or some other type of yielding in the clamped zone. This is not of general importance in calculating the deflection of a loaded bar. It can, however, be of very

Figure 7.2. (*a*) Elastic bending of a beam encastered at one end and subjected to a force F at the other. The line XX is the neutral axis where there is neither extension nor compression of the original length. (*b*) The bending moment has a minimum at the free end and a maximum at the point where the beam enters the encastered zone. Simple theory assumes that it then drops at once to zero; in practice there must be some transition region HK. (*c*) The bifurcation region of a mica specimen resembles the encastered zone of a bent beam.

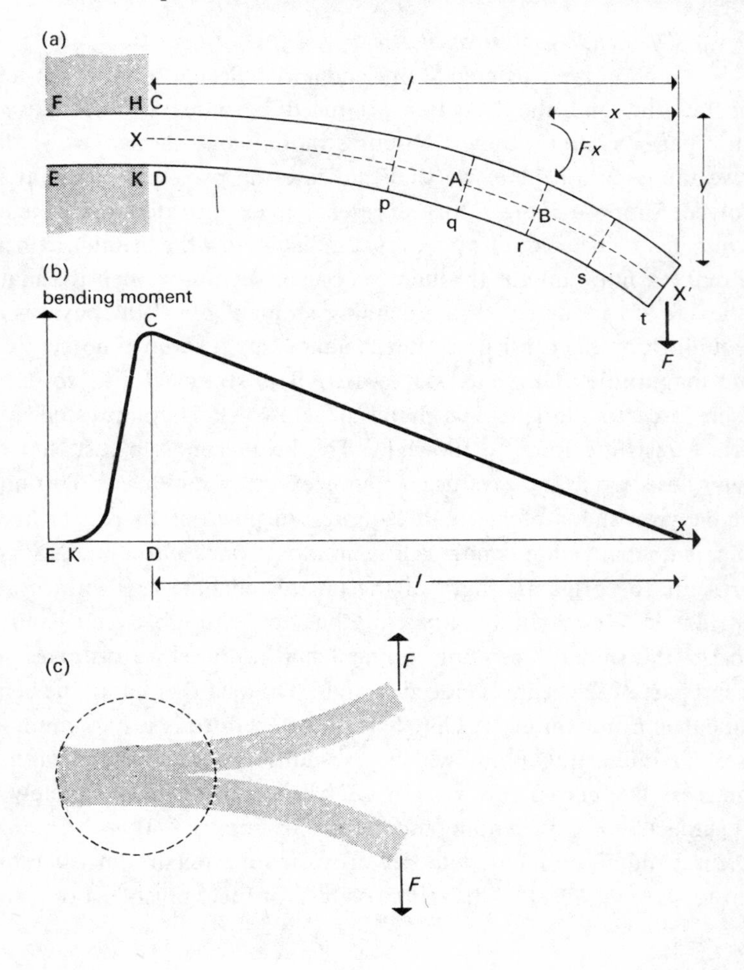

great importance; e.g. in the propagation of a crack, in fretting, or in the splitting of mica described earlier as a means of determining the surface energy. The conditions at the region of bifurcation resemble those of the encastered region in a bent beam (figure 7.2(*c*)). The analysis of the forces in this region is very difficult indeed.

If the second moment of area of the section of the beam about the vertical axis is I and the length of the beam is l, the deflection y produced at the end by a force F is given by

$$y=\frac{F}{EI}\frac{l^3}{3}. \tag{7.8}$$

For a beam of rectangular section, thickness a, width b, this becomes

$$y=\frac{F}{E}\times\frac{4l^3}{ab^3}. \tag{7.9}$$

7.1.4 *Thermal fluctuations in the length of an elastic bar*

Consider a bar of length L, cross-section A, Young's modulus E. If we apply a force F, the stress is $\sigma=F/A$ and the strain ε is then $(1/E)\times$ stress. Hence

$$F=EA\varepsilon. \tag{7.9a}$$

If an incremental force $\mathrm{d}F$ produces an incremental strain $\mathrm{d}\varepsilon$, the incremental extension is $L\,\mathrm{d}\varepsilon$ so that the force does work

$$FL\,\mathrm{d}\varepsilon=EAL\varepsilon\,\mathrm{d}\varepsilon. \tag{7.9b}$$

In stretching the bar from its initial state to a final extension ε_0 the work W done is the integral of equation (7.9*b*)

$$W=\tfrac{1}{2}EAL\varepsilon_0^2. \tag{7.9c}$$

This is the elastic strain energy and will be positive whether the bar is extended ($\varepsilon+ve$) or compressed ($\varepsilon-ve$).

If as a result of thermal fluctuations the bar undergoes extension or contraction equivalent to a strain ε_0 the bar will possess potential energy $\frac{1}{2}EAL\varepsilon_0^2$ over and above its average thermal energy. This is proportional to the square of a parameter (ε_0) and may therefore be treated as a single degree of freedom. The average value of $\overline{\varepsilon_0^2}$ is thus given by

$$\tfrac{1}{2}EAL\overline{\varepsilon_0^2}=\tfrac{1}{2}kT$$

$$\overline{\varepsilon_0^2}=\frac{kT}{EAL}. \tag{7.9d}$$

For a bar where $A=100\ \mathrm{mm}^2$, $L=1\ \mathrm{m}$, $E=10^{11}\ \mathrm{N\ m^{-2}}$ (say brass) we find that at $T = 300$ K, $\overline{\varepsilon_0^2} = 4 \times 10^{-28}$ m^2. The r.m.s. extension or contraction due to thermal fluctuations is thus of order 2×10^{-14} m or less than a ten-thousandth of the diameter of an atom.

7.1.5 *Young's modulus in terms of interatomic force constant*

In the previous chapter we showed how the elastic modulus is related to the force–displacement curve. We shall now make this more quantitative. Consider a solid composed of a single species of atom and let us assume a simple cubic structure, the separation between each atom being a. Let the bar have unit cross-sectional area and consider three neighbouring planes I, II and III (figure 7.3). We assume that there are only nearest-neighbour interactions; then if an atom q in plane II is displaced in a direction normal to the plane, it experiences a restoring force only from atom p in plane I and atom r in plane III. If a tensile force f is applied to atom r the spacing between the atoms in that row will each be increased by a small distance x (where $x \ll a$). The restoring force which each atom in this row experiences will be proportional to x. Let the restoring force be kx, where k is, in fact, a measure of the slope of the force–displacement curve shown in the previous chapter (figure 6.8(a)).

Since there are $1/a^2$ atoms per m^2, the force per m^2 (the stress) to achieve a separation of the planes of this amount will be $f/a^2 = kx/a^2$.

Figure 7.3. Simple cubic array of atoms, (a) undeformed, (b) when a force f is applied to each atom in plane III all the planes undergo an increased separation x, (c) a single force f acts in a similar way on a single atom q.

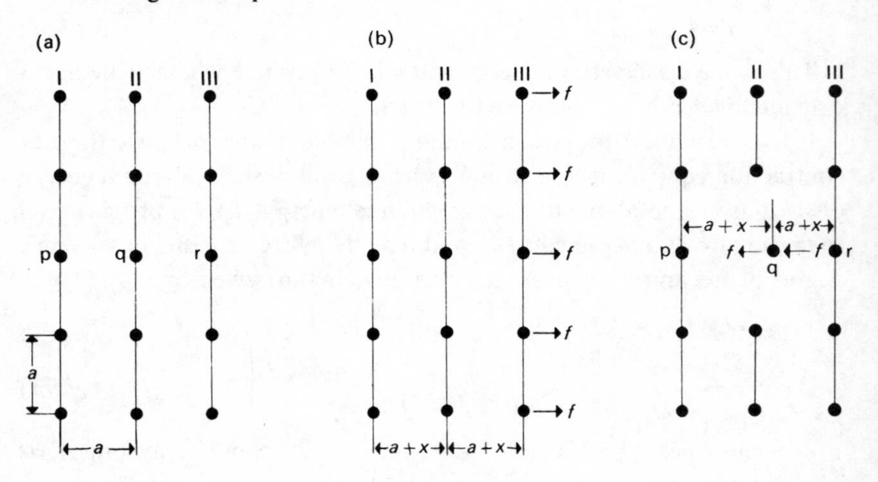

This displacement is equivalent to a linear strain of amount x/a. Then we may write for Young's modulus

$$E=\frac{\text{tensile stress}}{\text{strain}}=\frac{kx/a^2}{x/a}=\frac{k}{a}. \tag{7.10}$$

7.2 **Propagation of longitudinal waves along an elastic bar**

We extend this to a very simple description of wave propagation. Our major (and weakest) assumption is that we may isolate the interaction of one atom with its next neighbour. Suppose we hold planes I and III fixed (see figure 7.3(c)) and displace atom q relative to atoms p and r by an amount x. Then p exerts an attractive force kx on q and r exerts a repulsive force of the same magnitude. The resulting restoring force is $2kx$. If the atomic mass is m the equation of motion is

$$m\ddot{x}+2kx=0.$$

This gives simple harmonic motion of period

$$T=2\pi\left(\frac{m}{2k}\right)^{\frac{1}{2}}. \tag{7.11}$$

Suppose now we give an impulse to the free end of the bar and displace the row of atoms in plane I to the right (figure 7.3). These then act on the row of atoms in plane II which in turn are pushed to the right and so act on the atoms in plane III. In this way the impulse is propagated through the solid. The time taken for each plane to respond to the push of the neighbouring plane is of the order of one-quarter of the vibrational period T. Thus the impulse jumps from one plane to the next in a time

$$t\approx\frac{2\pi}{4}\left(\frac{m}{2k}\right)^{\frac{1}{2}}\approx\frac{2\pi}{5.7}\left(\frac{m}{k}\right)^{\frac{1}{2}}. \tag{7.12}$$

For our purpose we may take $2\pi/5.7$ to be unity. The disturbance thus travels a distance a in time $(m/k)^{\frac{1}{2}}$. The velocity of propagation is therefore

$$v=\frac{a}{t}=a\left(\frac{k}{m}\right)^{\frac{1}{2}},$$

or

$$v=\left(\frac{a^3}{m}\frac{k}{a}\right)^{\frac{1}{2}}. \tag{7.13}$$

We recognize m/a^3 as the density ρ, and the quantity k/a as the Young's modulus E. Hence

$$v=\left(\frac{E}{\rho}\right)^{\frac{1}{2}}. \tag{7.14}$$

This is the same as the equation for the velocity of a longitudinal wave in a bar in terms of its macroscopic properties. It shows that the vibrating atom is the 'messenger' for wave propagation.

The frequency of vibration is

$$\nu=\frac{1}{T}=\frac{1}{2\pi}\left(\frac{2k}{m}\right)^{\frac{1}{2}}=\frac{\sqrt{2}}{2\pi a}\,v,$$

$$\nu=\frac{1}{2\pi a}\left(\frac{2E}{\rho}\right)^{\frac{1}{2}}. \tag{7.15}$$

For, say, iron we have $E=2\times10^{11}\ \mathrm{N\,m^{-2}}$, $\rho=7.8\times10^{3}\ \mathrm{kg\,m^{-2}}$, $a=2.5\times10^{-10}$ m. Then $\nu\approx5\times10^{12}$ vibrations per second. This is the right order of magnitude.

The main defect in this model is that the vibrations of the atoms are not isolated but are coupled. A rigorous treatment does not, however, substantially alter the nature of this analysis.

7.3 Bulk moduli

7.3.1 *Bulk modulus of an ionic solid*

We consider here an ionic solid of the NaCl type in which the Na^{+} and Cl^{-} ions are distributed on a cubic lattice as shown in figure 7.4. We consider a single ion and its coulombic interaction with all the charged ions around it. As the figure shows there are:

6 neighbours of opposite sign at distance x

Figure 7.4. The structure of the NaCl crystal. (*a*) General view; (*b*) 6 neighbours of opposite sign at distance x; (*c*) 12 neighbours of same sign at distance $\sqrt{2}x$; (*d*) 8 neighbours of opposite sign at distance $\sqrt{3}x$.

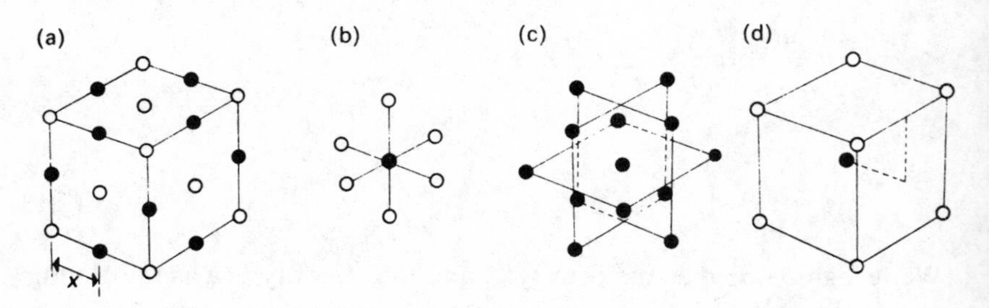

12 neighbours of same sign at distance $\sqrt{2}x$

8 neighbours of opposite sign at distance $\sqrt{3}x$, etc.

If each ion has charge e (or $-e$) the potential energy of one ion in relation to all its neighbours is

$$\frac{-e^2}{x}\left(6-\frac{12}{\sqrt{2}}+\frac{8}{\sqrt{3}}\ldots\right)\frac{1}{4\pi\varepsilon_0}=\frac{-e^2}{x}(2.1\ldots)\frac{1}{4\pi\varepsilon_0}. \tag{7.16}$$

Proper summation to infinity shows that the correct value, known as the Madelung constant A, is 1.748, say 1.75, for the rock-salt structure.

Assuming a repulsive potential energy term between each ion pair of the form A/x^9, the summation of all the repulsion terms will give B/x^9, where B is an appropriate constant. The potential energy of one ion allowing for both attraction and repulsion is

$$u=\left(-1.75\frac{e^2}{x}+\frac{B}{x^9}\right)\frac{1}{4\pi\varepsilon_0}. \tag{7.17}$$

If there are N positive and N negative ions per mole the potential energy for a mole of NaCl is

$$U=\tfrac{1}{2}(2N_A u)=N_A\left(-1.75\frac{e^2}{x}+\frac{B}{x^9}\right)\frac{1}{4\pi\varepsilon_0}, \tag{7.18}$$

where the $\frac{1}{2}$ is introduced so that we do not count each bond twice.

The equilibrium spacing σ occurs when U is a minimum.

Putting $\partial U/\partial x$ at $x=a$ equal to zero we find that B has the value $1.75e^2a^8/9$. Equation (7.18) becomes

$$U=-1.75N_Ae^2\left(\frac{1}{x}-\frac{a^8}{9x^9}\right)\frac{1}{4\pi\varepsilon_0}. \tag{7.19}$$

The next step appears complicated but is basically an attempt to connect the internal energy with an elastic modulus, in this case the bulk modulus K. If we apply a hydrostatic pressure P this produces a volume change $-\mathrm{d}V$, and the work done by the external forces on the crystal is then $-P\,\mathrm{d}V$. The ions are squeezed closer together and if the whole of this work goes solely in increasing that part of U which arises from interatomic forces, we have $-P\,\mathrm{d}V=\mathrm{d}U$ or

$$P=-\frac{\partial U}{\partial V}. \tag{7.20}$$

But

$$K=-\frac{\mathrm{d}P}{\mathrm{d}V/V}=\frac{\partial^2 U}{\partial V^2}V. \tag{7.21}$$

We need to express this in terms of the ionic spacing x. We have

$$\frac{\partial^2 U}{\partial V^2}=\frac{\partial}{\partial V}\left(\frac{\partial U}{\partial V}\right)=\frac{\partial}{\partial V}\left(\frac{\partial U}{\partial x}\times\frac{\mathrm{d}x}{\mathrm{d}V}\right)=\frac{\partial}{\partial x}\left(\frac{\partial U}{\partial x}\times\frac{\mathrm{d}x}{\mathrm{d}V}\right)\frac{\mathrm{d}x}{\mathrm{d}V}$$

$$=\frac{\partial^2 U}{\partial x^2}\left(\frac{\mathrm{d}x}{\mathrm{d}V}\right)^2+\frac{\partial U}{\partial x}\left(\frac{\mathrm{d}^2 x}{\mathrm{d}V^2}\right)+\ldots. \tag{7.22}$$

At the equilibrium separation $(x=a)$, $(\partial U/\partial x)=0$; consequently the second term vanishes so that

$$K=V\frac{\partial^2 U}{\partial x^2}\left(\frac{\mathrm{d}x}{\mathrm{d}V}\right)^2. \tag{7.23}$$

We now relate V to the atomic spacing x. For one mole

$$V=2N_{\mathrm{A}}x^3.$$

Therefore

$$\frac{\mathrm{d}x}{\mathrm{d}V}=\frac{1}{6N_{\mathrm{A}}x^2}. \tag{7.24}$$

To obtain K from equation (7.23) we differentiate equation (7.19) twice to obtain $\partial^2 U/\partial x^2$ and use equation (7.24) for $(\mathrm{d}x/\mathrm{d}V)^2$. We obtain

$$K=VN_{\mathrm{A}}1.75e^2\left(\frac{-2}{x^3}+\frac{10a^8}{x^{11}}\right)\frac{1}{36N_A^2x^4}\frac{1}{4\pi\varepsilon_0}. \tag{7.25}$$

At equilibrium when $x=a$ and $V=2N_{\mathrm{A}}a^3$ we find

$$K=\left(\frac{7.0}{9}\times\frac{e^2}{a^4}\right)\frac{1}{4\pi\varepsilon_0}=0.78\frac{e^2}{a^4}\frac{1}{4\pi\varepsilon_0}. \tag{7.26}$$

For rock-salt, assuming complete charge separation on the ions, e is the electronic charge, 1.6×10^{-19} C. The equilibrium spacing $a=2.82\times10^{-10}$ m. Substituting in equation (7.26) we obtain

$$K=2.8\times10^{10}\ \mathrm{N\ m^{-2}},$$

whereas the experimental value is about 2.4×10^{10} N m^{-2}. Taking into account the basic simplicity of the model this is surprisingly good agreement. It implies, of course, that the bonding is fully ionic. In most 'ionic' crystals only part of the bonding is ionic, the rest is covalent.

We may also express K in terms of the bond energy U_0 of the lattice in its equilibrium state. Substituting $x=a$ in equation (7.19) we have

$$U_0=-N_{\mathrm{A}}\frac{14.0}{9}\frac{e^2}{a}\frac{1}{4\pi\varepsilon_0}. \tag{7.27}$$

Hence from equation (7.26)

$$K=\left(\frac{7.0}{9}\times\frac{e^2}{a^4}\right)\frac{1}{4\pi\varepsilon_0}=\frac{-U_0}{N_A a^3}=\frac{-U_0}{V}, \tag{7.28}$$

where V is the volume of one mole. This is shown in Table 7.1 and we shall see that a similar result is obtained for van der Waals solids.

It is interesting to note that, for a free NaCl molecule in the vapour state, the distance between the Na^+ and Cl^- ions is 2.36×10^{-10} m compared with 2.82×10^{-10} m in the crystal lattice. This indicates the important part played by the surrounding ions of opposite sign in opening up the ionic spacing. It is relatively easy to make a rough quantitative estimate of this. For a single ionic pair in the free molecule we may write

$$u=\left(\frac{-e^2}{r}+\frac{A}{r^9}\right)\frac{1}{4\pi\varepsilon_0}. \tag{7.29}$$

The equilibrium spacing r_0 occurs when $\partial u/\partial r=0$. This gives the value of $A=r_0^8e^2/9$. Hence

$$u=-e^2\left(\frac{1}{r}-\frac{r_0^8}{9r^9}\right)\frac{1}{4\pi\varepsilon_0}. \tag{7.30}$$

In the crystal the coulombic interaction of the ionic charges gives alternative positive and negative energies; as we saw above, the final amount for each ion is $-1.75e^2/r$, where 1.75 is the Madelung constant. The repulsion term always involves a positive energy; however, the potential falls off so rapidly with distance that we need only consider the 6 nearest neighbours at distance r for which the repulsion energy is $6A/4\pi\varepsilon_0 r^9$ or

Table 7.1. *Typical results for alkali halides, assuming repulsive potential* A/x^9

Halide	a 10^{-10} m	$V=2N_Aa^3$ 10^{-6} m^3	U_0 kJ	U_0/V N m^{-2}	K N m^{-2}	Ratio $K:U_0/V$
NaF	2.33	15.1	899	5.9	5.3	0.90
NaCl	2.82	27.0	752	2.8	2.4	0.86
NaBr	2.98	31.8	718	2.3	2.0	0.87
NaI	3.24	40.9	672	1.6	1.4	0.87

Data from Moelwyn Hughes, *Physical Chemistry*, Pergamon Press (1960).
The theoretical ratio is unity (see equation (7.28)). Details depend on whether the bond is 100 per cent ionic and on the assumed repulsive potential index.

$6r_0^8e^2/4\pi\varepsilon_0 9r^9$. The energy per ion is then

$$u=-e^2\left(\frac{1.75}{r}-\frac{6r_0^8}{9r^9}\right)\frac{1}{4\pi\varepsilon_0}. \tag{7.31}$$

The minimum occurs when $\partial u/\partial r=0$, i.e. when

$$\frac{1.75}{r^2}=\frac{6r_0^8}{r^{10}},$$

$$r^8=3.43r_0^8 \quad \text{or} \quad r=1.17r_0. \tag{7.32}$$

If r_0 is taken as 2.36×10^{-10} m this would give for r a value of 2.76×10^{-10} m which is close to the observed value. Even for a one-dimensional infinite chain molecule NaClNaClNaCl . . . , the ionic spacing is already increased from 2.36×10^{-10} m to 2.69×10^{-10} m.

7.3.2 *Bulk modulus of a van der Waals solid*

The commonest type of van der Waals solid that we meet in everyday life consists of long-chain hydrocarbons. If the molecule contains less than say 50 carbons the chains pack together in a very nearly hexagonal array with the chains lying parallel to one another. It is possible, though rather tedious, to calculate the interaction between all the CH_2 groups in the crystal. If the chains are very long we enter the domain of rubbers or elastomers (see section 7.4) or polymers (see Chapter 13). In this section we shall deal with a much simpler type of van der Waals solid, though it is not one that we normally meet in everyday life, namely the solid rare gases, neon, argon, krypton and xenon.

Scattering experiments of rare gases (in the gaseous state) show that the potential energy between two atoms is well expressed by the relation

$$u=\frac{-A}{x^6}+\frac{B}{x^{12}} \tag{7.33}$$

and since the atoms are spherical we expect them to form a close-packed structure in the solid state. A detailed study shows that the lowest energy corresponds to the h.c.p. structure. However, the observed structure is f.c.c., the energy difference being less than 1 per cent. The reason for this discrepancy is that pairwise addition is not quite accurate enough since it ignores complications introduced by the presence of other atoms (the many-body problem). In what follows we shall assume that pairwise addition is sufficient for our purpose.

For the f.c.c. structure (see figure 7.5) each atom has 12 nearest neighbours distance $\sqrt{2}x$ apart and 6 neighbours $2x$ apart, etc. The contribution from the attractive forces is

$$-\left(\frac{12A}{(\sqrt{2}x)^6}+6\frac{A}{(2x)^6}+\cdots\right)=-\frac{A}{(\sqrt{2}x)^6}\left(12+\frac{6}{8}+\cdots\right). \tag{7.34}$$

Although the convergence is very rapid the contribution from even more distant neighbours is not entirely negligible. A full summation shows that the quantity in the bracket reaches a limiting value of 14.45. This is the Madelung constant for this structure and shows that taking all neighbours into account the *effective* number of nearest neighbours is increased from 12 to 14.45. Thus the contribution from the attractive forces becomes

$$-\frac{A}{8\times x^6}\times 14.45\approx-\frac{1.8A}{x^6}. \tag{7.35}$$

The repulsion term may be grouped in a single constant C so that the resultant potential energy may be written

$$u=-\frac{1.8A}{x^6}+\frac{C}{x^{12}}. \tag{7.36}$$

For N_A atoms the total energy will be $\frac{1}{2}N_A$ times this. As before we may differentiate U to find the equilibrium condition when $x=a$.

$$U=-0.9N_AA\left(\frac{1}{x^6}-\frac{1}{2}\frac{a^6}{x^{12}}\right). \tag{7.37}$$

Figure 7.5. The structure of an f.c.c. structure of a solid inert gas. Each atom has 12 neighbours at distances $\sqrt{2}x$ apart, 6 neighbours at distance $2x$ apart, etc.

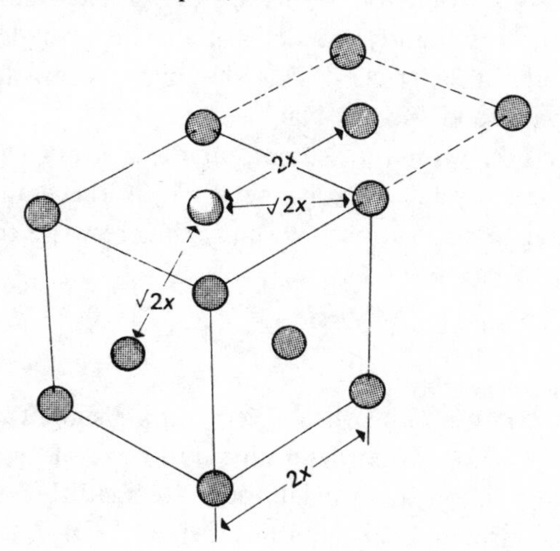

At the equilibrium separation

$$U_0 = -0.45\frac{N_A A}{a^6}. \tag{7.38}$$

The volume of the mole for this structure is

$$V = 2N_A x^3, \tag{7.39}$$

so that as before

$$\frac{dx}{dV} = \frac{1}{6N_A x^2}. \tag{7.40}$$

Repeating the previous procedure we find for the bulk modulus

$$K = \frac{1.8A}{a^9}. \tag{7.41}$$

As indicated in equation (7.35), for the attractive part of the potential (which falls off as x^{-6}) the Madelung constant is 14.45, i.e. the effective number of nearest neighbours is increased from 12 to 14.45. On the other hand, for the repulsive part where the energy falls off as x^{-12} we may ignore everything beyond second-nearest neighbours. This is not only because the numerical quantities are so small but because the repulsion arises from the overlap of orbitals so that it is physically unreasonable to think of repulsive interactions between more distant atoms. This gives an effective Madelung constant of 12.09. (Textbooks which sum over all neighbours give a value 12.13.) Thus the potential energy due to the attractive part gains from distant neighbours more than the energy due to the repulsive part. As a result the atoms are drawn closer together: and the atomic separation is slightly smaller in the solid crystal than in the simple diatom by about 3 per cent. This should be contrasted with the behaviour of an ionic solid such as NaCl.

As a first approximation we may assume that the nearest-neighbour distance of atoms in the crystal $\sqrt{2}x$ is the same as the equilibrium separation v_0 for a single pair of atoms. We may thus re-write equation (7.41) as

$$K = \frac{1.8}{a^3} \times \frac{8A}{r_0^6}. \tag{7.42}$$

The second factor in equation (7.42) has a very simple significance. By differentiating equation (7.33) we may determine A in terms of the equilibrium r_0 of one atom in relation to its neighbour. We shall then find the potential energy $\Delta\varepsilon$ of an isolated pair of atoms is simply $-A/2r_0^6$. Consequently from equation (7.42)

$$K=-\frac{28.8\Delta\varepsilon}{a^3}. \tag{7.43}$$

For the inert gases and for the smaller molecules such as H_2 and N_2, $\Delta\varepsilon$ is of order $(5 \text{ to } 25)\times 10^{-22}$ J, and a is about 3×10^{-10} m in the solid state. We obtain

$$K=(4 \text{ to } 20)\times 10^8 \text{ N m}^{-2}. \tag{7.44}$$

This is about one hundred times smaller than for ionic solids.

We may also express K in terms of U_0. Combining equations (7.38) and (7.41) to eliminate A we have

$$K=-4\frac{U_0}{N_A a^3}=-8\frac{U_0}{V}. \tag{7.45}$$

This is similar to the result for ionic crystals except that the numerical factor is appreciably larger. Table 7.2 shows reasonable agreement between the theoretical and calculated values of this ratio.

7.3.3 *Bulk modulus of metals*

We give here a simplified account (taken from Kittel) of the bulk modulus of those metals which, in the solid state, consist of positive ions and one free electron per ion. It would thus be applicable to the alkali metals (Li, Na, K) which are b.c.c. It might also be applicable, with some modification, to f.c.c. metals such as copper and aluminium.

The first problem is to establish the interaction between the constituent parts of the metal. For this purpose, it is useful to divide the solid into a

Table 7.2. *Solid inert gases: temperature ~0 K (experimental values)*

	Mol vol V 10^{-6} m^3	Bond energy U_0 KJ	U_0/V 10^8 N m^{-2}	K 10^8 N m^{-2}	Ratio $K:U_0/V$
Ne	18.9	1.88	1.0	11	11
Ar	23.7	7.74	3.3	29	9
Kr	38.4	11.15	2.9	34	11
Xe	49.1	18.16	3.7	3.6	10

(i) U is the latent heat of sublimation.
(ii) The simple theory suggests that the ratio of K to U_0/V should be 8: The observed ratio lies between 9 and 11.
(iii) From data in *Rare Gas Solids*, eds M. L. Klein and J. A. Venables, Academic Press 1975. (Especially the paper by J. A. Barker.). I thank Professor David Buckingham for bringing this to my attention.

series of contacting polyhedra, known as Wigner–Seitz cells, each of which contains a positive ion, i.e. the nucleus and the residual orbital electrons. If the volume of the solid increases or decreases, the cells will increase or decrease in size correspondingly. These cells may, for convenience, be replaced by spheres of radius r which have the same volume. How are we to estimate the coulombic interaction between the free electrons and the cells, since other models show that the electrons are free to wander throughout the lattice? We assume that, statistically, the *mean* distance between the electron and the ion is r. The interaction then consists of three parts:

(a) The electrostatic attraction between the valency electron and the positive ion in its cell: this will be of the form $-a/r$.

(b) The electrostatic repulsion between the orbital electron clouds in neighbouring cells: this will be of the form $+b/r$. This turns out to be numerically smaller than the preceding term so that the resultant electrostatic interaction is $-A/r$. We assume that the repulsion between the positive nuclei is negligible because of strong screening.

(c) The third part of the interaction arises from the kinetic energy of the electrons which tends to expand the lattice. We shall not derive this but make use of the fact that the Fermi energy, which is a measure of the kinetic energy, is proportional to (number of electrons per unit volume)$^{2/3}$ i.e. to $1/r^2$. The bond energy may thus be written:

$$U_0 = -\frac{A}{v} + \frac{B}{r^2}. \tag{7.46}$$

Thus the bond energy is of the form $-A/r^n + B/r^m$ where $n=1$ and $m=2$. Consequently, from equation (7.45) we may write for the bulk modulus:

$$K = -\frac{mn}{9}\frac{U_0}{V} = -\frac{2}{9}\frac{U_0}{V}. \tag{7.47}$$

The bond energy for metals is often quoted as the cohesive energy E_c (or the sublimation energy L_s), that is the energy required to separate all the atoms to infinity. If the solid metal consisted of neutral atoms, all somehow stuck together, this would be acceptable. But the metal consists of ions and electrons and the energy involved is that required to separate the constituent parts to infinity. A fair approximation to this is $L_s + I$ where I is the ionization potential of the metal, as illustrated in figure 7.6. We have already discussed a similar issue in relation to the bonding energy of NaCl.

Table 7.3. *Bond energies and bulk modulus of some alkali metals* (1 eV = 96.5 kJ mol^{-1})

Metal	L_s eV	I eV	$U_0 = L_s + I$ eV	$U_0 = L_s + I$ kJ mol^{-1}	V 10^{-6} m^3 mol^{-1}	U_0/V 10^{10} J m^{-3}	K 10^{10} N m^{-2}	Ratio $K: U_0/V$
Li	1.59	5.3	6.89	665	13	5.10	1.15	0.23
Na	1.13	5.1	6.23	601	24	2.50	0.64	0.26
K	0.98	4.3	5.28	510	45	1.13	0.28	0.25
Rb	0.82	4.1	4.92	474	56	0.85	0.19	0.22
Cs	0.82	3.9	4.72	455	70	0.65	0.14	0.22

Some typical results for alkali metals are given in Table 7.3. There is good agreement between the theoretical ratio of K to U_0/V, namely $\frac{2}{9}$, and the calculated ratio.

7.3.4 *Bond energy, or lattice energy and heat of sublimation*

We make a few additional comments on the bond or lattice energy and the heat of sublimation.

The total bond energy U_0 of a lattice (the lattice energy) is a fundamental quantity deriving from the forces between the constituent atoms and ions. It is not easily open to direct measurement. The nearest physical quantity is the heat of sublimation L_S, which in practice is a little greater than the sum of the latent heat of fusion L_F and the latent heat of vaporization L_V. We may easily form an estimate of L_S for an ionic crystal.

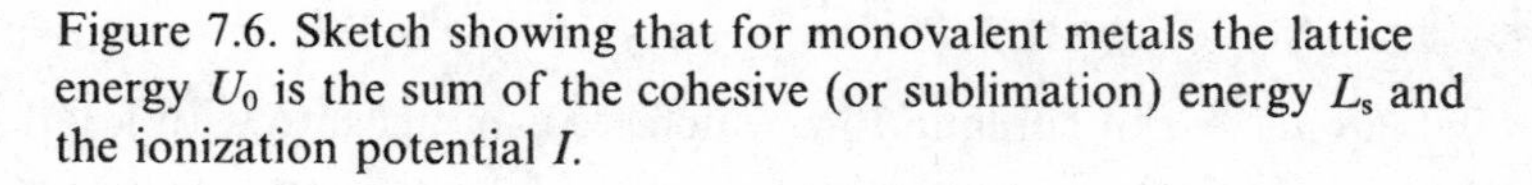

Figure 7.6. Sketch showing that for monovalent metals the lattice energy U_0 is the sum of the cohesive (or sublimation) energy L_s and the ionization potential I.

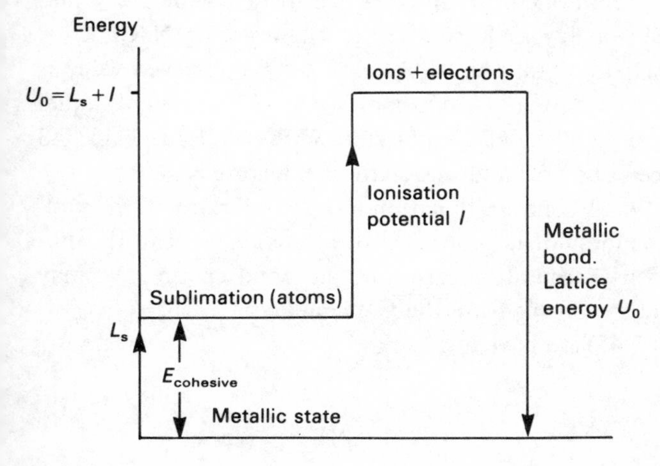

When sodium chloride is evaporated, the process does not involve pulling the positive and negative ions apart to infinity; this would be the true lattice energy. Vaporization generally produces individual NaCl molecules. Thus L_S is less than U_0 by the value of the interaction energy between Na^+ and Cl^- in the individual NaCl molecule. From equation (7.31) the lattice energy per ion in the solid state is obtained by putting $r = 1.17r_0$, where r is the ionic spacing in the crystal, r_0 the spacing in the free NaCl molecule. The value is

$$u_{\text{lattice}} = -\frac{e^2}{r_0}\frac{1.33}{4\pi\varepsilon_0},$$

and for one mole

$$U_0 = -\frac{N_A e^2}{r_0}\frac{1.33}{4\pi\varepsilon_0}.$$

The energy per ion pair in the free molecule is obtained from equation (7.30) by putting $r = r_0$. We obtain for the free NaCl molecule

$$u_{\text{molecular}} = -\frac{e^2}{r_0}\left(\frac{8}{9}\right)\frac{1}{4\pi\varepsilon_0},$$

and for one mole

$$U_{\text{molecular}} = -\frac{N_A e^2}{r_0}\left(\frac{8}{9}\right)\frac{1}{4\pi\varepsilon_0}.$$

The molar heat of sublimation is therefore

$$L_S = U_0 - U_{\text{molecular}} = \frac{N_A e^2}{r_0}\frac{0.44}{4\pi\varepsilon_0} \approx -\frac{1}{3}U_0. \tag{7.48}$$

The latent heat of sublimation, which we can measure with a fair degree of accuracy, is thus about one-third of the lattice energy. Tables of values support this conclusion for ionic solids. For example for NaCl, $U_0 =$ 750 kJ mol^{-1} so that L_S should be 250 kJ mol^{-1}. The observed value is *ca.* 240 kJ mol^{-1}. The difference (510 kJ mol^{-1} = 5.3 eV) is roughly equal to the ionization energy of the NaCl molecule. With metals, as Table 7.3 shows, L_S is between one-fifth and one-sixth of the lattice energy.

With a van der Waals solid, sublimation merely separates all the molecules (or atoms) as individual molecules (or atoms). The heat of sublimation is thus a fairly accurate measure of the bond energy. We may easily form an estimate of this from the above analysis.

From equation (7.45) we have

$$U_0 = -\frac{KV}{8}$$

and from equation (7.43), this becomes

$$U_0 = \frac{3.6\Delta\varepsilon V}{a^3}. \tag{7.49}$$

But from equation (7.39) we see that when the crystal is at its equilibrium separation ($x=a$), $V=2N_A a^3$. Substituting in equation (7.49) we have

$$U_0 = 7.2 N_A \Delta\varepsilon.$$

This is the same as the value obtained in equation (6.2) except that the numerical factor is slightly different. This arises from the fact that in the present model we have included the effect of more distant neighbours.

7.3.5 *Generalized relation between bulk modulus K and lattice energy U_0*

The previous sections dealing with ionic solids, van der Waals solids and metals may now be put into a more concise and general framework. Suppose the potential energy between constituent atoms or ions is of the form

$$u = -\frac{A}{x^n} + \frac{B}{x^m}. \tag{7.51}$$

Then the lattice energy can be written

$$U = -C\left(\frac{1}{nx^n} - \frac{a^{m-n}}{mx^m}\right), \tag{7.52}$$

where C is a suitable constant including the Madelung constant and other parameters and a is the equilibrium spacing. We see that this is the correct form for U since if we differentiate equation (7.52) with respect to x, we find that $\partial U/\partial x = 0$ at $x=a$ confirming that the energy is a minimum at the equilibrium spacing. Thus

$$\frac{\partial U}{\partial x} = C\left(\frac{1}{x^{n+1}} - \frac{a^{m-n}}{x^{m+1}}\right) \tag{7.53}$$

and

$$\frac{\partial^2 U}{\partial x^2} = -C\left(\frac{n+1}{x^{n+2}} - \frac{(m+1)a^{m-n}}{x^{m+2}}\right).$$

Consequently at $x=a$

$$\frac{\partial^2 U}{\partial x^2} = -C\left(\frac{n-m}{a^{n+2}}\right). \tag{7.54}$$

For any regular crystalline array the molar volume can be written

$$V = zN_A x^3,$$

where z is a constant determined by the crystal structure. Then

$$\frac{dx}{dV}=\frac{1}{3zN_A x^2}.$$

Consequently at equilibrium

$$\left(\frac{dx}{dV}\right)^2_{x=a}=\left(\frac{1}{9(zN_A x^2)^2}\right)_{x=a}=\frac{a^2}{9V^2}.$$

Hence

$$K=V\left(\frac{\partial^2 U}{\partial x^2}\right)\left(\frac{dx}{dV}\right)^2$$

$$=-VC\left(\frac{n-m}{a^{n+2}}\right)\left(\frac{a^2}{9V^2}\right)=C\left(\frac{m-n}{a^n}\right)\frac{1}{9V}. \quad (7.55)$$

But from equation (7.47) the equilibrium lattice energy U_0 at $x=a$ is

$$U_0=-C\left(\frac{1}{na^n}-\frac{1}{ma^n}\right)=-C\left(\frac{m-n}{mna^n}\right). \quad (7.56)$$

Thus equation (7.55) becomes

$$K=-\left(\frac{mn}{9}\right)\frac{U_0}{V}. \quad (7.57)$$

For rock-salt, $n=1$; $m\approx 9$; the factor is unity. For a van der Waals solid, $n=6$; $m=12$; the factor is 8. For monovalent metals $n=1$, $m=2$; the factor is 2/9.

7.3.6 *Bulk modulus of covalent solids*

The covalent bond involves the pairing of electrons (see p. 15). It extends over a few multiples of the bond length and, if stretched beyond this limit, is broken. This is quite unlike ionic forces, van der Waals forces and electrostatic forces which extend to infinity: they can be represented by a power law which applies over the whole distance. With covalent bonds, no power law can be applied. We may, however, pursue the general idea that the bulk modulus is related to the bond energy. For a single carbon–carbon bond, the energy is 59×10^{-20} J or 355 kJ mol^{-1}. In diamond, each carbon atom is attached to four neighbours. The energy per atom is thus $\frac{1}{2}(4\times 3.55)$ kJ mol^{-1}, the half being introduced to prevent us counting each bond twice. This gives a value of $U_0=710$ kJ mol^{-1}. The molecular (atomic) volume is 3.42×10^{-6} m^3 so that $U_0/V=2.08\times 10^{11}$ J m^{-1}. The bulk modulus of diamond is 5.45×10^1 N m^{-2}. Results for this and for silicon and germanium, which have the diamond

Table 7.4. *Bulk modulus of covalent solids*

Substance	U_0 kJ mol^{-1}	V 10^{-6} m^3	U_0/V 10^{11} J m^{-2}	K 10^{11} N m^{-2}	Ratio $K:U_0/V$
Diamond	710	3.42	2.08	5.45	2.6
Silicon	446	12.06	3.70	9.9	2.7
Germanium	372	13.56	2.76	7.7	2.8

structure, are given in Table 7.4. If we were bold enough to apply a power law of the form $-A/x^n+B/x^{12}$ we should obtain for n a value of about 2 implying an attractive force varying as $1/x^3$. We may compare this with the footnote to Table 1.6.

We should note that in these calculations for ionic solids, metals and covalent solids, we have ignored the contribution due to van der Waals interactions. It is, by comparison, very small.

7.4 Elastic properties of a rubber molecule

The elastic properties of rubber are very different from those of most other solids such as ionic solids, covalent solids and metals. In tension or shear the modulus is 1000 to 10 000 times smaller. On the other hand Poisson's ratio is almost $\frac{1}{2}$, so the bulk modulus is relatively high; it approaches that of a liquid. The two most striking elastic features are (*a*) rubber has enormous extensibility – up to 400 per cent compared with perhaps 1 per cent for ordinary solids, (*b*) the elastic modulus increases with increasing temperature. This is the only solid known which shows such an effect. We recognize that these two properties are very similar to the behaviour of a gas in compression.

Since rubber is known to consist of long-chain organic molecules we realize at once that its enormous extensibility and low modulus cannot arise from the stretching or bending of the C–C bonds, since these are extremely strong and difficult to deform. The rubbery behaviour is due to easy *rotation* of the carbon atoms about each C–C bond. As a result the chains are coiled up in higgledy-piggledy fashion and the result of applying a tension is to pull the chains into partial alignment. Thermal energy attempts to increase the disorder in the chain and therefore resists the aligning action of the tensile force. Consequently the modulus in tension increases with temperature.

If the molecular chains were completely separate the material would be a very viscous liquid and indeed natural latex is such a liquid. In order to become a rubbery solid the molecular chains must be cross-linked at a relatively small number of points. We now consider briefly the properties of a typical segment lying between two cross-links. Suppose the C–C bond has a length λ and there are N such bonds in a segment; because, as mentioned above, *rotation* about the C–C bond, in contrast to stretching or bending, is extremely easy, as a first approximation we may ignore any stiffness which arises from this factor. The segment then consists of N freely pivoted bonds. We may then ask: what is the most probable length of a segment of N units each of length λ?

Assuming that each link is completely free to take up any position relative to its neighbour we recognize this as the problem of a 'random walk' in three dimensions or a 'random flight'.†

We indicate briefly here how this may be calculated. Consider first a random walk in one dimension. Suppose a man takes N steps, each of length λ, along a line; each step may be forwards or backwards. What is the probability that he will have travelled a distance x from his starting point? If out of N steps A are forwards and B are backwards

$$A+B=N \tag{7.58a}$$

and

$$A\lambda-B\lambda=x. \tag{7.58b}$$

Hence

$$A=\frac{1}{2}\left(N+\frac{x}{\lambda}\right), \tag{7.59a}$$

$$B=\frac{1}{2}\left(N-\frac{x}{\lambda}\right). \tag{7.59b}$$

Any way of taking N steps such that A are forwards and B backwards will give the resultant distance x. The number of ways W in which this can be achieved is

$$W=\frac{N!}{A!B!}. \tag{7.60}$$

For simplicity assume N to be even so that A and B (from equations (7.58) and (7.59)) are integers. We use the Stirling approximation $N!=(N/e)^N$ and take $x \ll N\lambda$ (in the rubber molecule this is equivalent to

† This ignores the requirement of fixed bond-angles between the carbon atoms. If this is taken into account, and rotations about the bonds are permitted, the value of β in equation (7.66) is merely changed by a small numerical factor. A more serious conceptual difficulty is that in the random walk any given step may be covered many times, whereas in a real molecule the location of a molecular link can only be occupied once.

assuming that the distance between the two ends of the segment is very much less than the fully extended length of the segment). Then

$$\ln W = N \ln 2 - \frac{x^2}{2N\lambda^2} \tag{7.61}$$

or

$$W = B \exp(-\beta_1^2 x^2), \tag{7.62}$$

where $\beta_1^2 = 1/2N\lambda^2$. The probability $P(x)$ of the distance x occurring is directly proportional to W so that

$$P(x) = C \exp(-\beta_1^2 x^2). \tag{7.63}$$

The constant C may be determined by observing that

$$\sum_{x=-N\lambda}^{x=N\lambda} P(x) = 1 \approx \int_{-\infty}^{\infty} P(x)\,\mathrm{d}x. \tag{7.64}$$

This gives $C = \beta_1/\pi^{\frac{1}{2}}$.

We now turn to the three-dimensional random walk (figure 7.7(*a*)). If the coordinates of the starting point are (0, 0, 0) and those of the final point are (x, y, z) on orthogonal axes, the probability of finding the final position is

$$P(x, y, z) = \frac{\beta^3}{\pi^{\frac{3}{2}}} \exp[-\beta^2(x^2 + y^2 + z^2)]. \tag{7.65}$$

Normalizing for three dimensions it turns out that the β^2 in equation (7.65) is 3 times β_1^2 for the one-dimensional random walk, i.e.

$$\beta^2 = \frac{3}{2}\frac{1}{N\lambda^2}. \tag{7.66}$$

We are only interested in the exponential factor. We may therefore write for the most probable length of a segment

$$P(r) = A \exp(-\beta^2 r^2). \tag{7.67}$$

If one end of the segment is fixed, the probability of finding the other end in an element of volume $\mathrm{d}\tau$ is

$$P\,\mathrm{d}\tau = A \exp(-\beta^2 r^2)\,\mathrm{d}\tau. \tag{7.68}$$

We see that the probability density P is a maximum when $r = 0$, i.e. the segment makes its greatest effort to have zero distance between its ends. This is to be expected since if all steps in the random walk are equally probable a large number will contain as many forward as backward ones. This does not mean that the most probable distance between chain ends is zero. If we are interested only in the length r and not the direction then (as in our treatment of molecular velocities in a gas) we may consider one end of the segment fixed and determine the number of the other ends lying in a shell of volume $4\pi r^2\,\mathrm{d}r$. Then the number Π with lengths between r and $\mathrm{d}r$ is

$$\Pi\,\mathrm{d}r = A4\pi r^2 \exp(-\beta^2 r^2)\,\mathrm{d}r.$$

The result is shown in figure 7.7(*b*); the maximum occurs for a value of r given by

$$r_{\mathrm{m}} = \frac{1}{\beta} = \lambda\left(\tfrac{2}{3}N\right)^{\frac{1}{2}}. \tag{7.69}$$

Figure 7.7. (*a*) Probability density P as function of distance between the ends of a long chain assuming a three-dimensional 'random walk', (*b*) most probable distance r between the ends of a chain with N segments each of length λ occurs for $r_m = 1/\beta = \lambda\left(\tfrac{2}{3}N\right)^{\frac{1}{2}}$, (c) random arrangement of a chain containing 1000 segments, each segment set at the appropriate bond angle, but being given a random choice of 6 equally-spaced angular positions; this is a reasonably close approximation to complete randomness (from L. R. G. Treloar, *Physics of Rubber Elasticity*, 2nd edn (Oxford University Press, 1958)).

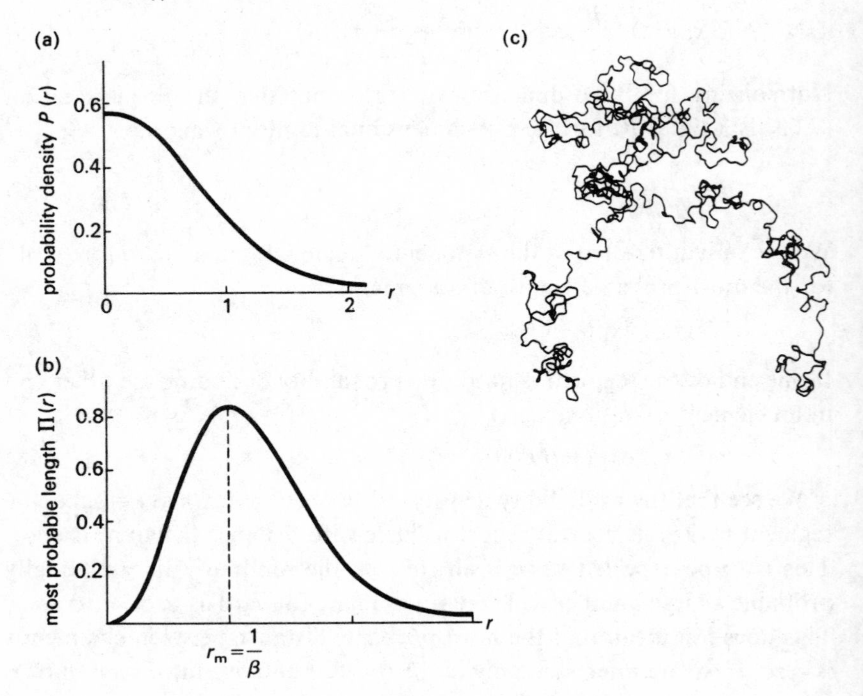

A simple analogy is to consider the concentration of bees around a drop of honey. The greatest density occurs near the honey but the greatest number of bees may well be found between 5 and 6 cm from the drop, rather than in the first centimetre.

Reverting to equation (7.67) the probability density is P and this is related to the (configurational) entropy of the segment by the relation

$$S = k \ln P = \text{constant} - k\beta^2 r^2. \tag{7.70}$$

If we apply a tensile force f to one end of the segment and extend it in the direction of r by an amount dr, the external work done is $f\,\mathrm{d}r$. Assuming that this involves no volume change we may equate it to the change in the Helmholtz free energy (equation (2.22*b*)).

$$f\,\mathrm{d}r = \mathrm{d}A = \mathrm{d}(U - TS). \tag{7.71}$$

For an isothermal extension

$$\begin{aligned} f &= \frac{\partial U}{\partial r} - T\frac{\partial S}{\partial r} \\ &= \frac{\partial U}{\partial r} + 2kT\beta^2 r. \end{aligned} \tag{7.72}$$

The term $\partial U/\partial r$ refers to the change in *internal* energy with extension, that is, with uncoiling of the segment by rotation about the C–C bond. As there is very little energy change (in our model no change at all) in this we may neglect this term and write

$$f = 2kT\beta^2 r = \frac{2RT\beta^2 r}{N_A}. \tag{7.73}$$

We see at once that (*a*) f is proportional to r so that the segment has Hookean properties, (*b*) f is proportional to T, (*c*) f is proportional to β^2, i.e. to $1/N$: the smaller the number of bonds in the segment, the greater its resistance to elastic extension.

So far we have only discussed the force that the segment exerts in resisting extension; whatever its length it is always attempting to contract. In bulk rubber which consists of a network of segments and chains it turns out that even if the rubber is subjected to uniaxial compression the overall effect, because of the large Poisson's ratio, is to extend the chains. The specimen therefore resists both extension and compression in a manner resembling the behaviour of a single segment discussed above.

7.5 The elastic properties of bulk rubber

7.5.1 *The tensile modulus*

Bulk rubber consists of long hydrocarbon chains such as

$$\text{polybutadiene} \quad \left(\begin{matrix} H & H & H & H \\ C- & C= & C- & C \\ H & & & H \end{matrix} \right)_n$$

or

$$\text{polyisoprene} \quad \left(\begin{matrix} H & H & CH_3 & H \\ C- & C= & C-- & C \\ H & & & H \end{matrix} \right)_n$$

The double bond in the monomer provides a convenient place for bonding agents such as sulphur to link one polymer to its neighbour, thus forming a large three-dimensional network. The length of segment between each pair of cross-links will cover a fairly wide range but we shall assume that we can use an average value of N for the number of links in each segment. All the segments will be pulling on one another in an attempt to contract, but in equilibrium the end-to-end length will be that given by r_m in equation (7.69). The main factor governing the resistance to deformation will be the entropic factor described in equation (7.73) and, as we shall see below, the number of segments per unit volume.

The segments have a random distribution of directions so that the elastic properties are isotropic. We shall simplify the structure by treating the rubber as an orthogonal network, each unit of the network being a cube of side r_m. The cubes are not empty! If we consider a typical rubber in which the segment contains 250 links ($N=250$) each of length λ the end-to-end length of r_m is only about 13λ. The cubes are in fact packed with random coils of the 250 links as well as with other segments which intrude from other parts of the other polymer chains in the rubber.† These all press on one another and account for the fact that rubber is highly incompressible possessing a Poisson's ratio of nearly $\frac{1}{2}$. In some ways one may regard rubber as a network of segments immersed in an incompressible fluid.

Because of the stiffening effect of double bonds in the greater part of each chain it would be misleading to treat each C–C bond as a freely rotating link. It is more realistic to treat the whole monomer as the link. We write v for the volume of each monomer.

† The open structure of the chain shown in figure 7.7 is the theoretical configuration of a single polymer chain. In the liquid or solid state the 'empty' spaces are filled with other chains showing similar random configurations. This is discussed more fully in Chapter 13.

Then if r_m^3 is the volume of the cube the number of links in the cube is

$$r_m^3/v. \tag{7.74}$$

Since each segment contains N links the number n of segments in the cube is

$$n = \frac{1}{N}(r_m^3/v). \tag{7.75}$$

On average, one-third point in each of the three orthogonal directions. Thus if we apply a stress σ in the x direction the number of segments bearing the stress is

$$\frac{n}{3} = \frac{1}{3N}(r_m^3/v). \tag{7.76}$$

The total force on the face of the cube is σr_m^2. The incremental force experienced by each segment is then

$$\Delta f = \frac{\sigma r_m^2}{\text{number of segments}} = \frac{\sigma r_m^2}{\dfrac{1}{3N}\left(\dfrac{r_m^3}{v}\right)} = \frac{3\sigma N v}{r_m}. \tag{7.77}$$

For each individual segment using equation (7.73) the incremental force is

$$\Delta f = 2kT\beta^2\,\Delta r = \frac{2kT}{r_m^2}\,\Delta r = \frac{2kT}{r_m}\left(\frac{\Delta r}{r_m}\right). \tag{7.78}$$

We see that $(\Delta r/r_m)$ is the strain in each segment. Combining equations (7.77) and (7.78) we have

$$\Delta f = \frac{3\sigma N v}{r_m} = \frac{2kT}{r_m}\left(\frac{\Delta r}{r_m}\right). \tag{7.79}$$

Hence the elastic modulus of the material in the cube is

$$E = \sigma\Big/\frac{\Delta r}{r_m} = \frac{2}{3}\frac{kT}{Nv} = \frac{2}{3}\frac{RT}{N_A} \times \text{number of segments/unit volume}. \tag{7.80}$$

This applies to all the cubes in the rubber specimen and is thus the tensile modulus of the rubber. For a typical rubber cross-linked with a few per cent by weight of sulphur N is of order 250 and (as deduced from the density of the rubber and its molecular weight) $v \approx 100 \times 10^{-30}\ \text{m}^3$. At room temperature $kT = 4 \times 10^{-21}$ J. Hence

$$E \approx 10^5\ \text{N m}^{-2}.$$

This is of the right order. For a more heavily cross linked rubber (say 8 per cent sulphur) the value of E might be ten times larger. But for more than 30 per cent sulphur the segments are too small to show flexibility. The material becomes relatively rigid (vulcanite) and the modulus is determined by the forces required to stretch or bend C–C bonds in the chains. This corresponds to the internal energy $(\partial U/\partial r)$ in equation (7.72).

7.5.2 *Effect of strain*

As pointed out above for a segment containing, say, 250 monomers the relaxed segment length, $(\frac{2}{3}N\lambda^2)^{\frac{1}{2}}$, is about 13λ. In principle it should be possible to stretch the segment until it is fully extended with a length of order 250λ. This implies a 20-fold extension in length. In practice a rubber cannot usually be extended elastically more than four- to five-fold. Beyond this the deformation becomes noticeably irreversible. This is known as hysteresis, the energy being lost in overcoming the attraction between the interacting chains. For even larger extensions the deformation force increases extremely rapidly. At this stage chain entanglements prevent the free movement of segments. More particularly portions of the chain segments become aligned, come into close contact and form crystalline regions which are very rigid. The residual fragments of free segments may still exhibit rubber elasticity before they become fully extended. The modulus is no longer explicable in entropic terms. Internal energy dominates the behaviour.

7.5.3 *Effect of temperature*

As we see from equation (7.80) the modulus of (unfilled) rubber is proportional to the absolute temperature, i.e. its modulus increases with temperature. This may be demonstrated in a very simple way. A rubber band is suspended vertically with a weight at the end sufficient to extend its length by, say, a factor 2. If the rubber is gently heated with a hair dryer the band will contract. Entropy is at work!

This may be made more quantitative. If a rubber specimen is extended and allowed to relax until the tensile force reaches a constant value the equilibrium force may be studied as a function of temperature. If the thermal expansion of the rubber is subtracted from the observed extension to give the true elastic extension, the force f is found to be linearly proportional to T and extrapolates fairly well back to $f=0$ at $T=0$ (see figure 7.8).

As we shall see in Chapter 13, the freedom of rotation about C–C is greatly restricted when the temperature falls below about 250 K. The free rotation assumed in the theory of rubber elasticity no longer applies.

The material is in the glassy state and deformation involves the bending or stretching of the C–C bond. The modulus is 10^3 or 10^4 times greater than the material in the rubbery state. (See Chapter 13).

7.6 Conclusions

In this chapter we have shown that the elastic properties of crystalline solids can be understood in terms of the forces between the atoms. Using simple pairwise addition we are able to relate the bulk modulus to the bond energy and to deduce appropriate relationships for van der Waals, ionic and metallic solids and even to some extent for covalent solids. There is reasonably good agreement between theory and experimental data.

With rubber we find that the elasticity arises from the entropy of the chain segments. It is for this reason that it resembles the elasticity of an ideal gas in compression where the pressure may be considered as arising from the reduction in entropy associated with a reduction in volume. Indeed one notices the presence of the 'gas constant' R as a fundamental

Figure 7.8. Tensile stress of an unfilled rubber specimen as a function of temperature. The rubber is extended by a certain amount and allowed to relax. The thermal expansion of the rubber is subtracted from the observed extension to give the true elastic extension shown as percentage strain. The tensile stress (or force) is seen to be roughly proportional to the absolute temperature.

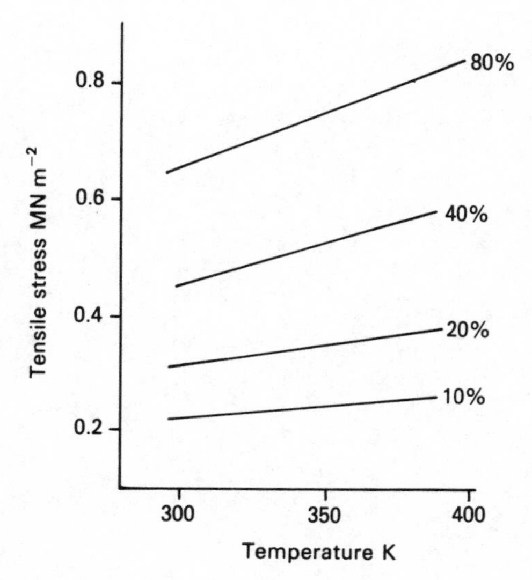

term in the expression for rubber elasticity in equation (7.73) and in the expression for the elastic modulus of bulk rubber in equation (7.80). In view of this one might well designate R as the 'rubber constant' in 'ideal rubber' elasticity.

8

The strength properties of solids

8.1 Deformation properties

8.1.1 *Ductile properties*

The strength properties of solids are most simply illustrated by considering the behaviour of a homogeneous specimen of uniform cross-section subjected to uniaxial tension (figure 8.1(*a*)). If we plot the true stress σ against the linear strain ε (i.e. the fractional increase in length) we may obtain a curve as illustrated in figure 8.1(*b*). The portion OA represents elastic deformation. The strain is proportional to the stress and the deformation is reversible. If the material is ductile, elastic deformation will proceed until at some critical stress Y the onset of permanent or

Figure 8.1. (*a*) Tensile specimen and (*b*) typical stress-strain curve, showing elastic deformation along OA, plastic yielding at Y and work-hardening along YZ. A brittle solid fails at a tensile stress S.

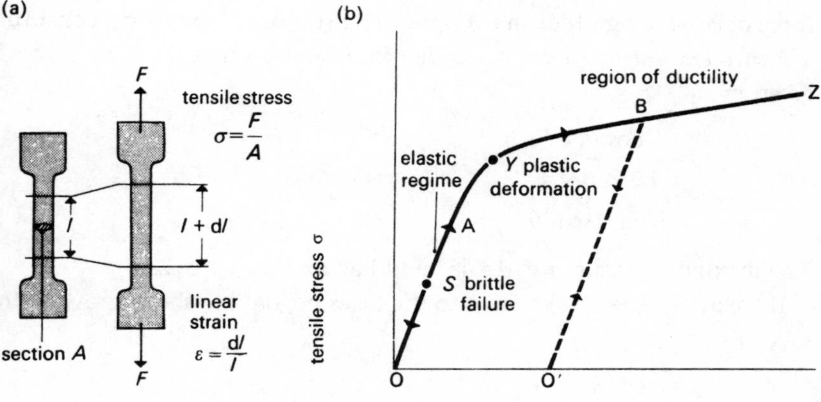

plastic deformation occurs. If we continue along the plastic curve there is generally an increase in yield stress with deformation. This is known as work-hardening. If at some point B we reduce the stress the material recovers elastically along BO′, where BO′ is very nearly parallel to OA. The displacement OO′ is the permanent plastic extension produced in the specimen. On reapplying the stress the deformation follows the curve O′BZ.

Experiments show that under simple uniaxial compression a cylinder will first deform elastically and then yield plastically at the same compressive stress Y.

Some insight into the yield criterion is provided by considering the effect of hydrostatic pressure, i.e. a stress situation in which the compressive stress on the material is the same in all directions. If we subject the specimen to a hydrostatic pressure P we find that plastic flow does not occur even if P exceeds Y. We must still apply a uniaxial stress (either in tension or compression) and its magnitude if plastic flow is to occur is still Y. Now analysis shows that the only part of a stress field which is unaffected by hydrostatic pressure is a shear stress. We conclude that plastic flow is associated with a critical shear stress. This is fully supported by microscopic studies which show that plastic deformation is always accompanied by slip of atomic planes over one another.

8.1.2 *Shear stress*

Consider a rectangular bar of uniform cross-sectional area A (figure 8.2(*a*)). Suppose we apply a tensile force F, so that the stress is $\sigma = F/A$. Consider a thin slice of material making an angle θ to the direction of F. On one side of the slice there is a force $F\cos\theta$, on the other side a force of equal magnitude in the opposite direction. These forces constitute a shear. The surface area of the slice is $A/\sin\theta$ so that the shear stress is given by

$$\tau_1 = \frac{F\cos\theta}{A/\sin\theta} = \frac{F}{A}\cos\theta\sin\theta$$

$$= \sigma\cos\theta\sin\theta. \tag{8.1}$$

The maximum occurs for $\theta = 45°$ and has a value $\tau = \sigma/2$.

If we apply a tensile stress σ to the faces bc, da, the shear stress on the slice is

$$\tau_2 = \sigma\cos\phi\sin\phi$$

$$= \sigma\sin\theta\cos\theta. \tag{8.2}$$

This stress is equal to τ_1 and in the opposite direction. If the two systems of tensile stresses σ are superposed (to constitute a two-dimensional hydrostatic tension), the two shear-stress systems completely annul one another. This is the basis for the statement that hydrostatic stresses do not in any way change the existing shear stresses in a system.

Finally we note that the conventional method of representing a shear stress by two equal and opposite parallel stresses τ is misleading (figure 8.3(*a*)). Such a stress system would produce a couple which would produce continuous rotation of the stressed element. There must be an opposing set of shear stresses s if static equilibrium is to be achieved. Consider a rectangular element length x, width y, depth z. On the upper and lower faces we apply shear stresses τ; on the side faces shear stresses s. For equilibrium take moments about O. Shear stress τ implies a shear force $\tau \times xy$; its couple is $\tau \times xy \times z$. Similarly shear stress s implies a couple $s \times yz \times x$.

Then

$$\tau xy \times z = s \times yz \times x,$$

$$\tau = s. \tag{8.3}$$

We see that a shear stress always involves a pair of equal orthogonal stresses. In practice, of course, one is often unaware of this. For example in shearing a long thin slice as in figure 8.3(*c*), the horizontal shear force $F_1 = \tau A_1$ may be very large and the vertical shearing force $F_2 = \tau A_2$ may

Figure 8.2. Shear stresses produced by tensile stresses: (*a*) vertical stress $\sigma = F/A$, (*b*) horizontal stress σ, (*c*) two-dimensional hydrostatic stress σ; this produces a resultant zero shear stress.

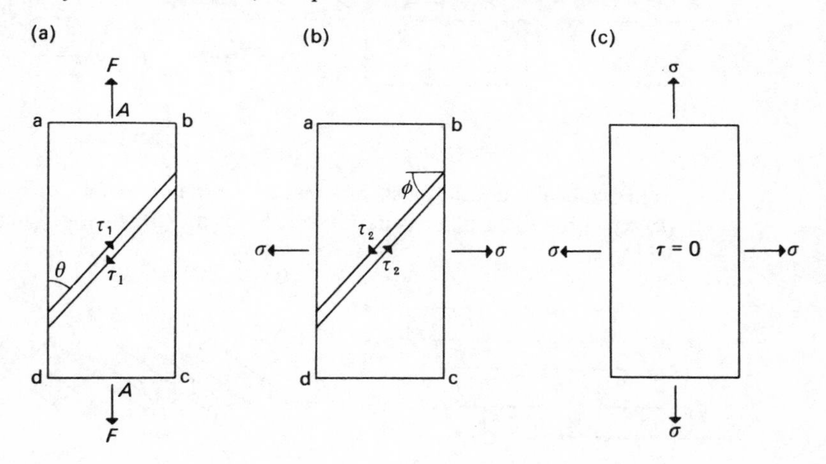

be vanishingly small if A_2 is very small, although the value of the shear stress τ must be the same in both cases.

A tensile stress Y produces a maximum shear stress $Y/2$ at 45° to the direction of Y. (The same applies to a uniaxial compressive stress Y.) For an isotropic material, shear will therefore occur in slip directions at 45° to the direction of the applied stress. If the material is not isotropic shear may occur more easily in some directions than others; if the shear stress is exceeded in this direction slip will occur in these more favourably oriented directions. This is often observed with single crystals, and is shown schematically for a simple shear stress in figure 8.4.

We may now explain the behaviour shown in figure 8.1(*b*). When the stress is first applied the atoms are displaced from their equilibrium positions; the resistance to deformation is determined by the interatomic forces. When the tensile stress reaches a critical value the shear stress is sufficient to produce slip along an appropriate plane and plastic yielding occurs. However, the whole stress must still be supported by the displaced atoms. Consequently when the stress is removed there is elastic recovery and the modulus, which arises from the interatomic forces, is essentially the same as originally.

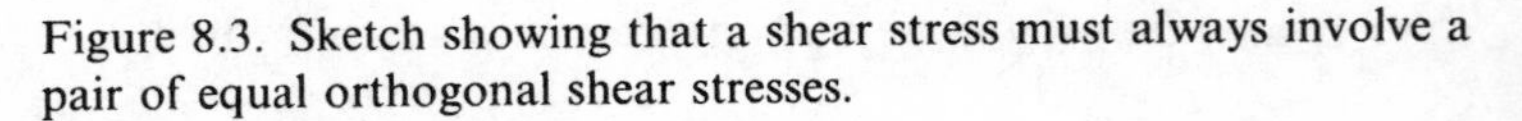

Figure 8.3. Sketch showing that a shear stress must always involve a pair of equal orthogonal shear stresses.

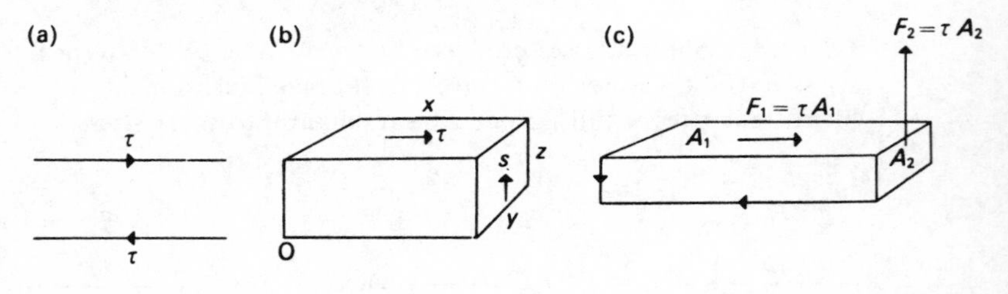

Figure 8.4. The effect of a shear stress on the yielding of a single crystal (*a*) when the shear plane is parallel to the shear, (*b*) when it is inclined.

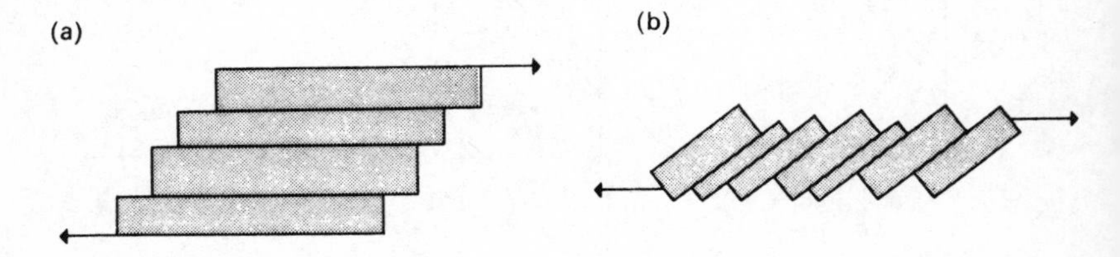

8.1.3. *Indentation hardness of ductile solids*

Engineers and metallurgists often wish to determine the strength properties of their materials without going to the trouble of preparing tensile specimens and carrying out full-scale tensile tests. A very convenient way of doing this is to measure the identation hardness. A very hard indenter (a hard steel sphere in the Brinell test, a diamond pyramid in the Vickers test) is pressed under a load W into the surface of the material to form a plastic indentation. When the indenter is removed the diameter of the indentation is measured and its area A determined. The mean pressure over the indentation is then

$$p = \frac{W}{A}. \tag{8.4}$$

In the industrial test procedure A is the surface area of the indentation; from the point of view of making p physically meaningful, the projected area is more appropriate. The difference, however, is generally small. A study of the stress situation around the indenter shows that almost two-thirds of p is in the form of a hydrostatic component, and therefore plays no part in producing plastic flow. Consequently only one-third of p is active in producing the indentation. Thus as a first approximation

$$p = 3Y, \tag{8.5}$$

where Y is the uniaxial yield stress of the material. This relation is well substantiated in practice.

Further, if the material does not appreciably work-harden after yielding in tension its yield stress Y will be very nearly the maximum stress the material can support before it pulls apart, i.e. its ultimate tensile strength (u.t.s.). The latter is therefore about one-third the indentation hardness, if they are measured in the same units. The conversion ratio

$$\text{u.t.s.} = \tfrac{1}{3}p \tag{8.6}$$

is widely used for polycrystalline homogeneous materials and is reliable to a few per cent.

In engineering hardness measurements the load W ranges from about 10 to 3000 kg and the indentation diameter lies between 1 and 5 mm. Because the early instruments used dead loading (either directly or through levers) the hardness was expressed as kg mm^{-2} and this is still the common practice. It would however be more correct to express it as kgf mm^{-2} (1 GPa $\simeq$ 100 kgf mm^{-2}).

Typical values for the indentation hardness in kgf mm^{-2} are 0.25 for krypton at 50 K, 0.8 for ice at 260 K, 1 for indium, 4 for lead, 40 for polycrystalline copper, 120 for mild steel, 900 for ball-bearing steel, 2000 for sapphire and over 10 000 for diamond at room temperature.

8.1.4 *Microhardness of solids*

Indentation hardness can be used as a research tool to study the strength properties (including the creep properties) of solids where other methods are not easily available. It is widely applied in metallurgical research to determine the properties and hence identify the various constituents of alloys or to estimate the amount of deformation produced by a particular metal working process. For this purpose the indentations must be small enough to resolve the various features under examination. In microhardness measurements the loads generally lie between 10 and 1000 gf (0.1 to 10 N) and the indentation diameters are of order 10–100 μm. These dimensions are easily measured with accuracy by optical microscopy. The depth of the indentation is about one-seventh of the diameter, i.e. 1–10 μm. This type of microhardness technique has also been applied to the determination of the fracture toughness of brittle solids when they are available only as very small specimens.

If we wish to study the properties of thin surface layers of metals deposited on hard substances or the effect of ion implantation where the layer of interest may be less than 1 μm thick much smaller indentations are needed. For a film 1 μm thick the indentation must not penetrate more than 0.2 μm (=200 nm) so that the diameter of the indentation will be of order 1 μm. Optical microscopy can resolve about 0.2 μm so that the boundary of the indentation cannot be determined with great precision. Electron microscopy may be used but it is not convenient. A different approach is to measure the *depth* of penetration of the indenter using microdisplacement transducers of great accuracy. From the geometry of the indenter the area of the indentation can be calculated for any specified load, though corrections have to be made for elastic yielding of the specimen and other topographical changes around the indentation. Loads range from 0.01 gf to 1 gf (0.1 mN–10 mN). Some typical results are shown in figure 8.5 for electrolytically polished (annealed) nickel where it may be assumed that any surface contamination films are less than a few nm thick. It is seen that the hardness increases as the size of the indentation is reduced and for an indentation depth of 20 μm reaches a value about five times that of the macroscopic hardness. This is not high enough to be identified with the ideal theoretical yield strength of the metal. The high values are probably due to the limited range of dislocation movement available when very small volumes are involved (see below).

8.1.5 *Calculation of critical shear stress for single crystals*

We now consider the shear stress at which two neighbouring planes in a single crystal can be caused to slide over one another. Consider the arrangement in figure 8.6 for a typical f.c.c. crystal and apply a shear stress τ to the planes X and Y. If θ is the strain produced and G is the rigidity or shear modulus

$$\tau = G\theta.$$

Atom B in its initial site in the lattice is in a position of minimum potential energy. As it is displaced the potential energy increases until a position of unstable equilibrium is reached at B′ – the potential energy curve at this point is horizontal but is a maximum. The potential energy curve thus has the form shown in figure 8.6(*b*). The force curve is found by differentiating the potential energy curve: it is drawn in figure 8.6(*c*) and shows that the maximum force occurs when B has reached some intermediate position B″ between B and B′. The angle of strain at this point is about $\frac{1}{4}$. If G remained constant we could say that B is sheared over to B″ for a shear stress of magnitude

$$\tau_{\mathrm{m}} \approx \frac{G}{4}. \tag{8.7}$$

The lattice could not resist a greater stress than this so that plane X would glide over plane Y for this value of τ.

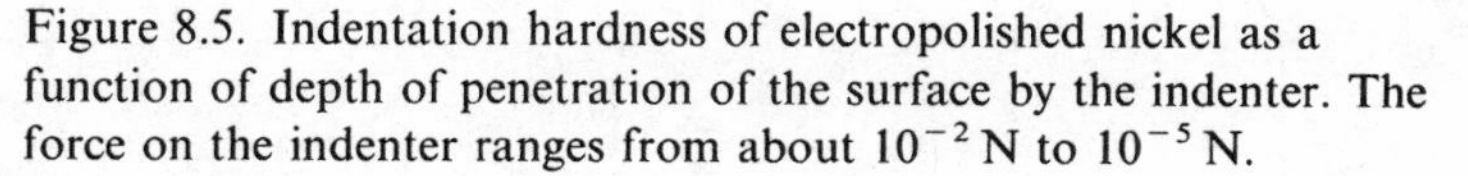

Figure 8.5. Indentation hardness of electropolished nickel as a function of depth of penetration of the surface by the indenter. The force on the indenter ranges from about 10^{-2} N to 10^{-5} N.

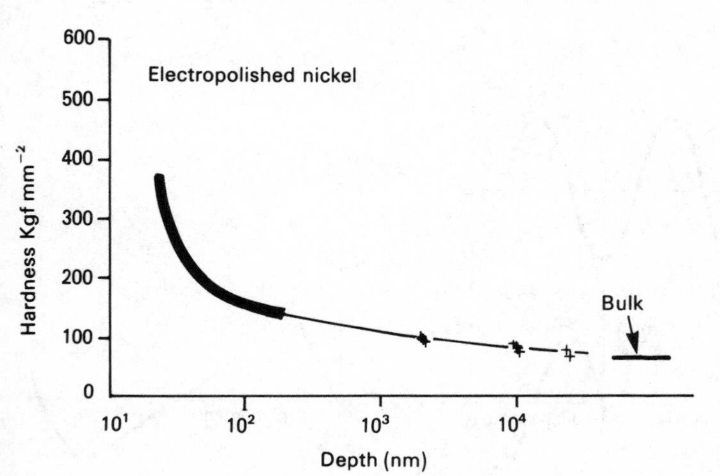

This is an overestimate since it assumes that G is a constant and has the same value as for small strains. This is rather unrealistic since for larger distortions we expect the modulus to decrease. We may form a better estimate of τ_m by assuming that the shear stress is a sine function of the displacement. It must clearly be zero for displacements $x=0$, $x=a/2$ and $x=a$, where a is the atomic spacing. We may thus use a function of the form

$$\tau = A_0 \sin\left(\frac{2\pi}{a}x\right), \tag{8.8}$$

where A_0 is a suitable constant. We may determine A_0 by specifying that the initial slope near $x=0$, where the strains are small, corresponds to the shear modulus G of the material. For small x

$$\tau = A_0 \sin\left(\frac{2\pi}{a}x\right) = A_0 \frac{2\pi}{a}x = G\theta, \tag{8.9}$$

Figure 8.6. Shearing of a perfect crystal. (*a*) Neighbouring planes of atoms at X and Y, (*b*) the potential energy curve as the top plane is slid over the bottom, (*c*) the equivalent force–displacement diagram, (*d*) a more realistic form of the force–displacement curve.

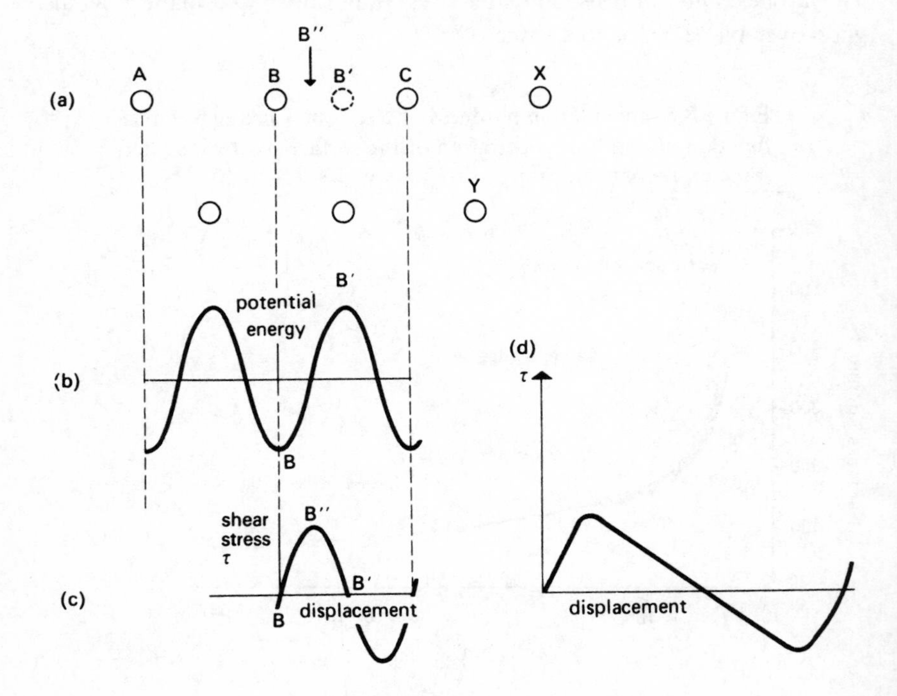

where θ is the shear angle x/a. Consequently

$$A_0 \frac{2\pi}{a} x = G \frac{x}{a} \quad \text{or} \quad A_0 = \frac{G}{2\pi}. \tag{8.10}$$

The shear stress reaches its maximum when the displacement is midway between B and B′; here $\sin(2\pi x/a)$ is unity so that

$$\tau_m = A_0 = \frac{G}{2\pi}. \tag{8.11}$$

A stress of this value will carry atom B over to position B″ and then for a slight increase in τ it will flick over to B′ and then to C. Thus slip along the atomic plane will take place for a shear stress of about $G/6$. A more realistic study of the atomic force field suggests a curve like that shown in figure 8.6(*d*). This reduces the critical shear stress τ even further but no amount of adjustment can reduce it to a value less than about $G/30$.

We conclude that the maximum shear stress τ_m the lattice can withstand is of the order of $\frac{1}{10}G$ to $\frac{1}{30}G$. Before discussing the relation of (τ_m) theoretical to (τ_m) experimental we may ask a very simple but pertinent question. Once we have dragged atom B to the top of the potential hill at B′ (figure 8.6(*b*)) why does it not glide over the rest of the potential hills until planes X and Y have completely slid off one another? Is not the position analogous to a 'frictionless' helter-skelter, where once the carriage has reached its highest point it can proceed indefinitely over all hills which do not exceed that height? The simple answer is that such a model would be appropriate if the atoms in row Y were absolutely rigid in space. Because they are themselves part of an elastic system the behaviour is much more like the interaction of two (frictionless) combs, the teeth of one being dragged through those of the other (figure 8.7(*a*)). Up to a certain tangential stress both sets of teeth are distorted reversibly (figure 8.7(*b*)); beyond this point the teeth from the upper comb escape from interaction with the lower comb and flick over into their new equilibrium position; whilst the teeth of the lower comb flick back into their original position (figure 8.7(*c*)). The whole of the distortional energy has disappeared as vibration of the teeth. This has a close analogue in the slip of crystals. Practically the whole of the work of plastic deformation is dissipated as vibrational energy in the lattice, i.e. as heat. Not more than a few per cent of the energy is retained as strain energy in the lattice. This has an interesting corollary in an unexpected area. The friction of metals is largely due to adhesion, shearing and deformation within and around the regions of real contact; it is, in fact, a process which is often dominated by plastic flow around the contact zones. Virtually all the plastic work appears as heat. This is the reason for Joule's observation that in frictional heating there is quantitative agreement between the work done against friction and the heat liberated. If large amounts of energy could be stored in the lattice this equivalence would not be observed.

8.2 Dislocations

8.2.1 *Observed shear stress, need for dislocations*

Experiments on pure single crystals show that minute shear stresses are sufficient to produce permanent, i.e. plastic, deformation. In general (τ_m) observed is of the order of one thousand times smaller than (τ_m) theoretical. The theoretical value cannot be wrong to this extent. This discrepancy has led to a search for sources of weakness in the lattice, such as flaws or imperfections: the simplest type is that known as the edge dislocation shown in figure 8.8(*a*). It involves, essentially, the insertion of an extra half-plane of atoms AB into the lattice. If a shear stress is applied as shown, atom B flicks over and joins with atom F for a very small value of τ. Then C joins up with G and D with H so that for a very small stress we have produced slip by one atomic spacing DE (figure 8.8(*b*)). In effect what we are saying here is that the theoretical value of τ_m assumes that slip ocurs simultaneously across the whole plane KE; in the dislocation model slip is able to occur one row of atoms at a time and this leads to an enormous reduction in shear stress.

We can already see that there are two basic problems associated with the above model. First, how are the dislocations originally produced in the crystal? Secondly when the edge dislocation has slipped out of the

Figure 8.7. The displacement of the teeth of two sets of engaging combs (*a*) initially, (*b*) after a small stress has been applied, (*c*) after the stress has been sufficient to cause slip: the elastic strain energy in the teeth is lost by vibrations which degrade into heat.

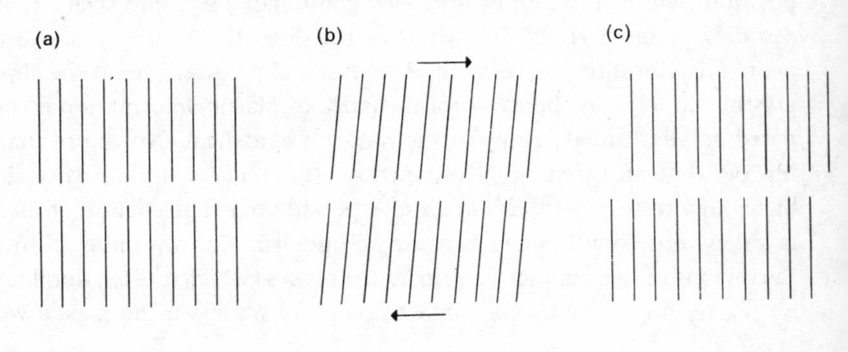

crystal as in figure 8.8(*b*), the crystal has become perfect so that it has its ideal theoretical strength. As this is not generally observed there must be some process for the generation of further dislocations whilst slip is occurring. Both these questions are part of a very large subject which we shall not pursue further in detail. We mention, however, a few further points of interest.

8.2.2 *Direct experimental evidence for dislocations*

The first and most direct evidence for the existence of dislocations is due to Dr J. W. Menter, who studied, with an electron microscope, very thin specimens of platinum phthalocyanine. This is a metallo-organic compound which forms well-defined crystals in which the platinum atoms lie on crystal planes 12 Å apart. In the electron microscope the organic part of the crystal scarcely absorbs electrons whilst the platinum atoms strongly absorb. In transmission, therefore, a microscope with a resolution of say 10 Å (1 nm) is able to resolve the planes of the platinum atoms. Typical electron micrographs reveal the presence of a defect strongly resembling a classical edge dislocation.†

In pure metals the atomic spacing is of the order 3 Å (0.3 nm) and it is only in recent years that electron microscopes have been developed with

Figure 8.8. (*a*) An edge dislocation, (*b*) the displacements which occur when a shear stress is applied and the dislocation at B moves out to E.

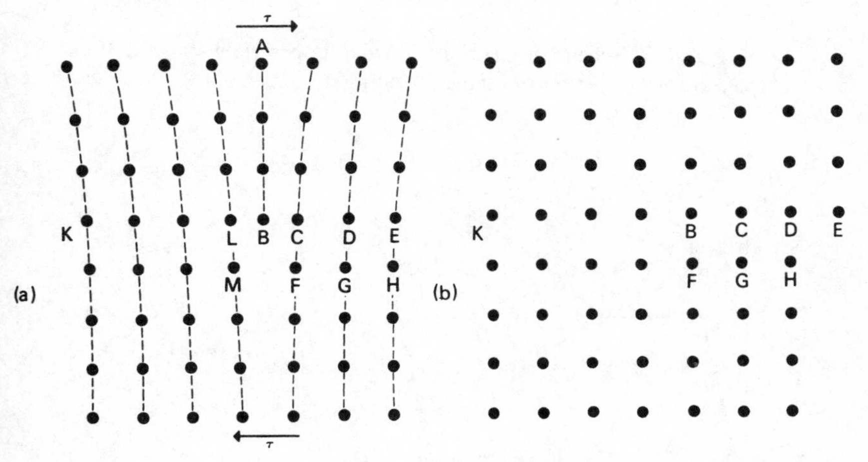

† See J. W. Menter, *Proceedings of the Royal Society*, A236 (1956) p. 119, and F. P. Bowden & D. Tabor, *Friction and Lubrication of Solids* (Clarendon Press, 1964), plate III.

sufficient resolution to 'see' dislocations directly in metals. This has led to a very extensive application of electron microscopy and diffraction to the study of dislocations in metals and other crystalline solids. Unfortunately, because of the relatively poor penetrating power of electrons, the specimens must be in the form of rather thin films and the results obtained may not be representative of the materials in bulk.

8.2.3 *Upper yield point*

The edge dislocation is a region of large elastic stress. At LBC in figure 8.8(*a*) the material is heavily compressed, at MF strongly extended. If impurity atoms are present in the lattice they will tend to concentrate at regions where they can relieve the stress. For example a *small* impurity atom will tend to take the place of the original atom B. Quite often the impurity atom forms a strong bond with the neighbouring atoms. As a result it may be more difficult to move the dislocation and detach it from the impurity. The impurity at B will thus tend to 'pin' the dislocation and a larger stress than normal will be required to make it move. Once the dislocation has moved across to position C or D it is in its normal environment and is now much easier to move. This explains the behaviour, for example, of certain mild steels where the dislocations are pinned by interstitial carbon or nitrogen. When such materials are tested in tension or compression or shear they show an 'upper yield point' (see figure 8.9 point B), then a region of fairly constant lower yield stress (CD) followed

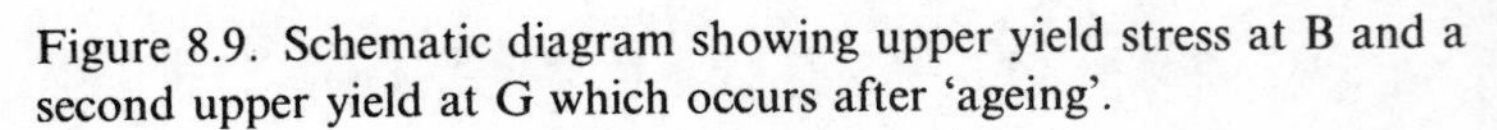

Figure 8.9. Schematic diagram showing upper yield stress at B and a second upper yield at G which occurs after 'ageing'.

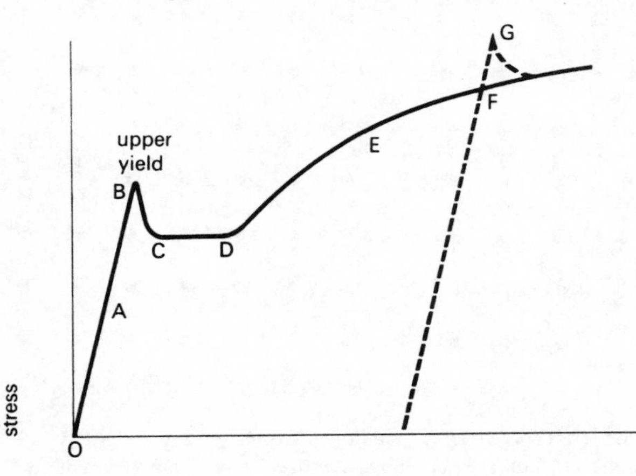

by a conventional work-hardening curve. If at some point F the stress is removed and the metal is 'aged' the carbon and nitrogen can diffuse to the dislocations and again pin them. The further deformation of the specimen may then show a second 'upper yield point' (G).

8.2.4 *Work-hardening*

Dislocations interact with one another. Continued deformation produces interlocking of dislocations and makes it more difficult for them to move. Although the details are still the subject of discussion this is the basic reason for work-hardening as a result of deformation.

8.2.5 *Screw dislocations*

Apart from edge dislocations there are 'screw' dislocations as shown in figure 8.10(*a*). If we follow a sequence of atoms, say, in the top plane starting from A and moving to BCDEF we end up at the point F which is an integral number of atomic spaces below A. This is essentially a screw movement. If the dislocation line XY moves towards CD the slipped region AXYZ increases in area.

Figure 8.10. (*a*) Screw dislocation, (*b*) crystal growth on an edge providing two points of attachment at E and F, (*c*) crystal growth on the edge of a screw dislocation which provides a self-perpetuating growth step.

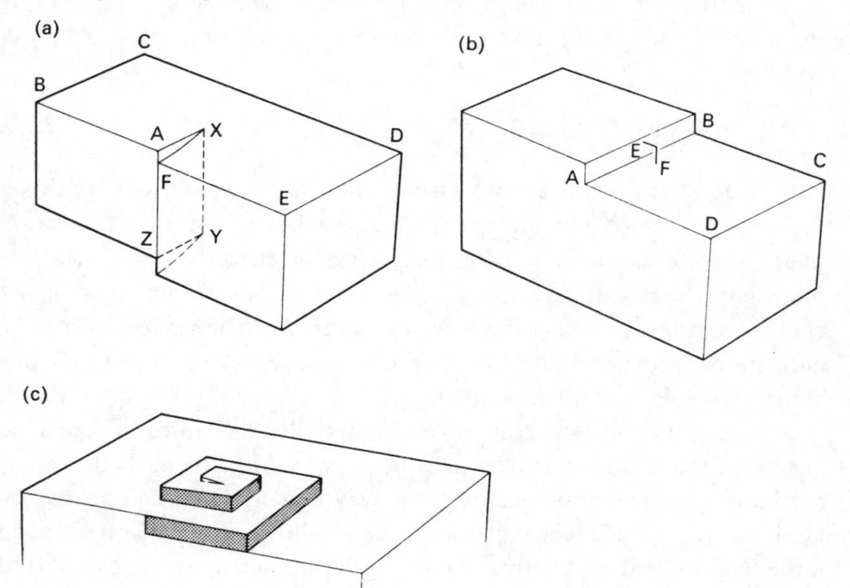

8.2.6 *Crystal growth*

The presence of sites for nucleation is very important in crystal growth. On a perfectly smooth atomic plane an atom can find only one point of attachment. On a step, each atom can find two points of attachment (figure 8.10(*b*)). Growth will therefore occur very much more easily until the whole of the step ABCD is filled, when the surface is then plane. A screw dislocation, however, provides a self-perpetuating growth step; as further atoms become attached to the step they build up a spiral 'staircase' and so provide further steps for further condensation. This mechanism accounts for the growth spiral that is often observed on the face of crystals (see figure 8.10(*c*)).

8.3 Vacancies, diffusion and creep

8.3.1 *Vacancies*

Before leaving the ductile properties of solids we refer briefly to a few related themes which are involved in dislocation movement.

A vacancy is a place in the lattice where an atom is missing. To create a vacancy in a crystal a number of bonds have to be broken: thus a vacancy implies a site where the atom has sufficient energy to escape. Suppose the energy required to create a vacancy is ε_v. If there are n_0 atoms in the bulk with their normal number of neighbours (i.e. ignoring the atoms in the free surface) the number n possessing extra energy ε_v will be

$$n = n_0 \exp(-\varepsilon_v/kT). \tag{8.12}$$

(For n_0 to remain constant we assume that the escaped atoms settle on the free surface of the specimen). We see that vacancy-concentration increases with temperature, thus increasing the energy of the crystal. At first sight it seems strange that a stable situation exists involving a higher energy than one in which there are no vacancies. The reason is that the entropy or disorder is increased by the presence of vacancies and this compensates for the increase in energy.

It is easy to see how this arises. If initially the lattice contains no vacancies the n_0 atoms have n_0 sites to accommodate them. If the atoms are indistinguishable there is only one way in which they can be distributed, i.e. $W_1 = 1$. If n vacancies are created, the n_0 atoms involved being ultimately located on the free surface of the specimen, we have a total of $(n_0 + n)$ sites. The number of ways in which n_0 indistinguishable atoms and n indistinguishable vacancies can be distributed is then simply

$$W_2 = \frac{(n_0+n)!}{n_0!\,n!}. \tag{8.13}$$

Thus the increase in entropy is

$$\Delta S = k \ln W_2/W_1 = k(\ln(n_0+n)! - \ln n_0! - \ln n!). \tag{8.14}$$

By Stirling's approximation for large numbers, $N! = (N/\mathrm{e})^N$ or

$$\ln N! = N \ln N - N. \tag{8.15}$$

Equation (8.14) thus gives

$$\Delta S = k\left[n_0 \ln\left(\frac{n_0+n}{n_0}\right) + n \ln\left(\frac{n_0+n}{n}\right)\right]. \tag{8.16}$$

Since $n \ll n_0$ (even at the melting point n is usually less than $10^{-4}\,n_0$) we can put $(n_0+n)/n_0 \approx 1$ and $(n_0+n)/n \approx n_0/n$.

Equation (8.16) then gives

$$\Delta S = kn \ln\left(\frac{n_0}{n}\right). \tag{8.17}$$

The volume change is so small that the PV term in the Gibb's free energy may be ignored. We may then write (see equation (2.28))

$$\Delta G = \Delta U - T\Delta S = 0 \text{ at equilibrium}$$

or

$$T\Delta S = \Delta U = n\varepsilon_{\mathrm{v}}, \tag{8.18}$$

since ε_{v} is the energy to create one vacancy. Combining equation (8.18) with equation (8.17) gives

$$kTn \ln\left(\frac{n_0}{n}\right) = n\varepsilon_{\mathrm{v}}$$

or

$$\ln\left(\frac{n_0}{n}\right) = \varepsilon_{\mathrm{v}}/kT. \tag{8.19}$$

This is the same as equation (8.12).

8.3.2 *Diffusion*

Atoms can move into a vacancy and leave their own site vacant. The vacancy has thus been displaced. This is the essential mechanism of diffusion in a solid. The energy for a neighbour to move into a vacant site will generally be less than ε_{v}, say ε_{d}. If we treat the crystal as an

Einstein solid with an atomic vibrational frequency ν (see Chapter 9, section 9.1.2), an atom has a chance of making

$$\nu \exp(-\varepsilon_d/kT) \tag{8.20}$$

jumps per second with this excess energy. If this is multiplied by the number of vacancies available for the atoms to jump into we obtain the number $\mathcal{N}$ of vacancy displacements per second.

$$\begin{aligned}\mathcal{N} &= \nu \exp(-\varepsilon_d/kT) n_0 \exp(-\varepsilon_v/kT) \\ &= n_0 \nu \exp\left(-\frac{\varepsilon_v + \varepsilon_d}{kT}\right).\end{aligned} \tag{8.21}$$

To convert this to the diffusion coefficient D we note that for a gas $D = \lambda c/3$, where λ is the mean distance between collisions. But $c = \mathcal{N}\lambda$, where $\mathcal{N}$ is the frequency of molecular collisions. Thus for a gas $D = \frac{1}{3}\mathcal{N}\lambda^2$ and this type of relation holds for all diffusion processes although the numerical constant may vary depending on the details of the diffusion mechanism. For the crystal $\lambda = a$. Hence

$$\begin{aligned}D &= \text{const} \times n_0 \nu a^2 \exp\left(-\frac{\varepsilon_v + \varepsilon_d}{kT}\right) \\ &= C n_0 \nu a^2 \exp(-q/kT),\end{aligned} \tag{8.22}$$

where q is the activation energy for self diffusion.

8.3.3 *Creep*

If a crystalline solid is deformed plastically under constant stress it will work-harden by the generation and interaction of dislocations until no further plastic flow occurs. If, however, the dislocations can escape from whatever obstacles impede their motion, flow at a vastly reduced rate will occur at constant stress. This is known as creep. The creep rate is usually a function of time, stress and temperature.

At temperatures above about 0.4 T_m (T_m = absolute melting point) the main mechanism which enables this to occur is the diffusion of atoms such that the dislocations can re-establish themselves in some slip plane where they are not blocked. In this case the creep rate, i.e. the rate of change of strain $\dot{\varepsilon}$ is constant with time. This is known as steady state or viscous creep and follows a law of the form

$$\dot{\varepsilon} = A(\text{stress})^n \exp(-q/kT). \tag{8.23}$$

In such a situation q turns out to be the activation energy for self-diffusion in the solid.

8.3.4 *Surface diffusion*

The free surface of a crystalline solid is not generally an ideally flat terrace. It usually contains some vacancies as in the bulk, some ledges or edges (see figure 8.10(*b*)) and atoms which have been able to detach themselves from these edges and sit individually on the surface (adatoms). The migration of vacancies (as in bulk diffusion) and the movement of adatoms are the two main factors involved in surface diffusion. Energy is required to activate these processes and one can write, as for bulk diffusion,

$$D = D_0 \exp(-q_s/kT)$$

where D, the diffusion coefficient, is expressed in $m^2\, s^{-1}$ and q_s is the activation energy. Since surface atoms have fewer neighbours than those in the bulk and are less tightly bound we would expect the activation energy to be less than for bulk diffusion. This appears to be the case to judge from the limited data available. For example for molybdenum and for tungsten, values of q_s are about 70 per cent of the bulk diffusion values although absolute values of q_s depend on the crystal face.

Surface diffusion is more difficult to study than bulk diffusion. A neat method which links atomic processes in surface diffusion with macroscopic changes in surface topography is the following. Sinusoidal grooves are etched chemically in the surface of the specimen. These grooves serve as an optical reflection diffraction grating and using a laser beam as the light source intensities of the diffracted beams provide a means of determining the depth of the grooves. The specimen is mounted in an ultrahigh vacuum system so that surface cleanliness can be monitored by electron diffraction: this is important since bulk impurities (especially carbon) diffuse to the surface and interfere with the surface diffusion process. At high temperatures the depth of the grooves gradually decreases and from measurements at different temperatures the activation energy can be calculated. It is clear that at a macroscopic level the surfaces should 'want' to become flatter since this reduces the surface area and hence the surface energy. Alternatively we may use the result obtained in section 10.11.6 that the vapour pressure over a convex surface is greater than that over a concave surface, implying that the atoms are freer to move. At the atomic level the process is less obvious since diffusion occurs in all directions both into and out of the hollows. We must assume that atoms at the top of the crests have fewer near neighbours inhibiting diffusion while those in the hollows are more tightly confined by their neighbours. Thus diffusion from the crests is favoured over diffusion from the hollows and the surface becomes less and less wavy. Surface diffusion is of practical importance in the sintering process.

Surface diffusion becomes more pronounced as the temperature is increased and a stage may be reached below the bulk melting temperature where the surface layer is mobile and virtually a liquid (see p. 158).

8.4 **Brittle solids**

8.4.1 *Brittle properties*

We now consider the behaviour of a solid which normally shows no ductility. If it is extended it stretches elastically and then snaps at some critical tensile stress S. If the material is homogeneous it will crack along a plane normal to the direction of the stress. If the specimen is subjected to a hydrostatic pressure P the tensile stress necessary to cause brittle failure is found to be $P+S$ (figure 8.11). This implies that brittle failure occurs when the resulting tensile stress across some appropriate plane exceeds a critical value, in this case S. In its idealized form this is the stress necessary to pull one plane of atoms completely away from a neighbouring plane.

We may estimate the theoretical brittle strength in a number of ways. If the bar has unit cross-section and γ is the surface energy, the act of snapping the specimen is to create two new unit areas of surface: the work done is 2γ. If the intermolecular forces are appreciable only over the distance of an atomic spacing, say 5×10^{-10} m, and the stress over

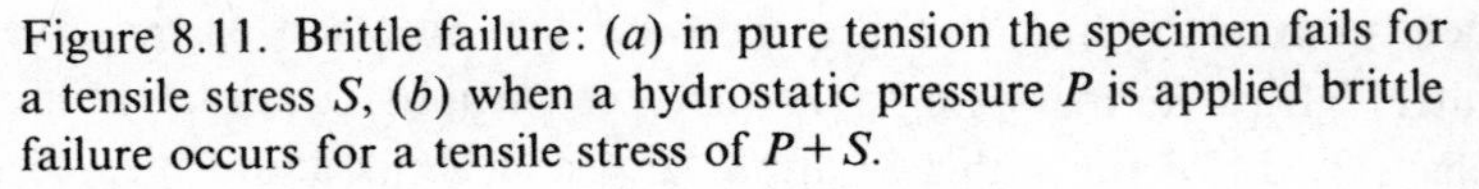

Figure 8.11. Brittle failure: (*a*) in pure tension the specimen fails for a tensile stress S, (*b*) when a hydrostatic pressure P is applied brittle failure occurs for a tensile stress of $P+S$.

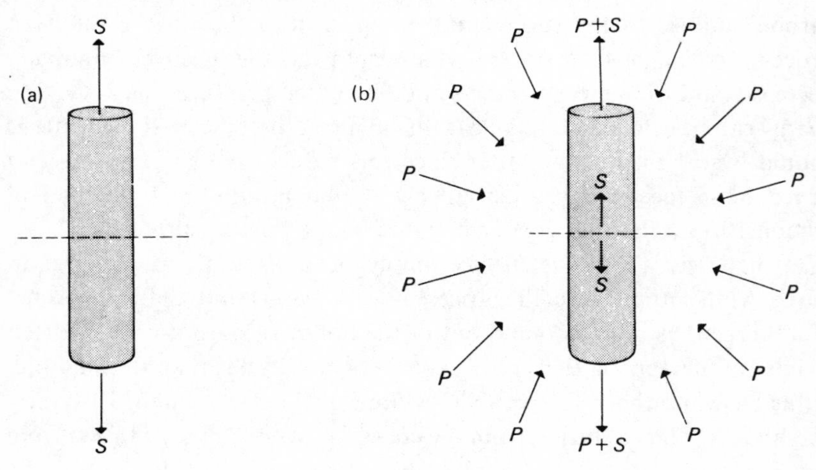

this distance is assumed constant, we may write

$$\text{work done} = S \times 5 \times 10^{-10} = 2\gamma. \tag{8.24}$$

For paraffin wax, where γ is of the order of 50 mJ m^{-2}, this gives a value of S of about 200 MN m^{-2} or 20 kgf mm^{-2}. For rock-salt, for which γ is about 500 mJ m^{-2}, the theoretical value is of order 200 kgf mm^{-2}. This is enormously larger than the observed strength.

We may also estimate the strength from the force–separation curve (figure 8.12(a)). When the force reaches the point B it is great enough to pull the atoms apart. This occurs when $\partial F/\partial x = 0$. If we express the energy U in terms of x we have $\partial U/\partial x = -F$. Hence failure occurs when

$$\frac{\partial^2 U}{\partial x^2} = 0. \tag{8.25}$$

We have not derived an expression for the energy of a crystal in terms of the atomic separation in one dimension. We may, however, estimate the hydrostatic tension necessary to pull a crystal apart. For an anionic crystal we have (from the previous chapter)

$$U = -\text{const} \times \left(\frac{1}{x} - \frac{a^8}{9x^9}\right). \tag{8.26}$$

Figure 8.12. (a) Force–displacement curve for calculating the theoretical brittle strength, (b) role of a crack, length l tip radius ρ, in facilitating brittle failure, (c) the formation of a crack by the 'pile-up' of dislocations.

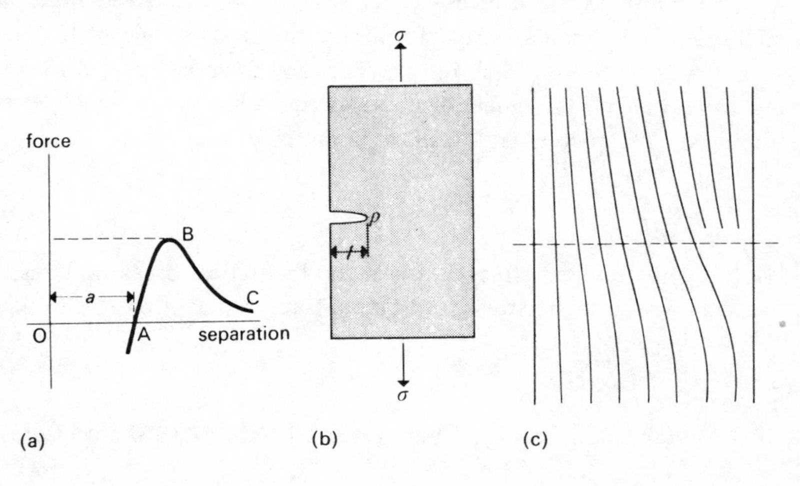

Ignoring the constant and differentiating,

$$\frac{\partial^2 U}{\partial x^2} = -\frac{2}{x^3} + \frac{10a^8}{x^{11}} = 0,$$

$$x = 5^{\frac{1}{8}} \times a = 1.23a. \tag{8.27}$$

This shows that the crystal can be expanded by a hydrostatic tension until when the ionic separation has increased by 23 per cent, i.e. the volume has increased by 85 per cent, it will disintegrate. If the bulk modulus K remained constant for these large strains it would imply a pressure equal to 0.85 K, but because of the large curvature of the force separation curve in this region the stresses are appreciably smaller. Simple calculation shows that this should be reduced by a factor of about 3.5 so that the corresponding pressure is about 0.24 K. Using the theoretical value of K from the previous chapter, this is 7000 MN m^{-2} or 700 kg mm^{-2}. This is of the same order as that deduced from the surface energy. In practice a hydrostatic tensile experiment would be difficult if not impossible to carry out; most estimates based on direct tension give a theoretical value for the brittle strength of the order of

$$S \approx \frac{E}{10}, \tag{8.28}$$

where E is Young's modulus.

8.4.2 *Observed brittle strength: need for cracks*

The observed brittle strength of solids is generally very variable but is usually 10 to 100 times smaller than the theoretical value. The source of this weakness is the presence of flaws or cracks in the solid, especially at the surface. These act as 'stress-raisers' or 'stress-multipliers'. Consider a rectangular strip of uniform thickness subjected to a tensile stress σ. If there is a crack in one edge, as shown in figure 8.12(b), and we can specify it by its length l and its tip radius ρ, the tensile stress at the tip of the crack is multiplied by a factor of the order

$$\left(\frac{l}{\rho}\right)^{\frac{1}{2}}. \tag{8.29}$$

Hence if the applied stress is σ and the theoretical breaking stress is S, we may write for the stress to produce fracture:

$$\sigma \approx S\left(\frac{\rho}{l}\right)^{\frac{1}{2}}. \tag{8.30}$$

If the crack radius is, say, 1 nm, a crack length of 1000 nm (1 μm) will increase the tensile stress at the tip of the crack by 30-fold. Thus an applied stress only one-thirtieth of the theoretical strength will be able to start the crack growing. Once this occurs the whole section will fail.

There are three general observations which support this explanation. First, brittle strengths are usually very variable; this is because of the large variety of sizes and shapes of surface cracks. Secondly, in recent years it has proved possible to prepare fine silica rods with very perfect surfaces which can withstand a tensile stress of $E/20$ before they fail. Thirdly, some fine crystalline fibres have been prepared which have tensile strengths approaching their theoretical value, presumably because their surfaces are free of surface flaws or growth steps.

Although this explanation is generally satisfactory there is an additional failure criterion which must be mentioned. Consider a flaw which is in the form of a sharp crack between two neighbouring atomic planes; the crack-tip radius tends to zero. Is then the stress-concentration factor infinite, and does the material now possess zero strength? The answer was given by A. A. Griffith in 1921. He showed that for a crack to grow it is not sufficient for the stresses at the crack-tip to exceed the theoretical strength; in addition sufficient elastic energy must be released from the system to provide the extra surface energy that a growing crack demands. By expressing the surface energy in terms of the elastic constants, the force-separation curve, and the atomic spacing, it is possible to estimate what this involves. It turns out that an infinitely sharp crack cannot grow with the application of a vanishingly small tensile stress.† In order to satisfy the surface energy criterion, the smallest stress σ_G capable of producing crack propagation, if the lattice spacing is a, is of order

$$\sigma_G \approx S\left(\frac{3a}{l}\right)^{\frac{1}{2}}, \tag{8.31}$$

that is to say, because of the surface energy criterion an infinitely sharp crack produces a reduction in strength equivalent to a tip radius of about three atomic spacings. For a crack of length l, the Griffith stress σ_G, is the smallest stress that will start the crack growing. This is the stress for a crack of tip radius $3a$ or less. For a more blunt crack ($\rho > 3a$) a larger stress is needed to start the crack moving. If the geometry of the tip is retained during propagation the failure stress falls slowly as the crack

† See A. H. Cottrell, *The Mechanical Properties of Matter* (Wiley, 1964), and E. Orowan, 'Fracture and strength of solids', *Reports on Progress in Physics*, vol. 12 (1949), pp. 185–232.

length increases. If, however, the crack grows into a sharp crack the failure stress falls to the Griffith value.

Equation (8.31) may also be expressed in terms of Young's modulus E and the surface energy γ of the solid in the following way. The work done in pulling an ideally brittle solid specimen of unit cross-section apart is equivalent to the work done in creating two new unit surfaces, i.e. 2γ. As we have seen, this pulling apart occurs when the applied stress S is of order $E/10$ (see equation (8.28)). The separating atomic planes are pulled apart at this stage by a distance of the order $a/10$. The work done is of order $(E/10)\times(a/10)=Ea/100$, which can be equated to 2γ. Hence γ is of order $Ea/200$. Putting $a=200\gamma/E$, equation (8.31) then becomes

$$\sigma_G=\frac{E}{10}\left(\frac{3\times 200\gamma}{El}\right)^{\frac{1}{2}}=\left(\frac{600}{100}\times\frac{\gamma E}{l}\right)^{\frac{1}{2}}.$$

A more rigorous analysis shows that the numerical factor inside the brackets is nearer unity than 6. The Griffith stress then becomes

$$\sigma_G\approx\left(\frac{\gamma E}{l}\right)^{\frac{1}{2}}. \tag{8.32}$$

The equilibrium spreading of a crack thus provides a means of determining the surface energy of a solid. We have already described one example of this in the splitting of a mica sheet. Attempts have also been made to apply this method to metals, polymers and other solids. Generally such experiments give extremely high values of the order of 100 J m^{-2}; this is because plastic deformation occurs at the tip of the crack and the plastic work involved completely swamps the much smaller true surface energy. Recently Gilman has shown that with some ionic crystals it is possible to 'freeze-out' the ductility by carrying out the experiments with the specimens immersed in liquid nitrogen. Under these conditions surface energies of the order of a few hundred mJ m^{-2} are obtained.

Edge dislocations are too small to act as effective stress-raisers from the point of view of crack propagation. However, if the free movement of edge dislocations is obstructed by some barrier so that several can pile up in a row (see figure 8.12(*c*)), an internal crack may be formed; this may be able to initiate crack propagation. In this way a ductile solid may become brittle.

8.4.3 *How brittle solids may be made ductile*

Consider a brittle solid which fails in a brittle manner for a tensile stress S (see figure 8.13). If we apply a hydrostatic pressure the tensile stress necessary for failure is $P+S$. Associated with this tensile stress is a shear stress equal to $\frac{1}{2}(P+S)$. If the critical shear stress of the material is less than this it will flow in a ductile manner before the tensile stress is large enough to produce brittle failure. This is one of the reasons why rocks below the earth can flow in a ductile way although they are normally very brittle materials. Indeed Bridgman has shown in the laboratory that under sufficiently high hydrostatic pressure even quartz can flow plastically. In this connection it is interesting to note that in indentation hardness experiments plastic indentation can often be made in relatively brittle materials (even though some cracking may also occur). This is because the large hydrostatic component of the stress field inhibits brittle failure. Furthermore the hardness values so obtained are a measure of the *plastic* properties of the brittle solid.

8.5 Conclusion

In this chapter we have discussed the strength and deformation properties of solids in terms of interatomic forces. The deformation characteristics fall into three main categories. When the stresses are below a certain level the strains are reversible and the deformation is said to be elastic. As we saw in the previous chapter elastic deformation involves the gentle distortion of the lattice and when calculations are possible there

Figure 8.13. Stress–strain curve for a brittle solid. Brittle failure occurs for a tensile stress S. In the presence of a hydrostatic pressure P this is increased to $P+S$. The associated shear stress may produce plastic yielding before brittle failure occurs.

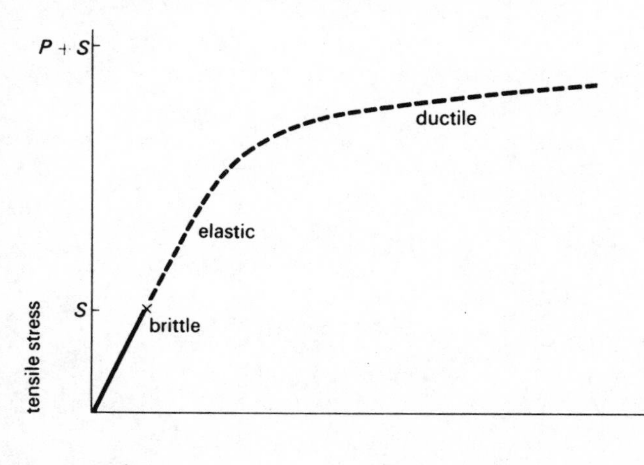

is good agreement between the observed behaviour and that derived from the interatomic forces. For larger stresses two other types of deformation occur, plastic and brittle: both are irreversible. These strength properties may also be calculated in terms of interatomic forces but observed values are generally very much smaller. We may say that solids owe their plastic and brittle strength to interatomic forces: their weakness to the presence of flaws or imperfections. The main characteristics are summarized in Table 8.1.

Table 8.1. *Plastic and brittle failure*

Mode	Criterion	Source of weakness	Effect of hydrostatic pressure
Plastic	Critical shear stress for planes to slide over one another	Dislocations, enable slip to occur row by row of atoms	None
Brittle	Tensile stress across a plane pulling planes of atoms apart	Cracks, act as stress-multipliers so that stress at crack-tip is much greater than the applied stress	Opposes applied tensile stress and so can inhibit brittle failure

9

Thermal and electrical properties of solids

In this chapter we shall discuss in terms of atomic mechanisms the specific heat capacity of solids, thermal expansion and thermal and electrical conductivity.

9.1 **Specific heat capacity**

9.1.1 *Definition of specific heat capacity*

The specific heat capacity is defined as the quantity of heat required to raise the temperature of unit mass of substance by 1 K. With solids the specific heat capacity at constant volume C_V is a little less than that at constant pressure C_P. For one mole the corresponding quantities are related by the thermodynamic function

$$C_P - C_V = \beta^2 TVK, \tag{9.1}$$

where β is the volume coefficient of expansion, K the bulk modulus and V the volume of one mole. In practice the specific heat capacity generally measured with solids is C_P, but the difference between this and C_V is small and can as a first approximation be neglected.

If we assume that the specific heat capacity is due to the change in internal energy we may easily calculate it, say, for one mole of the metal. This contains N_A atoms. The atoms have no translational energy, only vibrational. Each vibrational degree of freedom involves both potential and kinetic energy so that the average thermal energy is kT per degree of vibrational freedom. Each atom has three independent degrees of vibration so that its average vibrational energy is $3kT$. For all N_A atoms this gives a value of

$$U = N_A \times 3kT = 3RT. \tag{9.2}$$

If we keep the volume constant any additional heat goes solely in increasing the vibrational energy. Hence

$$C_V = \left(\frac{\partial U}{\partial T}\right)_V = 3R, \tag{9.3}$$

where C_V is the specific heat capacity per mole (atomic). Thus C_V should have a value of about 25 J K^{-1}, a conclusion reached empirically by Dulong and Petit in 1819. If one deals with a compound, say NaCl, then one would expect each ion-mole to possess a specific heat capacity of 25 J K^{-1}; so that for the mole the specific heat capacity would be 50 J K^{-1}; for a triatomic molecule it would be 75 J K^{-1}. Some typical results for C_P at room temperature for metals and diatomic and triatomic solids are given in Table 9.1. Ignoring the difference between C_P and C_V, it is seen that the specific heat capacities per mole are all indeed of the order of 25 J K^{-1} (except ice).

9.1.2 *The Einstein model*

Later measurements showed that the Dulong and Petit law holds only above a certain temperature. At lower temperatures C_P falls off markedly. For example with copper the specific heat capacity per mole falls off from 24 at room temperature to 21 at 170 K, to 11 at 70 K and to 1 J K^{-1} at 20 K. In the limit it approaches zero at the absolute zero.

The first simple explanation of this was due to Einstein in 1906 who made use of Planck's earlier discovery of the distribution of thermal energy in an oscillating system. If each atom behaves as an independent harmonic oscillator of frequency ν, its average energy (ignoring zero-point

Table 9.1. *Specific heat capacities of solids*

Substance	C_P (J kg^{-1} K^{-1})	Atomic or molecular mass (kg)	Specific heat capacity C_P mol^{-1} (J K^{-1})
Al	920	0.027	25
Cu	380	0.063	24
Ag	220	0.108	24
Pb	120	0.207	25
NaCl	880	0.0575	51 (diatomic)
CaF_2	920	0.0780	72 (triatomic)
SiO_2	1180	0.0600	71 (triatomic)
Ice	2100	0.018	38 (triatomic)

energy) is exactly the same as that given in our discussion of the vibrational energy of a diatomic molecule.

$$u = \frac{h\nu}{\exp(h\nu/kT) - 1}. \tag{9.4}$$

For the N_A atoms each of which has 3 independent degrees of vibrational freedom we obtain

$$U = 3N_A u. \tag{9.5}$$

For high temperatures this reduces to $3N_AkT$ in agreement with the Dulong and Petit law. For lower temperatures it diminishes and as T tends to zero, it also tends to zero. If an appropriate value of ν is assumed the shape of the Einstein theoretical curve agrees reasonably well with experiment except towards $T=0$ (see figure 9.1). In this region careful experiments show that C_V falls off as T^3, whereas in Einstein's theory it falls off as $\exp(-h\nu/kT)$, i.e. it falls off more rapidly than T^3.

9.1.3 *The Debye model*

There are two main defects in the Einstein model. First it assumes that all the atoms have the same single frequency ν, and secondly that the vibration of each atom is independent of its neighbour. In fact the atoms act as coupled oscillators and a whole range of frequencies is possible. Each value of ν (at a fixed temperature) contributes its own average thermal energy u (see equation (9.4)). These should all be added

Figure 9.1. Specific heat capacity of copper as a function of temperature: ● experimental values, —— theory given by Einstein assuming an atomic vibrational frequency of $\nu = 5\times10^{12}\ s^{-1}$. If one writes $h\nu_E/k\theta_E = 1$ this gives for the Einstein temperature, corresponding to $\nu = 5\times10^{12}\ s^{-1}$, a value of $\theta = 240$ K.

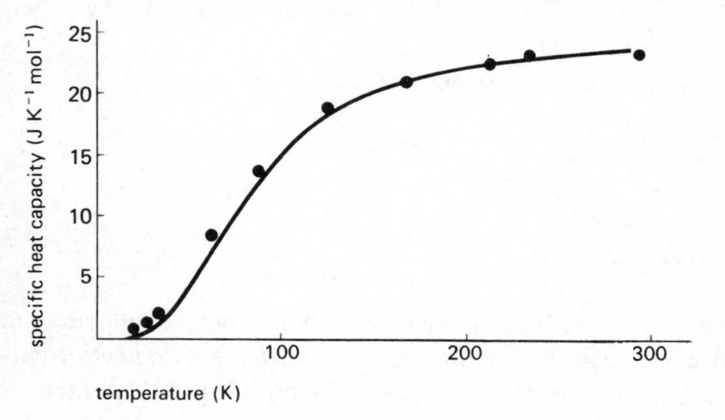

to determine the total energy U of the whole solid. An analysis along these lines was first developed by Born and von Karman in 1912. The main feature of such a treatment, which distinguishes it from the Einstein model, is that it reveals a whole spectrum of possible frequencies. The detailed analysis is, however, extremely complicated.

A completely different approach is that due to Debye (1912). The solid is treated as a continuum which, at first sight, seems a retrogressive step. However, its great merit is that it provides a frequency spectrum which is a close approximation to the true spectrum. Debye considers the way in which waves can travel through the solid. The most remarkable feature is that a *standing wave* of frequency ν behaves exactly like a quantum oscillator of frequency ν so that its thermal energy is again given by equation (9.4). This is surprising since the wave involves the vibration of all the atoms in the solid; yet its thermal energy is that of a *single* quantum oscillator. At first sight it would seem that this could not possibly give the same total energy. The difficulty, however, disappears when it is realized that the total vibration of each atom is the composite result of all the possible waves that can travel through the solid.

Suppose the solid is a crystal in the form of a cube of side L. To establish standing waves the free surface of the specimen must be either a node or an antinode and it may be shown that either condition leads to the same result. We shall treat the free surface as an antinode. Then the largest standing wave possible for a wave travelling normal to the faces of the cube has a wavelength $\lambda=2L$ (see figure 9.2). If the velocity of the wave is C, the frequency of the wave is $\nu=C/\lambda=C/2L$. The next possible standing wave has a wavelength $\lambda=\frac{2}{3}L$ and its frequency is $\nu=\frac{3}{2}C/L$. Thus the possible frequencies increase in the order:

$$\frac{C}{2L}, \quad \frac{C}{2L}\times 2, \quad \frac{C}{2L}\times 3, \quad \frac{C}{2L}\times 4, \quad \text{etc., or } \nu_1, 2\nu_1, 3\nu_1, 4\nu_1. \tag{9.6}$$

The corresponding thermal energies are

$$\frac{h\nu_1}{\exp(h\nu_1/kT)-1}, \quad \frac{2h\nu_1}{\exp(2h\nu_1/kT)-1},$$
$$\frac{3h\nu_1}{\exp(3h\nu_1/kT)-1}, \quad \text{etc.,} \tag{9.7}$$

and all that is now needed is to sum them. Very soon the possible frequencies become such large multiples of ν_1 that they become virtually continuous and the sum can be replaced by an integral. One needs to know the number of vibrations that are possible between ν and $\nu+\mathrm{d}\nu$. This is not difficult for a continuum. A simple approach is as follows.

For the waves discussed above

$$\nu_n = \frac{C}{2L} n,$$

where n is an integer. If we considered the more general case of a wave travelling in some arbitrary direction we would obtain the same result but n would have components n_1, n_2, n_3 (each of them integers) in the x, y, z directions.

Then

$$\nu = \frac{C}{2L} (n_1^2 + n_2^2 + n_3^2)^{\frac{1}{2}}. \tag{9.8}$$

We plot $n_1C/2L$, $n_2C/2L$, $n_3C/2L$ on x, y, z coordinates (see figure 9.3). Then the distance from the origin to any point is equal to ν. Since n_1, n_2, n_3 can only change by one unit at a time each possible point for ν occupies a volume of $(C/2L)^3$. The number of points available for frequencies between ν and $\nu+\mathrm{d}\nu$ is the volume of the spherical shell contained between radii ν, $\nu+\mathrm{d}\nu$ divided by the unit volume $(C/2L)^3$. Since ν can have only positive values we can only use one-eighth of the spherical shell.

Figure 9.2. Characteristic frequencies of standing waves in an elastic continuum of length L. We assume that the free ends are antinodes and that the velocity C of the waves is independent of frequency.

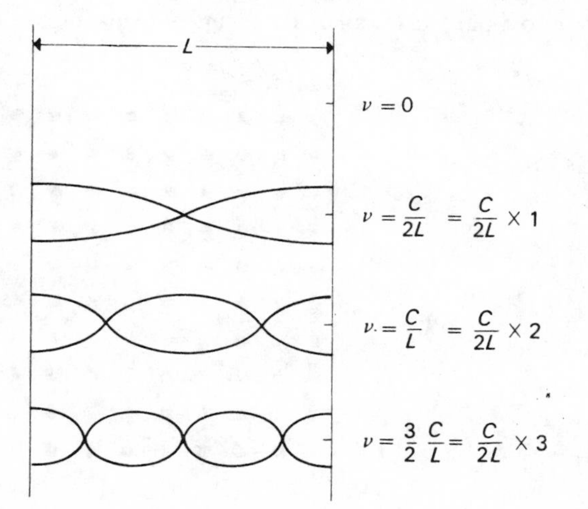

The resulting number is therefore

$$\mathrm{d}\mathcal{N} = (\tfrac{1}{8} \times 4\pi\nu^2\,\mathrm{d}\nu)\left(\frac{2L}{C}\right)^3$$

$$= \frac{4\pi L^3 \nu^2\,\mathrm{d}\nu}{C^3}, \qquad (9.9)$$

We allow for two transverse vibrations and one longitudinal vibration. These have different wave velocities but for simplicity we choose a suitable average velocity C. This triples the value of $\mathrm{d}\mathcal{N}$. Then the total thermal energy is simply

$$U = \int_0^{\nu_m} \left[\frac{h\nu}{\exp(h\nu/kT) - 1}\right] 3\,\mathrm{d}\mathcal{N}. \qquad (9.10)$$

We have ignored zero-point energy since it is independent of T and when we differentiate U to find C_V it will disappear.

The difficulty about equation (9.10) is that we have no way, so far, of specifying the upper limit ν_m of the integral. It is at this point that Debye ties his continuum model to a particulate model. If the crystal contains N atoms, Debye postulates that the total number of vibrations possible must be equal to $3N$ so that, in the limit, at high temperatures where each u has a value of kT the total thermal energy is $3NkT$. From equation

Figure 9.3. (*a*) Construction showing that the number of waves $\mathrm{d}N$ possessing frequency between ν and $\nu + \mathrm{d}\nu$ is proportional to the volume in the octet of the spherical shell lying between radii ν and $\nu + \mathrm{d}\nu$. (*b*) Two-dimensional figure showing that each point corresponding to a particular value of ν occupies a cube of side $C/2L$.

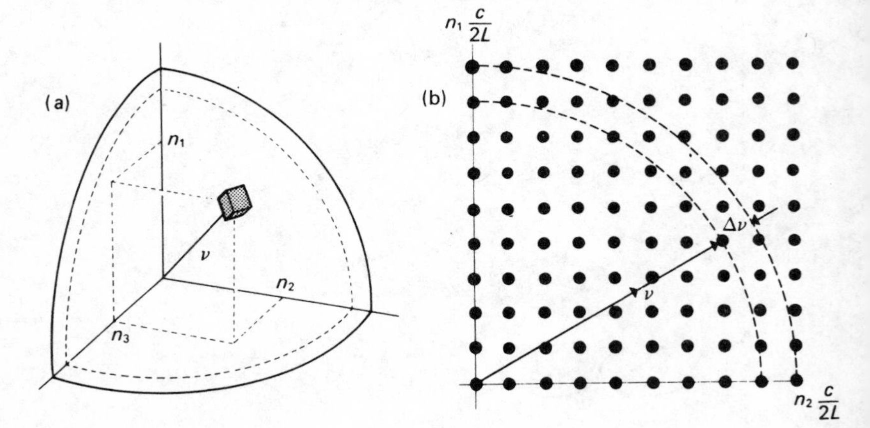

(9.9) this means that

$$\int_0^N 3\,\mathrm{d}\mathcal{N} = \int_0^{\nu_m} 12\pi \frac{L^3}{C^3} \nu^2\,\mathrm{d}\nu = 3N. \tag{9.11}$$

This gives

$$\nu_m^3 = \frac{3}{4\pi} \frac{C^3}{L^3} N, \tag{9.12}$$

where ν_m is the maximum frequency on the Debye model. Before we use this we may form some estimate of its magnitude. If the solid has a simple cubic structure with a distance a between each atom, Na^3 will be the volume of the solid, i.e. L^3. Then

$$\nu_m = \left(\frac{3}{4\pi}\right)^{\frac{1}{3}} \frac{C}{a} \approx \frac{C}{a}. \tag{9.13}$$

This corresponds to a minimum wavelength

$$\lambda_{\mathrm{min}} = \frac{C}{\nu_{\mathrm{m}}} \approx a. \tag{9.14}$$

Thus the limiting frequency on the Debye model corresponds to very short waves, the wavelength being comparable with the atomic spacing. This is reasonable since once we recognize the particulate nature of the solid, a wavelength less than the atomic spacing ceases to be meaningful. This result also corresponds rather closely to the very simple elastic model discussed in the previous chapter where we found that if the atoms have a natural (uncoupled) frequency ν the velocity of a wave is of order $2\pi\nu a$ compared with $\nu_{\mathrm{m}} a$ given in equation (9.13). Inserting equation (9.12) in (9.9),

$$3\,\mathrm{d}\mathcal{N} = \frac{9N}{\nu_{\mathrm{m}}^3} \nu^2\,\mathrm{d}\nu. \tag{9.15}$$

If this is substituted in equation (9.10) we have an explicit expression for U which can be evaluated. By differentiating, the specific heat capacity C_V may be found.

The Debye theory predicts that the heat capacity of a solid depends only on the characteristic frequency ν_{m}. This implies that if C_V for various solids is plotted against $kT/h\nu_{\mathrm{m}}$ they should all lie on a single curve. This is found to be very nearly true (see figure 9.4). The quantity $h\nu_{\mathrm{m}}/k$ is known as the Debye temperature θ_{D}. It is interesting to compare θ_{D} as calculated from equation (9.12) with the value of θ_{D} that gives the best fit of the Debye equation with the observed specific heat capacity measurements (Table 9.2). Although the theory is only really valid for isotropic solids containing a single type of atom it is seen that it holds extremely well for more complicated materials. It even holds approximately for ice, which is rather surprising. (An average wave velocity of 220 m s^{-1} for ice has been used to calculate θ_D). The specific heat capacity of ice at 200 K is about 1400 J kg^{-1} K^{-1} or 25 J mol^{-1} K^{-1}. It thus behaves as a monatomic solid, practically the whole of the thermal energy being due to oscillations of the molecule as a whole. Between 200 K and 273 K (the melting point) there is some increase in specific heat capacity to a value of about 40 J mol^{-1} K^{-1}. Presumably in this temperature range rotational degrees of freedom of the whole molecule contribute to the thermal energy

Table 9.2. *The Debye temperature*

	θ_D	
Substance	From specific heat capacity	From equation (9.12)
Pb	88	75
Cd	168	174
Ag	215	220
Cu	315	341
NaCl	281	305
CaF_2	474	510
Ice	220	200

Figure 9.4. Specific heat capacity of various solids plotted as a function of $kT/h\nu_m$, where ν_m is the characteristic Debye frequency. Writing $h\nu_m/k = \theta_D$, the quantity θ_D is known as the Debye temperature. ○, ●, ■, ×, experimental points for Cu ($\theta_D = 315$ K), Ag ($\theta_D = 215$ K), Pb ($\theta_D = 88$ K), C ($\theta_D = 1860$ K); —— theory given by Debye.

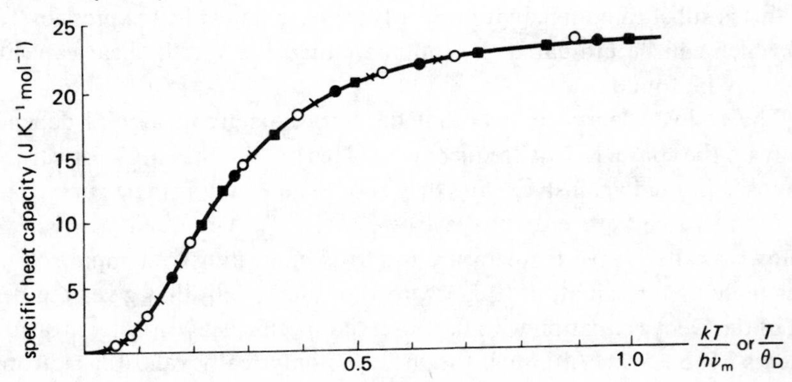

of the ice crystal. As we shall see in Chapter 14 this rotational freedom accounts for the rather large low-frequency dielectric constant of ice well below the melting point (see figure 14.10).

Although the Einstein model is in many ways a simpler and more direct one than the Debye model there are two features in the Debye treatment that make it preferable. First ν_m may be derived directly from bulk properties so that there are no assumed constants in the final calculation of U. By contrast the Einstein frequency (which turns out to be comparable with Debye's ν_m, in fact for many solids, $\nu_E \approx 0.75\nu_m$ or $\theta_E \approx 0.75\theta_D$) must be deduced empirically from the shape of the $C_V - T$ curve. Secondly, at low temperatures the Debye relation for the specific heat reduces to

$$C_V = \frac{12\pi^4 R}{5\theta_D^3} \times T^3, \tag{9.16}$$

so that it satisfactorily explains the observed T^3 dependence in this temperature range.

The main defect of the Debye treatment is that it considers waves of all frequencies to travel at the same speed. This is not true of coupled oscillators. For higher frequencies there is an appreciable drop in wave velocity. Fortunately the waves in this range generally contribute only a small part to the total vibrational energy.

9.2 Thermal expansion: Gruneisen's law

We may now derive a simple relation between thermal expansion and specific heat. It is based on the existence of an asymmetrical potential energy curve for the whole crystal in terms of the separation between atoms. For simplicity we consider here only the potential energy u between one atom and its neighbour and ignore the problem of coupled interactions.

The potential energy curve is shown in figure 9.5(a); we transpose it to axes as shown in figure 9.5(b) and express u in terms of the displacement x of the atoms from their equilibrium separation a. We assume that at 0 K the energy is zero (ignoring zero-point energy) and that the addition of thermal energy increases u according to a power law:

$$u = Ax + Bx^2 + Cx^3 \ldots . \tag{9.17}$$

Since O is the origin where $\partial u/\partial x = 0$ we obtain $A = 0$. Then

$$u = Bx^2 + Cx^3 + \ldots . \tag{9.18}$$

If we used only $u = Bx^2$ the curve would be symmetrical about the u axis and the restoring force for small displacements, $-\partial u/\partial x = -2Bx$, would be proportional to x, so that the atoms would oscillate with simple harmonic motion. We now assume that the cubic term is sufficient to match the real asymmetric potential energy curve.

$$u = Bx^2 + Cx^3. \tag{9.19}$$

Suppose we add energy u to the system by heating the solid. We see from figure 9.5(b) that there are two equilibrium positions P and Q between which x oscillates. In the first edition of this book we assumed that the mean position is the mid-point between P and Q. It occurs (see figures 9.5(b) and 9.6) at a displacement

$$e = -Cx^2/2B. \tag{9.20}$$

A more elegant and, physically, more attractive approach is the following due to Dr J. R. Waldram. From equation (9.19) the force between the two atoms is

$$F = -\frac{\partial u}{\partial x} = -2Bx - 3Cx^2. \tag{9.21}$$

We now observe that since there is no external force acting, the time average of F must be zero. Replacing x and x^2 by $\bar{x}$ and $\overline{x^2}$ (the time average of these quantities) we thus have

$$-2B\bar{x} - 3C\overline{x^2} = 0$$

Figure 9.5. (a) Potential energy u of an atom as a function of its separation x from its neighbours, (b) same curve transposed to different axes so that the minimum has coordinates (0, 0).

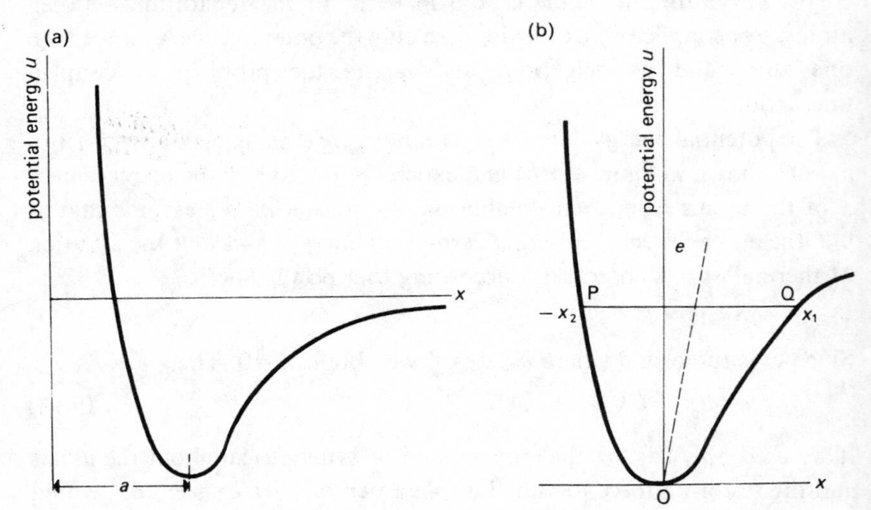

or

$$\bar{x} = \frac{-3C}{2B}\overline{x^2}. \tag{9.22}$$

The average displacement is three times as great as in the earlier derivation. This is because at Q the slope of the potential energy curve is less than at P, the restoring force is less and the rate of movement of the atom is less: it spends more time near Q than near P.

We now consider the energy as x oscillates between P and Q. At P and Q all the added energy is potential (u): there is no kinetic energy. At position O there is no additional potential energy, all the added energy is kinetic. For any system undergoing (approximate) simple harmonic motion the average kinetic energy = average potential energy = $\frac{1}{2}u$. The potential energy is given approximately by the first term of equation (9.19), the second term representing only that small part of the energy which accounts for the skew-shaped character of the potential energy curve. The time average potential energy may thus be written as

$$B\overline{x^2} = \tfrac{1}{2}u. \tag{9.23}$$

For a mole (N_A atoms) and for a solid in which the coordination number is n the increase U in lattice energy will, from equation (9.23), be

$$\tfrac{1}{2}U = B\overline{x^2}(\tfrac{1}{2}N_A n), \tag{9.24}$$

Figure 9.6. Force–separation curve for the situation described in figure 9.5(*b*); the expansion *e* is equivalent to an increase dx in the mean spacing.

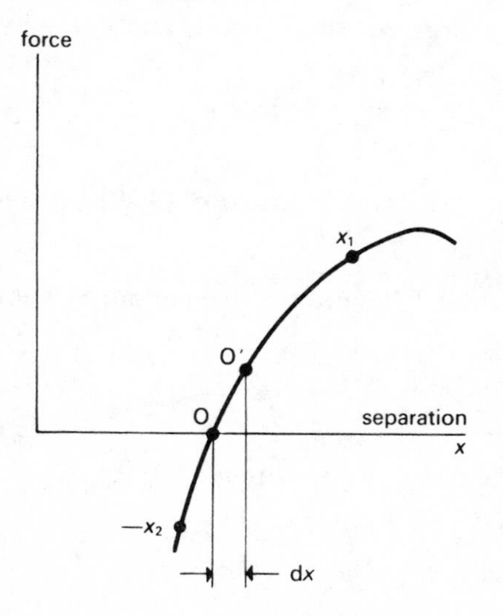

the half being introduced in the last term to avoid counting every bond twice. Then from equation (9.22) we have

$$\bar{x} = -\frac{3C}{2B^2}\frac{U}{Nn}. \tag{9.25}$$

Without any further elaboration we can see that the coefficient of linear expansion α which is equal to $\mathrm{d}(\bar{x}/a)/\mathrm{d}T$ will be proportional to $\mathrm{d}U/\mathrm{d}T$, i.e. to the specific heat capacity C_V of the solid. Thus

$$\frac{\alpha}{C_V} = \text{constant for a given material.} \tag{9.26}$$

This relation is due to Gruneisen and we shall now derive the value of the constant more quantitatively.

We shall find it convenient to introduce the bulk modulus K. We assume the solid to be isotropic. Let V be the molar volume. Suppose as a result of applying a hydrostatic pressure p on the solid we produce a volume change v corresponding to a mean change of x in the separation between each atom and its neighbour (v is negative).

$$K = \frac{p}{-v/V} = \frac{pV}{-v}. \tag{9.27}$$

The work done on the solid in increasing p from 0 to its final value is the gain in lattice energy U associated with changing the atomic spacing. The work done is

$$U = \tfrac{1}{2}p(-v) = \frac{1}{2}\frac{K}{V}(v)^2 = \frac{1}{2}KV\left(\frac{v}{V}\right)^2, \tag{9.28}$$

and since $U = \frac{1}{2}Nnu$ (u is the energy per single bond) and $v/V = 3x/a$ we have

$$\tfrac{1}{2}Nnu = \frac{1}{2}KV\left(9\frac{x^2}{a^2}\right). \tag{9.29}$$

From equation (9.19) we see that $u \approx Bx^2$. Equation (9.29) becomes

$$BNn = 9KV/a^2. \tag{9.30}$$

We can substitute this value of BNn into the denominator of equation (9.25)

$$\bar{x} = -\frac{3C}{2B}\frac{Ua^2}{9KV}.$$

The fractional volume change v/V can be written

$$\frac{v}{V} = \frac{3\bar{x}}{a} = -\left(\frac{Ca}{2B}\right)\frac{1}{K}\frac{U}{V}. \tag{9.31}$$

The temperature variation of this is the volume coefficient of thermal expansion β.

$$\beta = -\left(\frac{Ca}{2B}\right)\frac{1}{K}\frac{C_V}{V} = \gamma\frac{1}{K}\frac{C_V}{V}, \tag{9.32}$$

where C_V is the constant-volume molar heat capacity. The quantity $-(Ca/2B)$ is a pure number which depends only on the law of force between the atoms and is called the Gruneisen constant γ.

We may now take the final step. From equation (9.19) we see that

$$\frac{\partial^2 u}{\partial x^2} = 2B + 6Cx$$

so that

$$\left[\frac{\partial^2 u}{\partial x^2}\right]_{x=0} = \ddot{u}_0 = 2B.$$

Again

$$\frac{\partial^3 u}{\partial x^3} = \dddot{u} = 6C.$$

Consequently

$$\gamma = -\frac{Ca}{2B} = \frac{\dddot{u}}{\ddot{u}_0}\frac{a}{6}. \tag{9.33}$$

If we write out the full potential energy equation for an ionic solid as

$$u = -k\left[\frac{1}{a+x} - \frac{a^8}{9(a+x)^9}\right], \tag{9.34}$$

$\ddot{u}$ at $x=0$ is $-k(-8/a^3)$, $\dddot{u}$ is $-k(104/a^4)$. Hence $\gamma = 13/6 \simeq 2.2$.

For inert gases, assuming a 6–12 potential, the value predicted from equation (9.33) is $\gamma = 3.5$. The observed values for most solids appear to lie between about 1.5 and 3.

As an example we take an ionic solid such as rock-salt. For this $\gamma = 13/6$, $K = 2.8 \times 10^{10}$ N m^{-2}, $V = 4 \times 10^{-5}$ m^3 mol^{-1}, $C_P \approx C_V =$ 50 J K^{-1} mol^{-1}. Substituting in equation (9.32) we obtain a value

$$\beta \approx 10^{-4}\ \mathrm{K}^{-1}.$$

This is the right order of magnitude.

As has been pointed out at the end of Chapter 7 the above treatment avoids entropy since it treats the solid as though its behaviour were exactly analogous to that of a pair of atoms. By contrast, for an assembly of a large number of atoms, as in a solid crystal, an entropy factor arises.

This occurs in the following way. Because of the shape of the force–separation curve an expansion involves a decrease in the force constant between individual atoms. This leads to a decrease in the natural frequency ν of the atoms and this in turn implies a smaller energy quantum $h\nu$. Since at constant temperature, in the classical region, the total vibrational energy is constant ($u=3kT$ per atom) this means that the number of vibrational quanta in the material is increased by the expansion. The more quanta there are, the more ways there are of distributing them randomly, i.e. there is an increase in entropy. Consequently the entropy term favours expansion. On the other hand a change in atomic spacing necessitates an increase in the stored elastic energy and is therefore 'opposed' by the elastic bulk modulus of the material. A balance is struck at a certain value of the coefficient of thermal expansion β. It may be derived from the Helmholtz free energy which contains two terms, one being the stored elastic energy and the other the entropy term. When this analysis is carried out it is found that, to a very close approximation,

$$\beta=\gamma\frac{1}{K}\frac{C_V}{V},$$

as in equation (9.32), where $\gamma=-\mathrm{d}(\ln\nu)/\mathrm{d}(\ln V)$. The natural frequency is proportional to (force constant)$^{\frac{1}{2}}$, i.e. to $(\partial^2u/\partial x^2)^{\frac{1}{2}}$. The volume V is proportional to a^3 so that $\mathrm{d}(\ln V)$ is equal to $3\mathrm{d}(\ln a)$. Using equation (9.19) for u we find that

$$\gamma=-\frac{\mathrm{d}(\ln\nu)}{\mathrm{d}(\ln V)}=-\frac{Ca}{2B}.$$

This is identical with the result obtained above.

Finally we may note that even if the atomic vibrations were perfectly harmonic, i.e. if the potential energy curve were perfectly symmetrical, a notional increase in the atomic spacing would reduce the restoring force on a displaced atom and the vibrational frequency would be correspondingly reduced. This would result in a finite coefficient of thermal expansion (Guggenheim, 1965†). However, the thermal expansion due to this factor would be very small compared with that arising from the asymmetric potential energy curve considered above.

† E. A. Guggenheim, *Boltzmann's Distribution Law* (North-Holland, 1965).

9.3 Thermal conductivity

9.3.1 *Heat transfer*

There are three main forms of heat transfer.

(*a*) Radiation; this involves the emission of electromagnetic waves. All materials can radiate but radiation is generally most marked with solids.

(*b*) Convection; this involves the streaming of matter. With forced convection the streaming is imposed by some external driving mechanism; with natural convection it occurs as a result of density differences. Convection is observed only in liquids and gases.

(*c*) Conduction; this involves the transfer of thermal energy by some form of collision. It occurs in gases, liquids and solids. With poorly conducting solids the energy transfer is provided by lattice vibrations; with metals this is greatly augmented by electron collisions. In this chapter we shall deal solely with conduction in solids.

The rate at which heat flows through any element of a solid is proportional to the cross-section area $\mathrm{d}A$ of the element and the temperature gradient measured normal to $\mathrm{d}A$. We may make this proportionality into an equality by inserting a constant K which we call the thermal conductivity. Then

$$\frac{\mathrm{d}Q}{\mathrm{d}t} = -K\,\mathrm{d}A \times \frac{\mathrm{d}T}{\mathrm{d}x}. \tag{9.35}$$

The negative sign shows that a positive flow of heat takes place from a hotter to a cooler region, i.e. where $\mathrm{d}T/\mathrm{d}x$ is negative. K has the dimensions $MLT^{-3}\theta^{-1}$. It varies from the best thermal insulators to the best conductors by a factor of only about 1000. By contrast, electrical conductivities can vary by a factor of 10^{30}.

9.3.2 *Heat flow down a uniform bar*

As a simple example we consider the heat flow down a uniform bar under two extreme conditions. When it is:

(*a*) thermally insulated. In the steady state if no heat escapes from the surface of the bar the quantity of heat crossing every section per second must be constant. The temperature gradient must therefore be constant (see equation (9.35)). Knowing the temperature of the hot and cold ends, the cross-section of the bar and the rate of heat conduction, K may be determined directly. This is a convenient method of measuring K for good conductors. It may also be used with poor conductors if the bar is replaced by a thin disc of the material.

(*b*) exposed to a cooler constant temperature environment. Consider a very long bar of uniform cross-sectional area A held at a temperature T_0 at one end (figure 9.7). Let the environment be at a temperature T_e. Consider a small element of length $\mathrm{d}x$ at some point along the bar where its temperature is T. Express these temperatures as temperature excesses over the environmental temperature. The hot end is at θ_0, the element at θ and the environment at zero. In the steady state heat enters at A, some is lost by radiation and convection from the surface of the element, and the remainder passes through the element at B. If the circumference of the bar is c the surface area of the element is $c\,\mathrm{d}x$. If the heat lost is proportional to the surface area, the emissivity ε of the surface, and the temperature excess over the surroundings, the heat lost per second is $c\,\mathrm{d}x\varepsilon\theta$.

Figure 9.7. (*a*) Steady-state flow of heat along an exposed uniform bar. The heat flow in at A, less the heat flow out at B is equal to the heat lost by radiation and convection from the surface between A and B. (*b*) Heat flow through a thermally insulated element: the heat flow in at A, less the heat flow out at B is equal to the heat accumulated in the element.

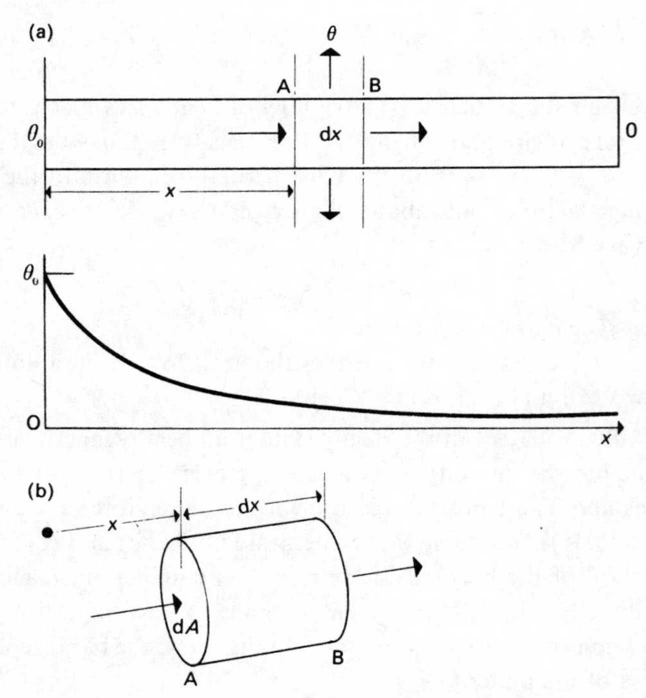

Heat flow per second in at $A = -KA\dfrac{d\theta}{dx}$.

Heat flow per second out at

$$B = -KA\frac{d}{dx}(\theta + d\theta) = -KA\left(\frac{d\theta}{dx} + \frac{d^2\theta}{dx^2}\,dx\right).$$

Heat flow per second at A exceeds outflows at B by

$$KA\frac{d^2\theta}{dx^2}\,dx = \text{heat lost from surface of element, so that}$$

$$KA\frac{d^2\theta}{dx^2} = c\varepsilon\theta. \tag{9.36}$$

The solution is of the form

$$\theta = B\,e^{wx} + C\,e^{-wx},$$

where $w^2 = c\varepsilon/KA$. At great distances from the hot end the temperature approaches the ambient, i.e. $\theta = 0$. Hence $B = 0$.

$$\theta = c\,e^{-wx}. \tag{9.37}$$

Since at $x = 0$, $\theta = \theta_0$ we have finally

$$\theta = \theta_0 \exp\left[-\left(\frac{c\varepsilon}{KA}\right)^{\frac{1}{2}} x\right]. \tag{9.38}$$

The temperature falls off exponentially with distance from the hot end.

9.3.3 *Thermal diffusivity*

Consider a body which is insulated so that no heat is lost to the surroundings. We heat one part of it and consider how the heat flows before steady-state conditions are reached. Clearly the body has to be warmed up (i.e. it must absorb heat) as the heat flows through it.

Consider a small element of cross-section dA at a point A (ordinate x) where the temperature is T. At the neighbouring point B (ordinate $x + dx$) the temperature is $T + dT$ (see figure 9.7(*b*)).

Heat flow per second in at $A = -K\,dA\dfrac{dT}{dx}$.

Heat flow per second out at $B = -K\,dA\dfrac{d}{dx}(T + dT)$.

The heat accumulated in the element per second $= K\,dA(\partial^2 T/\partial x^2)\,dx$. If the element has density ρ, specific heat capacity C, its heat capacity is

$\rho\, dA\, dxC$. If the temperature rise per second of the element is dT/dt we have

$$\rho C\, dA\, dx \frac{dT}{dt} = K\, dA \frac{\partial^2 T}{\partial x^2} dx,$$

$$\frac{dT}{dt} = \frac{K}{\rho C} \frac{\partial^2 T}{\partial x^2}. \tag{9.39}$$

The quantity $K/\rho C$ is known as the thermal diffusivity or *temperature* conductivity. It is of great importance in non-steady-state thermal problems.

9.3.4 *Theory of thermal conductivity*

The thermal conductivity of solids can be explained in terms of some type of collision process by means of which thermal energy is transported from the hotter to the colder regions. It is natural therefore to express it in a manner resembling the treatment for gases. For a gas where the transfer is due to molecular collisions we have

$$K = \tfrac{1}{3} nm\bar{c}\lambda c_V, \tag{9.40}$$

where c_V is the specific heat capacity per unit mass. We note that the quantity nmc_V is the specific heat capacity per unit volume s. Then

$$K = \tfrac{1}{3} s\bar{c}\lambda. \tag{9.41}$$

9.3.4.1 *Poor conductors.* In poor conductors the thermal energy is transferred by lattice vibrations. These travel through the specimen as waves which are scattered by the lattice as they progress through it. The wave is attenuated corresponding to the drop in temperature through the specimen. These thermal waves thus behave like particles possessing energy and momentum: they are known as *phonons*.

We may treat the waves as energy packets moving with the velocity c of the sound wave, and travelling a distance λ before they collide and communicate their thermal energy to the lattice. By exact analogy with gases we may write

$$K = \tfrac{1}{3}(\text{specific heat capacity per unit volume}) \times (\text{velocity of sound wave})\lambda.$$

This relation is found to be quite satisfactory for most poor conductors, the value of λ being of the order of 2 nm.

9.3.4.2 *Good thermal conductors: metals.* Good thermal conductors are generally good electrical conductors – their thermal conductivity is 100–

1000 times greater than that of insulators. This suggests that the major part of the thermal conductivity, like that of the electrical conductivity, arises from movement of electrons.

We make the following assumptions about the metallic state:

(*a*) The metal consists of fixed positive ions in a sea of electrons: in general there will be one or two free electrons per ion.

(*b*) The electrons behave as a perfect gas transporting thermal energy as in gas conduction.

(*c*) Each electron has thermal energy of translation equal to $\frac{3}{2}kT$, i.e. its thermal capacity is $\frac{3}{2}k$.

(*d*) Each electron travels a distance λ before colliding with a positive ion and giving up all its thermal energy.

We rewrite equation (9.40) in the form

$$K = \tfrac{1}{3} n \bar{c} \lambda \, (m c_V), \tag{9.42}$$

and note that mc_V is the thermal capacity of a single travelling particle. Replacing this by $\frac{3}{2}k$ for our electron gas, we obtain

$$K = \tfrac{1}{2} n \bar{c} \lambda k, \tag{9.43}$$

where $\bar{c}$ is the velocity of an electron assuming it to behave as a gas molecule. At room temperature its value is of order 10^5 m s^{-1}. If we consider a monovalent metal such as sodium, for which there is a single electron per ion, we have $n \approx 4 \times 10^{28}\, m^{-3}$, $k = 1.4 \times 10^{-23}$ J K^{-1}. The observed conductivity at room temperature is about 130 J m^{-1} K^{-1}. This gives $\lambda \approx 4$ nm or 40 Å. In general, equation (9.43) gives a good agreement with observation on the assumption that λ is of the order of a few atomic spacings. Although this seems satisfactory there is a major defect in the theory which we shall discuss later.

9.4 **Electrical conductivity of metals**

9.4.1 *Drude model*

The general picture that we adopt here, which is due to Drude (1912), is similar to that used in the preceding section. It is assumed that the electrons move like a perfect gas in all directions through the lattice and that their thermal motion is terminated by collisions with positive ions. Under normal conditions there is no net transfer of charge in any direction. When, however, a potential is applied across the metal the electric field imposes a drift velocity on the electrons and it is this which is responsible for the observed electrical conductance. The drift velocity is very small compared with the mean electron-gas velocity $\bar{c}$.

Consider a bar of uniform cross-section A, length L, across which a potential V is applied (figure 9.8). The electric field on the electrons is $X = V/L$. The force on an electron is eX and its acceleration is $\alpha = eX/m$. If λ is the mean free path of the electron the time between collisions is $t = \lambda/\bar{c}$. The drift velocity acquired during this period is αt so that the average drift velocity u between collisions is $\frac{1}{2}\alpha t$:

$$u = \tfrac{1}{2}\alpha t = \frac{1}{2}\frac{eX}{m}\frac{\lambda}{\bar{c}}. \tag{9.44}$$

The model still assumes that the electron gives up the velocity gained from the field by colliding with a positive ion. If n is the number of free electrons per m^3 the number travelling per second along the bar due to drift is the number in a volume Au, i.e. Aun. Each electron carries a charge e so that the charge carried per second, i.e. the current i, is

$$\begin{aligned} i &= eAun \\ &= \frac{1}{2}\frac{AX\lambda e^2 n}{m\bar{c}}, \end{aligned} \tag{9.45}$$

or

$$i = \frac{AV}{L} \times \frac{\lambda e^2 n}{2m\bar{c}}. \tag{9.46}$$

If ρ is the specific resistivity and κ the electrical conductivity we know that

$$i = \frac{AV}{L} \times \frac{1}{\rho} \quad \text{or} \quad \frac{AV}{L} \times \kappa. \tag{9.47}$$

Figure 9.8. Electrical conductivity model for metallic conductors.

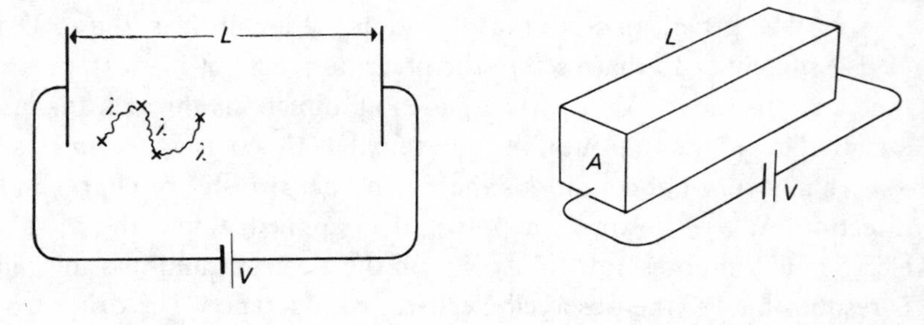

Comparison with (9.46) shows that†

$$\kappa = \frac{\lambda e^2 n}{2m\bar{c}} = \frac{\lambda e^2 n\bar{c}}{2m\bar{c}^2}. \tag{9.48}$$

If we ignore the difference between $\bar{c}^2$ and $\overline{c^2}$ we may, for the electron gas, write

$$\tfrac{1}{2}m\bar{c}^2 = \tfrac{3}{2}kT. \tag{9.49}$$

Hence, from equation (9.48)

$$\kappa = \frac{\lambda e^2 n\bar{c}}{6kT}. \tag{9.50}$$

For a monovalent metal assuming $\lambda = 0.3$ nm, $n \approx 4 \times 10^{28}\ \mathrm{m}^{-3}$, $e = 1.6 \times 10^{-19}$ C,

$$\bar{c} = 10^5\ \mathrm{m\ s}^{-1}, \quad k = 1.4 \times 10^{-23}\ \mathrm{J\ K}^{-1}, \quad T = 300\ \mathrm{K},$$

$$\kappa \approx 10^6\ \mathrm{ohm}^{-1}\ \mathrm{m}^{-1}. \tag{9.51}$$

This is in reasonable agreement with the resistivity values of most metals, $\rho \approx (2$ to $20) \times 10^{-8}$ ohm m, or conductivity $= 5 \times 10^6$ to $5 \times 10^7\ \mathrm{ohm}^{-1}\ \mathrm{m}^{-1}$.

9.4.2 *Wiedemann–Franz relation*

We may combine equations (9.43) and (9.50) to find the ratio of thermal to electrical conductivity. We have

$$\frac{K}{\kappa} = 3\left(\frac{k}{e}\right)^2 T. \tag{9.52}$$

Historically, Wiedemann and Franz observed that K/κ is the same for all metals at the same temperature. This was extended by Lorenz who observed that the ratio was proportional to the absolute temperature. The above analysis contains both 'laws'. The theoretical ratio is indeed close to the observed values over an appreciable temperature range for a large number of metals.

9.4.3 *Specific heat capacity of the free electron*

In the Drude model discussed above the electron gas is assumed to possess thermal energy. Like any monatomic gas it should therefore contribute a specific heat capacity of $\tfrac{3}{2}R$. The specific heat capacity of a

† There is a subtle distinction between the time average ($t = \lambda/c$) and the distance average of the mean free path λ. This distinction, if followed through, leads to a value of κ twice that given in equation (9.48). A similar factor applies to the thermal conductivity K so that the ratio (see equation (9.52)) is unchanged.

monovalent metal should therefore consist of $3R$ from its vibrational (lattice) energy and $\frac{3}{2}R$ from its free electrons, i.e. the specific heat capacity should have a value of about 38 J mol^{-1} K^{-1}. The observed value in fact does not exceed 25 J mol^{-1} K^{-1}.

This is a basic defect of the electron-gas model and the explanation can only be given in terms of the energy states of the electrons in a metal. The spacing of these states is independent of the size of the specimen: it depends only on the number of electrons per unit volume. Considering the energy distribution of electrons in unit volume we observe that no energy state can contain more than two electrons (of opposite spins), so that the electrons fill the available states from the lowest available energy level until they reach the topmost or Fermi level, E_F (see figure 9.9). This is the situation at absolute zero. If the metal is heated, only those electrons which can be moved into a higher vacant state above E_F can acquire further energy. In effect this means that only the electrons at the top of the Fermi level can acquire energy. Analysis shows that if there are n free electrons per m^3 the number of thermally excited electrons at temperature T is

$$\Delta n \approx n \times \frac{kT}{E_F}. \tag{9.53}$$

For a typical metal $E_F \approx 2$ eV and at room temperature ($T \approx 300$ K) kT is about 4×10^{-21} J ≈ 0.025 eV. Thus Δn is only about 1 per cent of n. If

Figure 9.9. Electronic energy levels in a metal shown as dn/dE plotted against the energy E so that the shaded area represents the number of electrons per m^3 with energy between E and $E + dE$. (*a*) At absolute zero all the energy states are filled up to a level E_F known as the Fermi level. (*b*) At higher temperature T a small number of electrons at the top of the Fermi level are able to acquire additional energy of order kT.

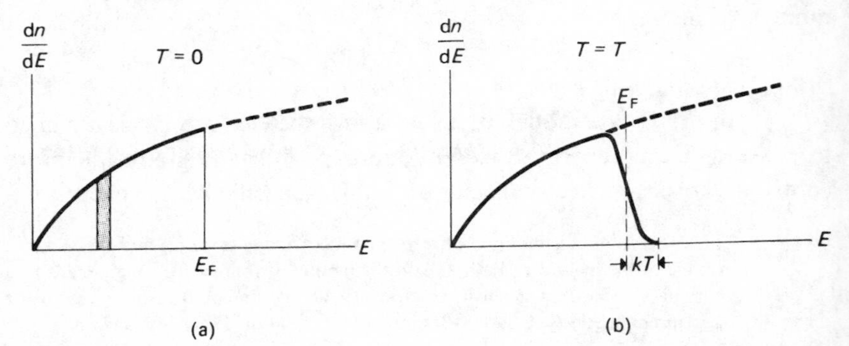

the thermal energy acquired by these electrons is of order kT we see that the specific heat capacity is only about 1 per cent of the value quoted above. It is for this reason that the electronic contribution to the specific heat capacity is so small.

We may take this a stage further. Since the specific heat capacity of a single electron is of the order k, we see from equation (9.53) that the specific heat capacity per m^3 is of order

$$k\,\Delta n = \frac{nk^2T}{E_F}, \tag{9.54}$$

so we may regard the *average* specific heat capacity per electron as

$$c_e = \frac{k^2T}{E_F}. \tag{9.55}$$

The expression from equation (9.42) for the thermal conductivity may be written

$$K = \tfrac{1}{3} n\bar{c}\lambda\,(c_e), \tag{9.56}$$

whilst the electrical conductivity, from equation (9.48), may be written

$$\kappa = \frac{\lambda e^2 n}{2m\bar{c}}. \tag{9.48}$$

The ratio is

$$\frac{K}{\kappa} = \frac{2}{3}\,\frac{m\bar{c}^2}{e^2}\,(c_e). \tag{9.57}$$

(At this point if we assume the electrons behave like a perfect gas $m\bar{c}^2$ is $\frac{3}{2}kT$ and $c_e = \frac{3}{2}k$; we then obtain precisely the classical Drude result given in equation (9.52).)

The quantity $\bar{c}$, the velocity of those electrons in the Fermi distribution which take part in the thermal transport process, is of order

$$\bar{c}^2 \approx \frac{E_F}{m}, \tag{9.58}$$

so that equation (9.57) becomes, after substituting for c_e from equation (9.55), and for $\bar{c}^2$ from equation (9.58),

$$\frac{K}{\kappa} \approx \frac{E_F}{e^2}\,(c_e) \approx \frac{k^2T}{e^2}. \tag{9.59}$$

A more rigorous treatment gives a constant $\frac{1}{3}\pi^2$ in front of this expression so that it is very nearly the same as equation (9.52) obtained from the simple Drude model.

The conclusions one derives from this, however, expose another issue of very great importance. The value of E_F of about 2 eV implies that the electrons have velocities (see equation (9.58)) about 10 times greater than the free electron-gas velocity. To explain the observed conductivities one must assume a mean free path λ of the order of 50–100 nm. On the Drude model λ has a value comparable to the distance between the positive ions in the lattice; this seems reasonable. A value of 100 atomic spacings or so seems strange. The explanation lies in the fact that electrons travelling through a perfect lattice experience no resistance whatsoever. If we treat the electrons as waves it is possible to show that electrons scattered by the lattice recombine to give a wave of undiminished amplitude; in other words there is no attenuation of the wave so that the resistance is zero. At normal temperatures the lattice vibrations render the lattice irregular so that the electron waves are scattered in an irregular way. It is this which accounts for the observed finite resistance. Any factor which increases the irregularity of the lattice increases the resistance. This is why the resistance increases with temperature and with impurity content.

10

The liquid state

The main characteristics of the gaseous state are random, translational motion of the molecules and relatively little molecular interaction. In the solid state the main characteristics are fixed sites for the individual atoms or molecules and very little translational movement. Both these states are well understood. By contrast the liquid state, in some ways, has 'no right to exist'. The molecular interaction must be quite strong since a given quantity of liquid, like a solid (and unlike a gas) occupies a definite volume. On the other hand the molecules have so much freedom that a liquid, unlike a solid, flows very readily and easily takes up the shape of the vessel it occupies.

The gaseous state has been thoroughly investigated and most of its problems have been satisfactorily resolved. The solid state is still being studied in hundreds of physics laboratories all over the world. By contrast the liquid state is a neglected step-child of the physical scientists. This is partly because the liquid state raises a number of very difficult theoretical problems. It is also partly a matter of fashion dictated by the economic value and engineering uses of solids as structural materials, electrical devices, etc. But for chemists, physical chemists and especially biologists, the liquid state is possibly of greater importance than the solid state. We may well expect, we should certainly hope for, marked advances in our understanding of the liquid state in the next two or three decades.

There are basically three approaches to the liquid state. The first is to treat it as a modified gas; the second is to treat it as a modified solid; the third is to treat it *sui generis* – this is the most difficult of all.

10.1 The liquid as a modified gas

A liquid in static equilibrium cannot withstand a shear stress. If we compress a column of liquid with a piston the situation is not like that of a solid where a uniaxial pressure or tension produces a shear stress in the medium. Unlike a solid the liquid must be held in a container; the walls exert pressures on the liquid in such a way that no resultant shear stress remains. This is equivalent to saying that the only stress the liquid can experience is a hydrostatic pressure or tension. It is for this reason that, at any point in a stationary liquid, the pressure is the same in all directions. In this sense a liquid closely resembles a gas and indeed both states are sometimes designated by the single generic term 'fluid'. In terms of resistance to pressure it therefore seems reasonable to consider how far the gas laws can be applied to explain the behaviour of liquids. Naturally we would not expect the ideal gas laws to be applicable but we might expect the imperfect gas laws to have some relevance.

We start by considering a liquid as a limiting case of a van der Waals gas. We ask what meaning may be attached to the van der Waals constants in the equation

$$\left(P+\frac{a}{V^2}\right)(V-b)=RT.$$

(*a*) The quantity b. In a gas $b=4v_m$ and at the critical volume $V_c=3b$ so that the quantity (V_c-b) is still positive and meaningful. At temperatures well below the critical temperature the density of the liquid increases, the volume occupied by the liquid is only perhaps $(1.5 \text{ to } 2)v_m$ so that $(V-b)$ is negative unless a different correction value is given to b.

(*b*) The quantity a/V^2. So long as the molar volume is large a/V^2 is small and is meaningful as a correction term. However, when V is comparable with or less than b the term a/V^2 has a value of the order of 1000–10 000 atmospheres. This internal pressure is so large compared with the atmospheric pressure that it cannot be treated as a correction factor. This is, of course, consistent with the fact that the external pressure P is not an important factor in determining the volume occupied by a liquid.

(*c*) The tensile strength of a liquid. In what follows we shall show that for low-temperature isotherms a liquid, according to van der Waals' equation, can actually withstand negative pressures. If we apply an increasing negative pressure (hydrostatic tension) the liquid will expand until at some critical value vapour will begin to form within the liquid, that is to say the liquid begins to cavitate or pull apart. The pressure at which this occurs provides, therefore, a measure of the tensile strength of the liquid.

Following Temperley (1947)† we use the reduced equation of state and write

$$\left(\pi+\frac{3}{\phi^2}\right)(3\phi-1)=8\theta, \tag{10.1}$$

where π, ϕ and θ are the reduced pressure, volume and temperature respectively. We may expand the liquid from D to B (see figure 10.1) by applying a hydrostatic tension. At B the liquid would become a vapour, i.e. it would pull apart. The pressure at which this occurs may be obtained by writing equation (10.1) as

$$\pi=\frac{8\theta}{(3\phi-1)}-\frac{3}{\phi^2}, \tag{10.2}$$

and determining the minimum at B by putting $\partial\pi/\partial\phi=0$. This gives

$$\frac{-3\times 8\theta}{(3\phi-1)^2}+\left(\frac{6}{\phi^3}\right)=0.$$

This defines the value of ϕ at B; we call this ϕ_m. Then

$$4\phi_m^3\theta=(3\phi_m-1)^2. \tag{10.3}$$

Figure 10.1. The tensile strength of a liquid treated as an extreme case of a van der Waals fluid. Using the reduced equation of state it is shown that the liquid fails when a negative pressure π_m is applied.

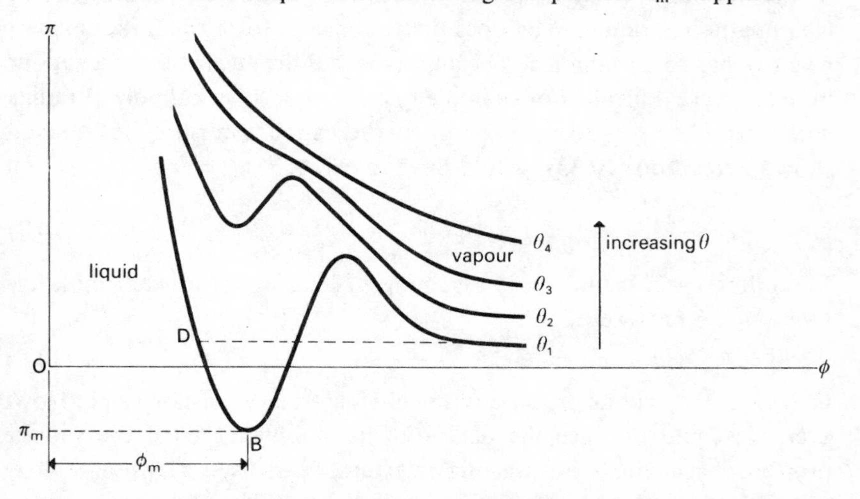

† H. N. V. Temperley, *Proceedings of the Physical Society*, vol. 59 (1947), p. 199. See also A. H. Cottrell, *The Mechanical Properties of Matter* (Wiley, 1964).

If we insert this in equation (10.2) we obtain the corresponding value of π which we call π_m.

$$\pi_m = \frac{3\phi_m - 2}{\phi_m^3}. \tag{10.4}$$

We see from this equation that $\pi_m = 0$ when $\phi_m = \frac{2}{3}$. This is the 'last' isotherm for which the liquid has no tensile strength; it occurs when $\phi = \frac{2}{3}$ or $V = \frac{2}{3}V_c$. If the liquid is less condensed than this it can exist only if a positive pressure is applied.

At lower temperatures $-\pi_m$ becomes larger. It reaches its maximum (negative) value, as equation (10.3) shows, when $\theta = 0$, i.e. when $3\phi = 1$. Putting $\phi = \frac{1}{3}$ in equation (10.4) we obtain as the maximum value

$$\pi_m = \frac{1-2}{(\frac{1}{3})^3} = -27. \tag{10.5}$$

The limiting hydrostatic tension is then

$$P = -27P_c.$$

For water $P_c = 218$ atmospheres, so that

$$P = -5900 \text{ atmospheres}. \tag{10.6}$$

(This is comparable with, though appreciably larger than the internal pressure quoted on p. 131). Before we leave this calculation we note that $\theta = 0$ corresponds to absolute zero temperature and $\phi = \frac{1}{3}$ corresponds to $V = \frac{1}{3}V_c = b$, i.e. the molecules are packed as close as the van der Waals model will allow. We may ask how such high negative pressures can be meaningful in a liquid. Why does the vapour not form when the smallest negative pressure is applied? The answer is that the nucleation of a vapour bubble is very difficult. For example to open up a spherical hole of radius r in a liquid of free surface energy γ, the vapour pressure, as we shall show in equation (10.33), would have to exceed

$$P = \frac{2\gamma}{r}. \tag{10.7}$$

For water $\gamma = 72$ mJ m^{-2} and assuming r is the size of a water molecule (=about 0.3 nm) we obtain

$$P = 4800 \text{ atmospheres}. \tag{10.8}$$

Generally the vapour pressure is a negligible fraction of this value. However, we could produce the nucleation of bubbles, even if the vapour pressure is very small, by applying an external hydrostatic tension of 4800 atmospheres. Cavities would form and the liquid would then rupture. This tension is very close to that calculated in equation (10.6).

Readers may note that we could, in fact, estimate the breaking strength of the liquid simply in terms of the surface energy. Consider a liquid column of unit cross-section. If we pull it apart we create two unit areas of surface, so that the work done is 2γ. The intermolecular forces are short range and if they become negligible for a distance greater than say x, the force F to pull the column apart will be of order $F \times x = 2\gamma$ or $F = 2\gamma/x$. If x is about 0.5 nm we again obtain as the breaking stress (force per unit area) a value close to that given in equations (10.6) and (10.8). Exact agreement is not to be expected; the main point we wish to make is that because of intermolecular forces liquids can, in theory, withstand large negative pressures. In practice, of course, liquids are generally much weaker because of dissolved gases which readily nucleate cavities as the negative tension is applied. This is particularly marked at the walls of the vessel where small amounts of gas may be trapped in the surface irregularities. Carefully outgassed liquids in smooth-walled containers can, however, possess tensile strengths up to one-tenth of the values calculated above.

10.2 The structure of liquids: the 'radial distribution function'

Before turning to a consideration of a liquid as a modified solid we mention one characteristic feature of a liquid which distinguishes it in a very basic way from a gas. A gas has no structure. If we study the diffraction of neutrons, electrons or X-rays by a monatomic gas we find that there is no coherent diffraction; the gas behaves like a diffraction grating in which the spacings between the lines are random. (For diatomic molecules one obtains diffraction maxima corresponding to the mean distance between the atoms in the molecule.) If, however, a monatomic gas is liquefied it shows marked coherent diffraction; the maxima are not sharp as with a solid crystal, but are diffuse.

The information obtained by such diffraction experiments is presented most graphically by means of the 'radial distribution function' which may be derived in the following way. We choose a molecule at random within the liquid and draw a series of concentric spheres around it such that the volume interval between two neighbouring spheres is a constant. The distribution function is the average number of molecules contained between each spherical shell. It is evident that the radial distribution function is a measure of the average density as a function of distance from some arbitrary origin. A typical result from X-ray diffraction is given in figure 10.2 and similar results are obtained from the diffraction of neutrons. As a matter of interest the radial distribution function for the crystalline solid is also shown; this brings out the marked difference between the liquid and the solid state. The solid behaves like a diffraction grating containing a very large number of regularly spaced lines: the liquid resembles the diffraction of light by a grating containing only a few lines. This suggests that there is short-range order in a liquid; any small group of atoms or molecules is in fairly regular array and resembles a crystal lattice but it is just a little out of register in a random way so that, as soon as one moves a short distance from the group, the neighbouring atoms (or molecules) no longer possess the long-range order characteristic of the solid crystal.

The contrast between the liquid and solid states may be expressed in another way.† Neutron diffraction experiments suggest that a molecule in a liquid has time to vibrate 10 to 100 times before the local structure

Figure 10.2. Radial distribution function plotted against radius measured from an arbitrary atom chosen as origin. (*a*) Liquid mercury, (*b*) solid mercury (from results given by A. Guinier, *X-ray Diffraction in Crystals, Imperfect Crystals and Amorphous Bodies* (Freeman, 1963) p. 70, figure 3.9). For the liquid, the vertical ordinate can be normalized such that for large distances the radial distribution function tends to unity (uniform density). For the solid this is not possible since in principle each peak is infinitely narrow and infinitely high. Instead the vertical ordinate has been drawn proportional to the number of atoms at the specified radial distance. It is seen that with the liquid the structure is lost beyond distances of 1 nm; with the solid it persists indefinitely.

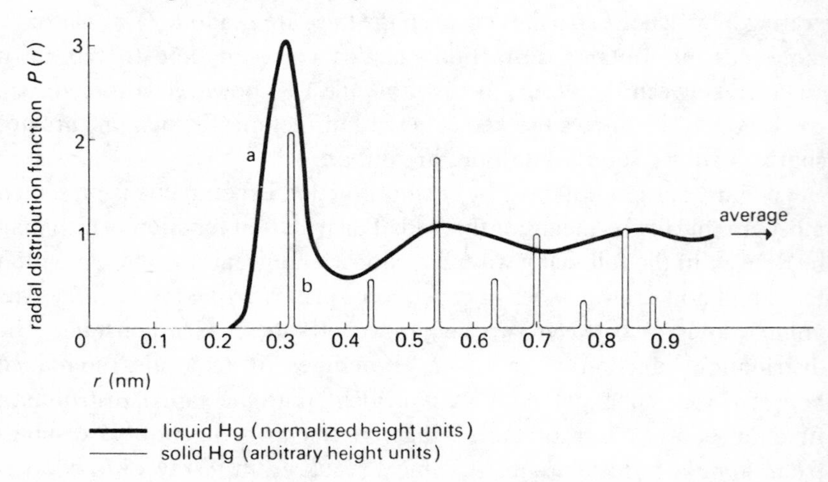

† See J. D. Bernal, *Scientific American*, no. 203 (1960), pp. 124–34; and *Proceedings of the Royal Society*, A280 (1962), pp. 299–322.

changes. During that time the structure of a liquid is physically, though not geometrically, similar to that of a crystal. The difference, however, is crucial. For example the solid crystal, because of its high degree of order, admits only a limited degree of variation of composition. Crystallization as a means of purifying substances is evidence of the particular architecture that marks each crystalline phase. By contrast, liquids mix much more readily either with other liquids or with solids or with gases (though of course they do not mix as readily as gases which are always perfectly miscible with one another). We may describe this situation by saying that 'atoms in a liquid do not occupy such closely specified positions as they do in solids. They have more elbow room, and they are not so particular about their partners.'

Nevertheless the structural resemblance to the solid over limited regions appears to be indisputable. This conclusion is supported from a completely different type of measurement. It is found that when metals melt their electrical resistance increases by a factor of only about two. As mentioned in a previous chapter the resistance is primarily due to non-coherent scattering of the electrons by the metal ions. This implies that, to the conductance electrons, the metal ions, in the liquid, appear to be oscillating about regular sites, very much as in the solid state.

There are two main differences between the solid and liquid state. First, as discussed above, there is the absence of long-range order in liquids: in this sense they closely resemble amorphous solids, glasses and polymers. It is significant that these materials can flow viscously like liquids although, as pointed out in Chapter 6, their viscosity is generally enormous compared with that of true liquids. On the other hand, with crystalline solids flow or 'creep' can occur, but it takes place extremely slowly and through an entirely different mechanism – through the movement of dislocations, the behaviour of which is determined by the highly ordered structure of the material. This brings us to the second major difference: compared with the molecules in the solid state (both crystalline and amorphous), the molecules in liquids have very much greater mobility. They can swap places or shuffle around to other positions with relative ease.

10.3 **The liquid as a modified solid**

When solids melt they expand generally by 5–15 per cent in volume. One way of describing this is to say that there is an increase in free volume between the constituent molecules or atoms. Another is as follows (cf. E. A. Moelwyn-Hughes, private communication).† If one thinks of a typical crystalline solid for which each molecule, or atom, has say 8 or 12 nearest neighbours, we may regard the liquid state as a solid from which one nearest neighbour has been subtracted. This leaves the material with the right amount of volume expansion and a reasonable degree of short-range order. What is the nature of this disorder? The X-ray diffraction evidence suggests that the mean intermolecular distance does not change as much as would be expected from the bulk volume expansion. For example when solid argon melts the mean argon–argon spacing increases by only about 1 per cent, whereas the volume change would lead us to expect a mean increase in separation of about 5 per cent. Such evidence may be used to justify the view that the melting process consists mainly in producing 'holes' in a structure which otherwise closely resembles the solid structure. This idea was first developed by J. G. Kirkwood and constitutes the simplest form of the 'hole-theory' of liquids. As we shall see later the modified 'hole-theory' introduced by H. Eyring provides a very convenient model for the study of viscous flow in liquids.

Attempts have been made to treat the disordered structure and the increase in free volume in the liquid state in terms familiar to the solid-state physicist. One approach is to concentrate on the role of vacancies or holes (A. L. G. Rees, 1966, private communication). As the temperature rises the concentration of vacancies increase. So long as vacancies are individual holes the behaviour is still that of a solid. As soon, however, as two or three vacancies coalesce a whole aggregate of molecules surrounding the hole becomes relatively mobile. The material is now very much more like a liquid. One advantage of this approach is that it stresses the cooperative nature of the melting process.

10.4 The liquid state *sui generis*

10.4.1 *The Bernal approach*

The most difficult approach is to tackle the liquid state without directly referring to the gaseous or solid state. The simplest method as pioneered by J. D. Bernal and his colleagues in England and by G. D. Scott in Canada, may be referred to as a random-packing model. Recognizing the disordered but relatively close-packed nature of the liquid structure Bernal took a large number of plasticine spheres, chalked them to stop them sticking, placed them in a football bladder, removed the air to avoid bubbles and then squeezed them together until they filled the whole

† See also E. A. Moelwyn-Hughes, *Physical Chemistry* (Pergamon Press, 1960).

of the space. On examining the aggregate it was found that the spheres had become polyhedra of various irregular shapes. The most common number of faces was thirteen (this was not the true coordination number since the centres of some of the neighbours were more than the average distance apart). More striking was the observation that the most common number of sides to a face was five. This is of great significance for the following reason. In crystallography molecules can be arranged with two-fold, three-fold, four-fold, or six-fold symmetry, never with five-fold symmetry. This is because 'one cannot form regular patterns with five-fold symmetry that will fill space solidly and extend indefinitely in three dimensions. It is like trying to pave a floor with five-sided tiles'. This is illustrated in figure 10.3 and shows how this type of arrangement would explain the presence of short-range order and the absence of long-range order.

In a later study Bernal emphasized the point that the attractive forces between molecules play little part in influencing the packing which occurs in the condensed state: the packing is almost entirely determined by the repulsive forces, i.e. the molecules behave very much like hard spheres. As Bernal remarks, the key word (originally used in a different sense) is 'the one used by Humpty Dumpty in *Alice Through the Looking Glass*, impenetrability'. We are familiar with this in the solid state where, for practically all crystalline structures which are not dominated by covalent or other directed bondings, the arrangement is rather like that arising from packing the molecules as close as their repulsive forces will allow. For simple atoms, indeed, the crystal structure is like that of close-packed spheres. Bernal therefore considered the random packing, not of deformable plasticine spheres as in the earlier work, but the packing of a large

Figure 10.3. The difference between five-fold symmetry and six-fold symmetry, after Bernal.

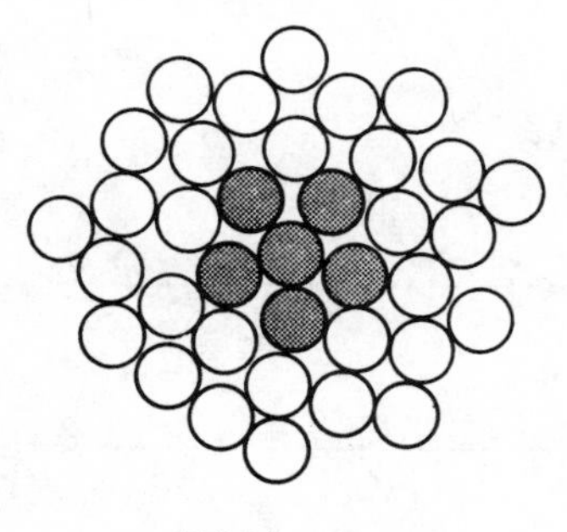

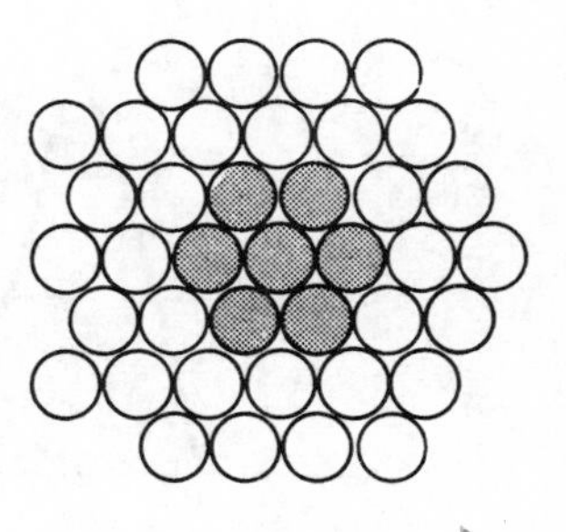

number of hard steel balls. After introducing black paint and letting it harden the aggregate could be broken open and studied. In this way the number of contacts and near contacts could be determined. The model gives a coordination number varying between four and eleven, and Bernal considers this variation in contact number from molecule to molecule to be the most significant feature of the irregular liquid structure. The packing very adequately reproduces the radial distribution function for liquid argon derived from neutron diffraction experiments (see figure 10.4), though this is probably not a very critical test. Further, the model does not allow vacancies between neighbours to be large enough to admit a third molecule.

The Bernal model aims to provide essentially an instantaneous snapshot (in say 10^{-15} s) of the structure of a liquid. If we now add to this the reality of thermal motion we see that the molecules are continuously changing the number of nearest neighbours, continuously shifting and shuffling around, and occasionally are able to allow one molecule to squeeze its way between neighbours without leaving a vacant hole. This aspect of the model may be open to criticism. For example with a solid, because it is rigid, vacant sites or holes can exist in the lattice. The concentration of holes increases as the temperature is raised and by the time the solid reaches its melting point (but is not yet melted) about one site in 10^4 or 10^5 is vacant. Once melting occurs there is a further increase in volume, but according to Bernal the free space is more uniformly spread

Figure 10.4. Radial distribution function for liquid argon plotted against distance from centre in units of sphere diameter. — results from neutron diffraction; ● ▽ calculated from random packing model (● Bernal, ▽ Scott).

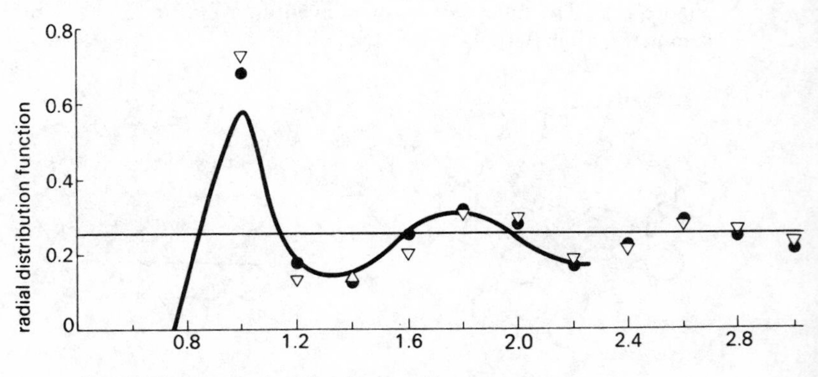

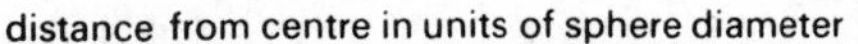

out. The X-ray evidence quoted above suggests that this is not possible, even at temperatures just above the melting point; in any case the thermal expansion of liquids is rather large so that at somewhat higher temperatures there must be vacancies or holes. Indeed, as Bernal himself points out, it is this increase in holes and the consequent drop in average coordination number that marks the transition from the coherent liquid phase to the incoherent gas phase (see figure 10.5). Bernal suggests that this probably occurs when the coordination number has fallen to an average value of three or four. Since the coordination number in the solid state (or in the liquid state near the freezing point) is about twelve, this means that the density of a gas at its critical point is between one-quarter and one-third of the density of the most highly condensed state. This is in fairly good agreement with observation for most *normal* liquids.

The Bernal approach is essentially that of a crystallographer; his model is a static one and concentrates on structural features. A far more analytical and difficult approach is that due to M. Born and H. S. Green. They carried out an extensive study of the mutual potential between molecules in the liquid state. They treat the molecules first in pairs, then in triplets and so on. The treatment is extremely complicated and the results are not easy to apply. In a sense their approach approximates to the theory of dense gases. A calculation is made of sets of coordinates of a random collection of hard spheres. As the density of spheres is increased the behaviour passes from that of a gas to that of a dense gas and then to some other state. In this condition, by an iterative process, the coordinate positions can be repeatedly perturbed until a state of minimum energy is

Figure 10.5. The transitions from the close-packed coherent liquid phase to the expanded, associated and finally free (incoherent gas) phase, as proposed by Bernal.

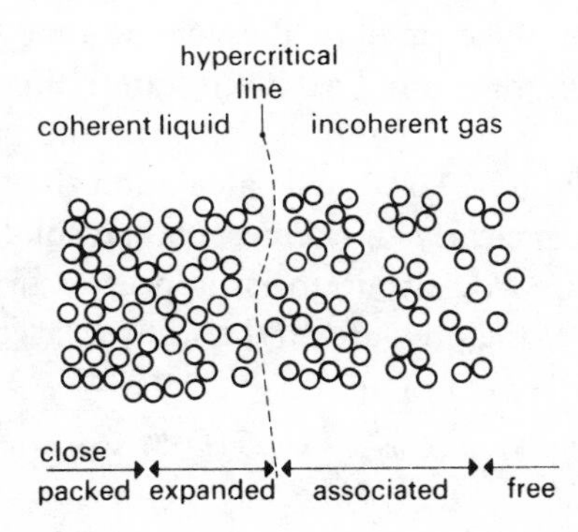

obtained. This corresponds over a critical density range to the liquid state. The computation here is formidable even for only 100 spheres.

10.4.2 *The Alder approach: molecular dynamics*

Yet another approach is that associated with the name of B. J. Alder; it is now referred to as the method of molecular dynamics. It involves elaborate computations but the basic concept is simple. The molecules are treated as hard smooth perfectly elastic spheres, although as a later refinement an attractive square-well potential with a rigid core has been used. The molecules are placed in a box and started off with equal speeds in random directions and their subsequent motion determined by applying Newton's laws of motion. The boundary conditions chosen are such that a particle which passes out of one side of the box re-enters with the same velocity through the opposite side. In this way the number of molecules in the box and the total energy remain constant. The collisions within the system are analysed by a computer. It is found that after relatively few collisions the velocity distribution becomes very nearly Maxwellian. The total energy is determined by the initial velocities and of course remains constant; it determines the effective temperature of the system.

The behaviour depends on the velocity and on the packing of the particles. At one extreme the particles oscillate about an equilibrium position; this corresponds to the solid state. At the other extreme the particles move around throughout the box with relative freedom; this corresponds to the vapour regime. At some intermediate stage the particles vibrate about equilibrium positions but are also able to change places and slowly diffuse from one region to another. This corresponds to the liquid state.

We quote some typical results obtained by Alder and Wainwright for 32 hard smooth particles in a box.† They were able to map out a view of the calculated particle trajectories as seen from one face of the box. If V is the volume occupied by the particles and V_0 the volume they would occupy in a close-packed situation, they observed a very interesting instability for a ratio $V/V_0 = 1.525$. For one group of collisions the particle trajectories appear as shown in figure 10.6(*a*) and calculation shows that the pressure on the walls is small; clearly this represents the solid state. In other groups of collisions the particle trajectories appear as in figure 10.6(*b*) and calculations show that the pressure on the walls is large. This

† B. J. Alder & T. E. Wainwright, *Nuova Cimento*, vol. 9 (1958), sup. 1, p. 16, and *Journal of Chemical Physics*, vol. 31 (1959), p. 459.

is interpreted as the liquid phase. It is interesting to note that in the 'liquid' state the particles oscillate about a mean position but occasionally swop places (see figure 10.6(*b*), particles A and B; particles C, D and E). The situation shown in these figures refers to only 32 particles in the box. Apparently the system is not sufficiently large for the solid and liquid phases to exist in equilibrium: the system is in just that density range where part of the time it is in the liquid phase and part of the time in the solid phase.

Although the results reproduced in figure 10.6 are very striking, some reserve is necessary since an assembly of particles which show no attraction for one another can hardly be considered as a very realistic model of a liquid or a solid. However, from the point of view of packing, the structure scarcely depends on intermolecular forces (see Bernal above).

If the molecules are less closely packed the results obtained are as shown in figure 10.7. Figure 10.7(*a*) shows the potential for particles with a square-well attractive field whilst figure 10.7(*b*) shows the particle trajectories for 108 such particles in a box; the behaviour evidently represents the liquid–vapour region. The Alder model brings out very clearly the dynamic features of the condensed state and the similarities and differences between the solid and liquid phases.

Since Alder's pioneering studies there have been very important developments in the technique of molecular dynamics as applied to both solids

Figure 10.6. End face of box showing calculated particle trajectories for 32 hard elastic spheres (zero attractive forces) for a packing $V/V_0 = 1.525$. (*a*) Behaviour resembling a solid. (*b*) Behaviour resembling a liquid. The particles spend most of their time oscillating within a *cage* of other particles, but are also to swap places. See particles A and B; C, D and E.

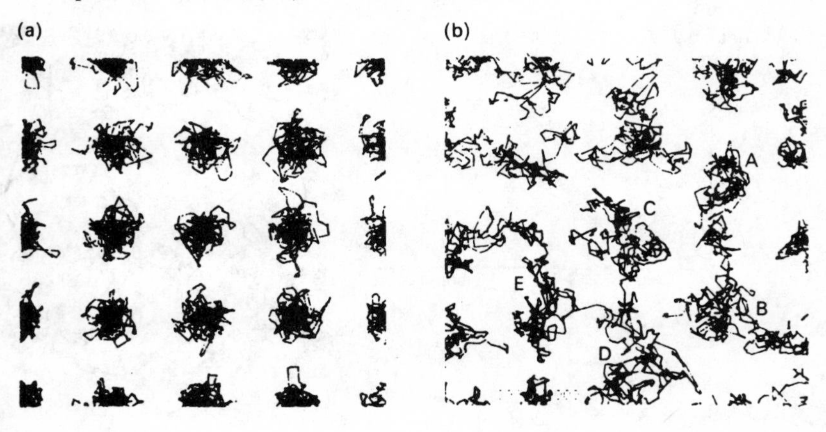

and liquids. A crucial factor has been the enormous increase in the capacity and speed of modern computers. For example it is now possible to model the behaviour in the liquid state of hydrocarbon molecles containing 20 carbons in the chain, allowing for realistic rotations about the C–C bonds.

10.5 Ice and water

We introduce here a short digression on the relation between ice and water. Water, which is the basis of all our ideas of liquidity, is in a sense an atypical liquid. The forces between the molecules are very strongly directional; they arise from the interaction between electron-deficient hydrogen atoms in one molecule and electron-rich oxygen atoms in another. In the solid state this hydrogen bonding leads to a very open structure in which every water molecule is surrounded symmetrically (in the form of a regular tetrahedron) by four other water molecules. Consequently in ice the coordination number is four. As the temperature is raised the coordination number remains constant but the mean distance between the molecules increases. There is thus a small decrease in density with increasing temperature as with all 'normal' solids. At the melting point, however, a radical change occurs. The thermal energy is sufficient to destroy the regular bonding and the coordination number actually

Figure 10.7. (*a*) Potential energy diagram of spheres possessing a square-well attractive field and a hard elastic core. The repulsive core has an effective diameter σ_1. The attractive forces operate at an effective separation of σ_2. (*b*) End face of box showing calculated trajectories of 108 such spheres. The behaviour of the particles resembles the vapour phase.

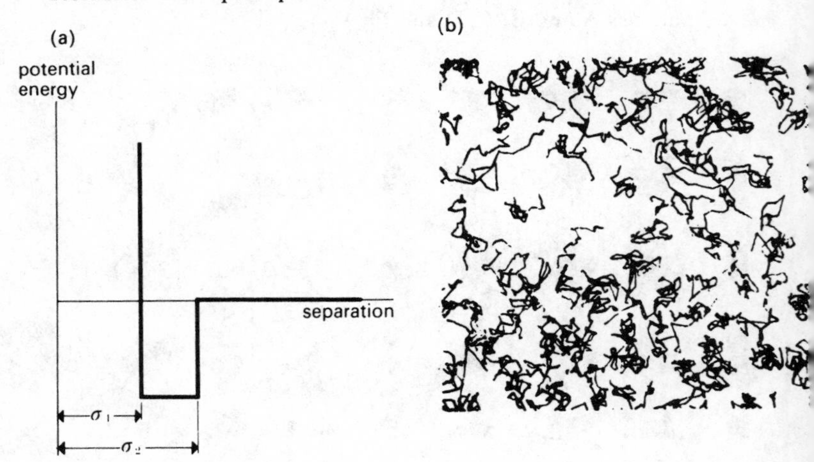

increases. At 0 °C the average coordination number appears to be between 4.5 and 5; with rising temperature this probably increases but soon reaches a constant value. This increase on its own must lead to a corresponding *rise* in density. However, an increase in temperature also leads to an overall increase in the mean distance between the molecules (thermal expansion). This expansion, on its own, would produce a *decrease* in density with increasing temperature. When combined with the change in coordination number it results in a 10 per cent jump in density on melting, a maximum density at 4 °C followed by a steady decrease in density at temperatures above about 10 °C. The behaviour is shown schematically in figure 10.8. The details are by no means definite but the general pattern is reasonably valid. Another way of looking at this is in terms of the 'mixture-models' of water (e.g. H. S. Frank models). These stress the idea that water may not simply be a random or an ordered arrangement of individual molecules with uniform characteristics and bonding. On the contrary water may consist of two or more different molecular species,

Figure 10.8. The melting of ice. Schematic diagram showing (*a*) the change in coordination number, (*b*) the change in molecular separation, (*c*) the corresponding change in density as ice is heated from −5 °C to +15 °C.

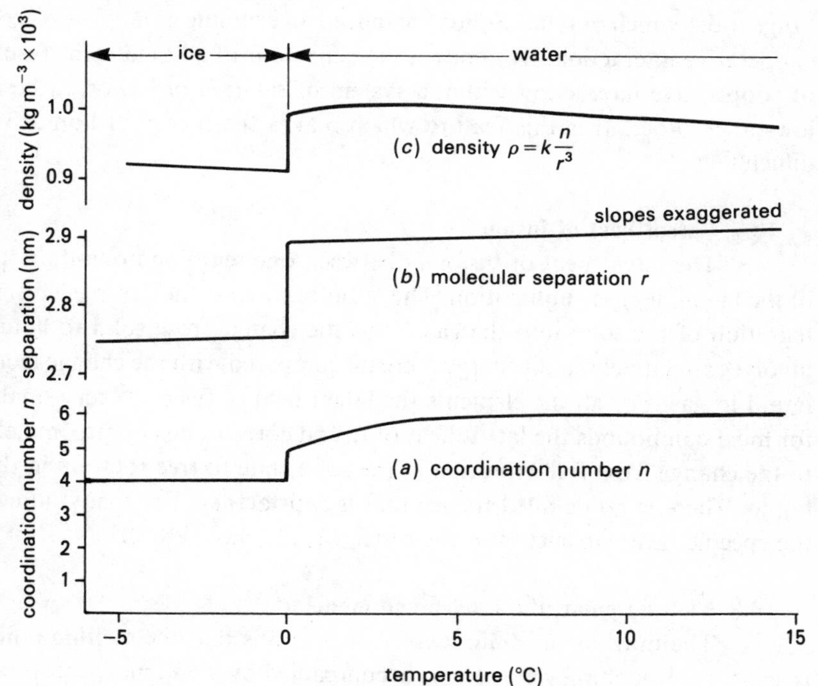

each species being structurally organized in its own distinctive way, with its own degree of coordination. The effect of temperature on this model is to increase the proportion of one species at the expense of the other. Whichever approach is adopted, the main point brought out by this discussion is the immense importance of the coordination number in determining the detailed properties of liquids.

10.6 General approach

In what follows we shall treat the liquid as a modified solid, particularly when we are concerned with energies. When we consider flow we shall treat the liquid essentially as a solid, the molecules of which have a fair degree of mobility because of the presence of holes or free volume. This is the Eyring theory of viscous flow in a liquid; it differs from the Bernal theory in concentrating on thermal fluctuations rather than on the structural nature of the liquid state. It is much closer to the Alder model of the liquid state.

Before leaving this discussion we may make the following point. In the gaseous state the molecules are, more or less, completely independent. In the solid state they interact but the interaction, at least as far as crystalline solids are concerned, is capable of analysis precisely because of the long-range order which obtains. Solids are indeed an example of highly ordered cooperative interaction. By contrast the behaviour of a liquid is the result of cooperative interaction within a system of random order, or, at best, low-range order. It is this feature which makes the theory of liquids so difficult.

10.7 Latent heat of fusion

The latent heat of fusion is between one-tenth and one-thirtieth of the latent heat of sublimation. The latter represents the complete disintegration of the solid into the gas. Thus the change from solid to liquid involves a relatively small energy increase compared with the change from liquid to gas. For all the elements the latent heat of fusion is very small; for most compounds the latent heat of fusion corresponds approximately to the change from free vibration in the solid state to free rotation in the liquid. There is no detailed theory that is satisfactory. For many liquids the specific heat capacity is of the order 12–25 J mol^{-1} K^{-1}.

10.8 Melting point: the Lindemann model

The most astonishing feature of fusion is that the melting point is so sharp. If melting were always accompanied by a volume increase we could understand (if not explain) the sharpness of the melting point in the following way. Melting represents essentially the 'loosening-up' of the solid structure. If local melting occurs molecules on the edge of melted regions are exposed to molecules which are already a little further apart than usual. It would not be unreasonable to see in this the basis of a runaway situation where a little bit of melting soon engulfs the whole solid. If a pressure were applied to the solid to prevent a volume increase we should, in fact, find that the melting point is increased (Clausius–Clapeyron equation). Thus melting could cover a whole range of temperatures depending on the extent to which the volume increase is inhibited; only if the solid were completely free to expand would melting occur at a single sharply defined temperature. This 'explanation' would, of course, be a little embarrassing if we attempted to apply it to the melting of ice since here melting is accompanied by a decrease in volume.

As we saw earlier it is not possible to explain melting in terms of our potential energy curve, since this can only describe basically the solid state or the gaseous state. However, Lindemann (1910) suggested that one could obtain a good idea of the melting point from the potential energy curve by assuming that melting occurs when the amplitude of vibration exceeds a critical fraction of the atomic spacing. At this stage numerous collisions occur between neighbours and the crystal structure is destroyed.

Let x_0 be the vibration-amplitude when melting occurs. If f is the force constant between one atom (or molecule) and its neighbour and the motion is simple harmonic, the mean total energy during vibration will be

$$u_v = \tfrac{1}{2} f x_0^2. \tag{10.9}$$

Since melting points are generally high we may assume the classical value for the internal energy, namely kT per atom or molecule for each principal direction of vibration. Thus at the melting point we write

$$u_v = kT_m = \tfrac{1}{2} f x_0^2. \tag{10.10}$$

But as we saw in an earlier chapter, for a simple cubic structure, $f = Ea$, where a is the atomic (or molecular) spacing and E is Young's modulus of the solid.

Hence

$$kT_m = \tfrac{1}{2} E a x_0^2.$$

We assume that melting occurs when x_0 is some fraction of the atomic

Table 10.1. *Melting point T_m (K) of some solids, assuming $\beta^2 = 1/50$*

Solid	E ($N\,m^{-2}$)	ρ ($kg\,m^{-3}$)	M (kg)	$T_m(K)$ Calculated	Observed
Lead	1.6×10^{10}	11.3×10^3	0.207	400	600
Silver	8.3×10^{10}	10.5×10^3	0.108	1100	1270
Iron	21.2×10^{10}	7.9×10^3	0.056	1800	1800
Tungsten	36.0×10^{10}	19.3×10^3	0.184	4200	3650
Sodium chloride	4.0×10^{10}	2.16×10^3	0.057	1200	1070
Quartz	7.0×10^{10}	2.6×10^3	0.060	1900	2000

spacing a. Let x_0 be βa where $0 < \beta < 1$.

$$T_m = \frac{Ea^3\beta^2}{2k} = \frac{Ea^3N\beta^2}{2R} = \frac{EV\beta^2}{2R},$$

where V = molecular volume, assuming a cubic structure.

Hence

$$T_m = \frac{E}{2\rho}\frac{M}{R}\beta^2, \tag{10.11}$$

where ρ = density of the solid in kg m^{-3}, M is the molar mass in kg, and $R = 8.3$ J mol^{-1} K^{-1}.

Some typical results are given in Table 10.1 where it is assumed that β^2 has the value of $\frac{1}{50}$, i.e. $x_0 \approx \frac{1}{7}a$. Note that this model has assumed simple cubic packing (so that $Na^3 = V$) and β a constant fractional amplitude of vibration at melting. Considering the crudity of the model and the range of materials to which it has been applied the agreement is surprisingly good.

This model, by its very nature, cannot be applied *quantitatively* to surface melting (see Chapter 6, section 6.2.1). The only satisfactory theoretical approach is that of molecular dynamics.

10.9 Vapour pressure

10.9.1 *Molecular interpretation of vapour pressure*

Suppose we place a quantity of liquid in a sealed container and evacuate the air above it. Some of the liquid will evaporate since a certain fraction of the molecules will have enough excess thermal energy to escape from the attraction of neighbours. Some of these will return and condense. Equilibrium will be reached when the rate of evaporation equals the rate of condensation. The pressure exerted by the vapour under these conditions is known as the saturation vapour pressure or, more usually, the vapour pressure. It increases as the temperature is raised. The equilibrium vapour pressure will be almost identical in the presence of air or another gas (see, however, the section below on the effect of external pressure), since the vapour and the alien gas exert their partial pressures independently.

If the pressure above the liquid is set at some specified level, for example 1 atmosphere, the vapour pressure may be increased by raising the temperature of the liquid, until it is equal to the external pressure. At this point the liquid can evaporate freely: bubbles of vapour form copiously within the bulk of the liquid and the liquid is said to be boiling. The boiling point is thus the temperature at which the vapour pressure equals the external pressure – usually one atmosphere. If the external pressure is reduced the boiling point is lowered. It is a matter of experience that the boiling point on the absolute scale, at atmospheric pressure, is approximately two-thirds of the critical temperature, but there is no simple explanation for this.

The molecular interpretation of vapour pressure is simple. The molecules in the liquid have a wide range of thermal energies. The only molecules which can escape are those whose thermal energy is sufficient to overcome the attraction of neighbouring molecules. If the latent heat of vaporization per mole is L, the energy for a molecule to escape is

$$\varepsilon = \frac{L}{N_A}. \tag{10.12}$$

It follows that the average potential energy of the molecules in the vapour phase is greater than the mean value in the liquid phase by an amount ε. Consequently if n_V and n_L represent the number of molecules per unit volume in the vapour and liquid phases respectively and if the Boltzmann distribution holds for the liquid as well as for the vapour, we have

$$\frac{n_V}{n_L} = \exp\left(-\frac{\varepsilon}{kT}\right) = \exp\left(\frac{L}{N_A kT}\right) = \exp\left(-\frac{L}{RT}\right). \tag{10.13}$$

As the temperature is raised the ratio n_V/n_L increases but n_L scarcely changes with temperature. Since the vapour pressure p is directly proportional to the number of molecules per unit volume n_V in the vapour phase and to the temperature T, since the vapour obeys the gas-laws, we may re-write equation (10.13) as

$$p = BT \exp\left(-\frac{L}{RT}\right). \tag{10.14}$$

We shall find in many expressions of this type that the variation of the exponential term with temperature is far greater than the variation of the pre-exponential term. For example if T is increased from 0 °C to 30 °C the pre-exponential term increases by a factor of 1.1, whereas, for say water, the exponential term increases by a *factor* of 8. Since equation (10.14) is, in any case, based on a very simple model we may simplify it further by ignoring the pre-exponential term in T. We may then write

$$p = A \exp\left(-\frac{L}{RT}\right), \tag{10.15}$$

where A is a suitable constant. This relation is almost exactly correct.

We may approach the vapour pressure relation in another way. Suppose we consider the liquid to resemble a very dense gas containing n_L molecules per m^3. The number per m^3 with velocities, normal to the surface, between u and $u+du$ is given by equation (4.45) in Chapter 4. The number striking each m^2 of surface per second is then

$$u\,dn = n_L\left(\frac{m}{2\pi kT}\right)^{\frac{1}{2}} \int_0^\infty \exp\left(-\frac{mu^2}{2kT}\right) u\,du.$$

If we now specify that of these molecules only those with energy greater than ε (where $\varepsilon = L/N_A$) can escape, the number per second crossing each m^2 of surface into the vapour phase is the same integral as above but the limits are from $mu^2/2 = \varepsilon$ to ∞. This gives us the number escaping per m^2 per second,

$$n_L\left(\frac{kT}{2\pi m}\right)^{\frac{1}{2}} \exp\left(-\frac{\varepsilon}{kT}\right).$$

But if n_V represents the number of molecules per m^3 in the vapour phase, the number striking each m^2 of surface per second will be (see Chapter 4, equation (4.56*c*))

$$n_V\left(\frac{kT}{2\pi m}\right)^{\frac{1}{2}}.$$

If all those molecules striking the liquid surface are adsorbed (this is the most dubious part of the treatment) then equilibrium occurs when the number escaping from the liquid exactly equals the number arriving from the vapour phase, i.e.

$$n_L \exp\left(-\frac{\varepsilon}{kT}\right) = n_V.$$

This is identical with equation (10.13).

10.9.2 *Effect of temperature on vapour pressure*

We use equation (10.13) to calculate the effect of temperature on vapour pressure. Since

$$p = A\exp\left(-\frac{L}{RT}\right),$$

we have by differentiating

$$\frac{dp}{dT} = A\frac{L}{RT^2}\exp\left(-\frac{L}{RT}\right) = \frac{L}{RT^2}\times p = \frac{L}{RT}\times\frac{p}{T}. \tag{10.16}$$

If the vapour behaves like a perfect gas and V is the volume occupied by one mole of vapour at T we have $pV = RT$, so that

$$p = \frac{RT}{V}.$$

Equation (10.16) becomes

$$\frac{dp}{dT} = \frac{L}{T}\times\frac{1}{V}. \tag{10.17}$$

The correct relation, according to thermodynamics, is

$$\frac{dp}{dT} = \frac{L}{T}\frac{1}{(V - V_L)}, \tag{10.18}$$

where V_L is the volume occupied by one mole of the liquid. This is known as the Clausius–Clapeyron equation. We may note that a plot of $\ln p$ against T^{-1} should give a straight line of slope equal to L/R. This is indeed found to be so.

10.9.3 *Effect of external pressure on vapour pressure*

If we apply an external pressure, for example by compressing with an inert gas, how is the vapour pressure p of the liquid affected? The answer is surprising; p is increased. This may be explained in terms of figure 10.9 which represents an idealized experiment. The top horizontal limb of the tube abcd contains a membrane permeable only to the vapour above the liquid AB. When the space above the liquid contains only vapour, A is level with B. If now an inert gas is pumped into the right-hand limb at a pressure π, since it cannot pass through the membrane, the level A rises to A′ and B falls to B′. If h is the difference in height, clearly

$$\pi = h\rho_L g. \tag{10.19}$$

Since the membrane is permeable to the vapour, the vapour over A′ must be in equilibrium with the vapour over B′. This implies that the vapour pressure at B′ is greater than over A′ by $h\rho_V g$. The increase in vapour pressure is thus

$$\Delta p = h\rho_V g = \pi\frac{\rho_V}{\rho_L}. \tag{10.20}$$

Another way of understanding this is to consider the arrangement in figure 10.9(*c*). The liquid is contained in a cylinder one end of which is closed by a movable piston. The bulk movement of the liquid is blocked by a membrane, which is impermeable to the liquid, but permeable to the vapour. The whole is maintained at a temperature T, so that heat is available for evaporation. If the piston is compressed with a pressure π until one mole of liquid has been forced to evaporate, the liquid diminishes in volume by amount V_L; the piston does work πV_L on the liquid. Consequently the thermal energy necessary for evaporation is reduced from L to $L-\pi V_L$. The vapour pressure then becomes

$$p = A\exp\left[-\frac{(L-\pi V_L)}{RT}\right] = \left[A\exp\left(\frac{-L}{RT}\right)\right]\exp\left(\frac{\pi V_L}{RT}\right) \tag{10.21}$$

$$= p_0\left(1+\frac{\pi V_L}{RT}+\cdots\right) = p_0 + \frac{p_0\pi V_L}{RT}.$$

Hence

$$\Delta p = p - p_0 = \frac{p_0\pi V_L}{RT}. \tag{10.22}$$

Figure 10.9. Effect of external pressure on the vapour pressure of a liquid. (*a*) Initial condition – the membrane is permeable only to the vapour. (*b*) Condition when external pressure is applied. (*c*) Alternative model.

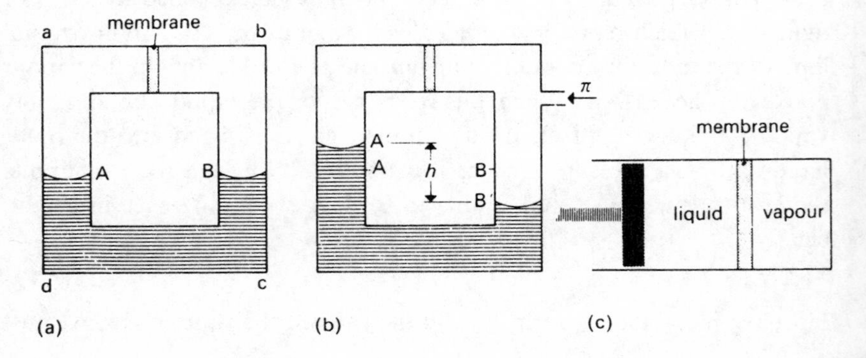

If the vapour behaves like a perfect gas $p_0 V_V = RT$ so that we may insert $p_0 = RT/V_V$ in equation (10.22). This gives

$$\Delta p = \pi \frac{V_L}{V_V} = \pi \frac{\rho_V}{\rho_L}. \tag{10.23}$$

Since at room temperature ρ_V/ρ_L is of the order of 10^{-3}, the increase in vapour pressure with external pressure is generally small. Descriptively one might say that the external pressure squeezes out the liquid molecules. It is not so easy to see why a negative pressure reduces the ease of escape of the liquid molecule. Nevertheless it is clear from the preceding experiment that a negative pressure will in practice reduce the vapour pressure.

10.10 Dilute ideal solutions

In this section we discuss briefly some of the physical properties of dilute solutions in terms of the concepts considered in sections 10.9. We shall restrict ourselves to ideal solutions; that is to say, if solute B is dissolved in solvent A the molecular interactions A–A, A–B and B–B are all considered equal. The solute is assumed to be non-volatile.

10.10.1 *Vapour pressure: Raoult's law*

The vapour pressure of a solution is less than that of the pure solvent. Consider a pure solvent A consisting of n_A mols in a volume V: the vapour pressure of the pure solvent is p_0. Now consider a solution occupying the same volume which contains n_B mols of solute B. The molar concentration of A is reduced to $(n_A - n_B)/V$. Since this is an ideal solution the tendency of solvent molecules to leave the solution and enter the vapour phase is unaffected by the presence of B. The vapour pressure p is reduced simply because the number of solvent molecules per unit volume has been reduced by the presence of B. Then

$$p = \left(\frac{n_A - n_B}{n_A}\right) p_0. \tag{10.24}$$

The reduction Δp is $p_0 - p$.

Hence

$$\Delta p = p_0 - \left(\frac{n_A - n_B}{n_A}\right) p_0 = \frac{n_B}{n_A} p_0 \tag{10.25}$$

or

$$\frac{\Delta p}{p_0} = \frac{n_B}{n_A}. \tag{10.26}$$

10.10.2 *Increase in boiling point*

Boiling occurs when the vapour pressure reaches atmospheric pressure. Since the solution has a lower vapour pressure than the pure solvent its temperature must be increased above its normal value to compensate for this pressure defect. We see from equation (10.16) that temperature and vapour pressure are related by the expression

$$\frac{\Delta p}{\Delta T}=\frac{\mathrm{d}p}{\mathrm{d}T}=\frac{L}{RT}\times\frac{p_0}{T} \tag{10.27}$$

where we have replaced p by p_0.

Thus the temperature rise ΔT needed to achieve an increment Δp in the vapour pressure and so produce boiling is

$$\begin{aligned}\Delta T&=\frac{\Delta p}{p_0}\frac{RT^2}{L}\\&=\frac{n_\mathrm{B}}{n_\mathrm{A}}\frac{RT^2}{L}\end{aligned} \tag{10.28}$$

where T is the boiling point of the pure solvent and L its latent heat of vaporization. An analogous relation may be derived for the depression of the freezing point where L represents the latent heat of fusion.

10.10.3 *Osmotic pressure*

During the nineteenth century many studies were carried out on the diffusion of solvent, through suitable membranes, into a solution. A simple example is shown in figure 10.10(*a*). A vertical tube contains a solution of B in A. Its lower end is sealed by a semipermeable membrane, that is a membrane which is permeable to solvent A but not to solute B. If the end is immersed in A, solvent will diffuse into the solution unless a pressure Π is applied to prevent it. The quantity Π is the osmotic pressure (osmosis = pushing).

Careful experiments by Pfeffer showed that for dilute solutions Π is proportional to the concentration C of B, i.e. the number of mols of B per unit volume (n_B/V), and to the absolute temperature

$$\Pi=\text{constant}\times CT=\text{constant}\times\frac{n_\mathrm{B}T}{V}. \tag{10.29}$$

In a classical paper published in 1886 van't Hoff showed that the constant in equation (10.29) is the gas constant R so that the relation for Π resembles that of a perfect gas. This led him to suggest that the solute behaved like a perfect gas consisting of n_B mols in a volume V. By a rather ingenious, if dubious, argument it could be shown that the impact of the solute molecules on the membrane caused the solvent on the other side of the membrane to exert the observed pressure Π.

A more direct approach is to treat osmosis in terms of vapour pressure. The following treatment is essentially that given by Lowry and Sugden†. Consider the arrangement shown in figure 10.10(*b*). The large reservoir of solution initially contains n_B mols of solute B in n_A mols of solvent A occupying a volume V. Osmosis occurs through the semipermeable membrane and solvent continues to enter the tube until the hydrostatic pressure is sufficient to prevent further osmosis. We assume that the reservoir is so large that the influx of solvent does not change the concentration.

Clearly

$$\Pi = h\rho_L g \quad \text{or} \quad hg = \Pi/\rho_L \tag{10.30}$$

where ρ_L is the density of the solution. The system is surrounded by a

Figure 10.10(*a*). Simple arrangement for observing and determining osmotic pressure. The membrane is permeable to solvent A but not to solute B. The solvent may be prevented from entering the solution by application of a pressure Π, the osmotic pressure.

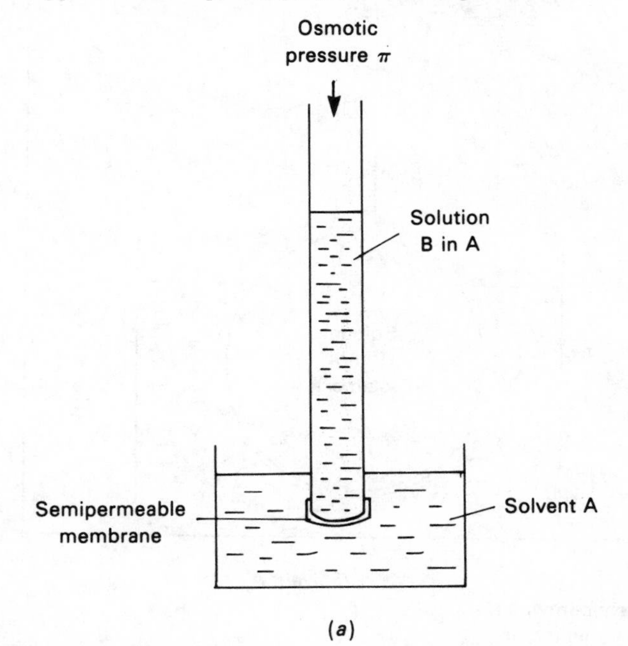

† T. M. Lowry and S. Sugden, *A Class Book of Physical Chemistry*" (Macmillan, 1949).

larger vessel which is initially evacuated of air and gradually the vapour establishes equilibrium. At the surface of the solvent the vapour pressure is p_0. At the top of the column the vapour pressure of the solution is p: for equilibrium the vapour pressure of the solution must be equal to the

Figure 10.10(*b*). Experiment for determining the osmotic pressure in terms of the difference in vapour pressure of the solvent and the solution. The large bell-jar is initially evacuated and the vapours allowed to achieve equilibrium.

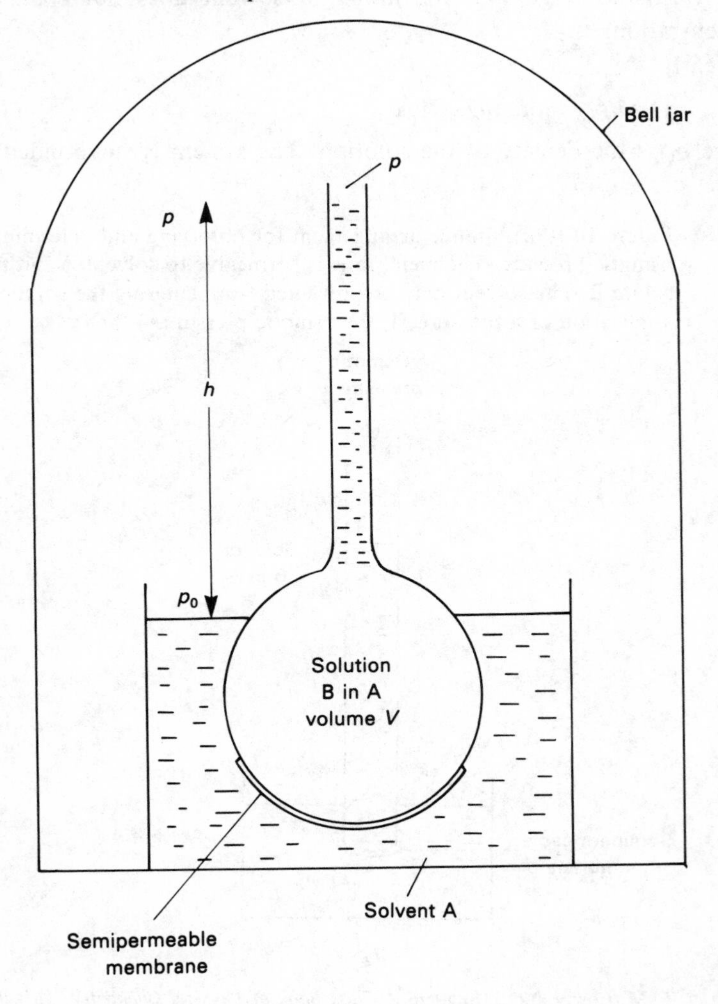

vapour pressure of the solvent at height h. Hence

$$\Delta p = p_0 - p = h\rho_V g \tag{10.31}$$

where ρ_V is the density of the vapour of the solvent.

If M is the molecular mass of one mole of the solvent the density of its vapour is

$$\rho_V = \frac{M}{\phi} \tag{10.32}$$

where ϕ is the volume occupied by one mole of the vapour at temperature T. Since the vapour behaves like a perfect gas

$$p_0\phi = RT. \tag{10.33}$$

Hence from equation (10.32)

$$\rho_V = \frac{Mp_0}{RT} \tag{10.34}$$

then from equation (10.31)

$$\Delta p = \frac{hgMp_0}{RT}. \tag{10.35}$$

Using the value of hg from equation (10.30) we have

$$\Delta p = \frac{\Pi}{\rho_L}\frac{Mp_0}{RT}$$

or

$$\Pi = \rho_L \frac{RT}{M}\frac{\Delta p}{p_0} = \rho_L \frac{RT}{M}\frac{n_B}{n_A}. \tag{10.36}$$

But, if n_B is small compared with n_A, the quantity Mn_A is the total mass of the solution so that

$$\rho_L/Mn_A = V \tag{10.37}$$

where V is the volume of the solution

Hence from equation (10.36)

$$\Pi = \frac{RT}{V}n_B \quad \text{or CRT}. \tag{10.38}$$

Thus the osmotic pressure is the pressure required to raise the vapour pressure of the solution to that of the pure solvent. Evidently the difference in vapour pressure between solvent and solution provides the driving force for the osmotic pressure. As an example we calculate the osmotic pressure of a sucrose solution, consisting of 1 g of sucrose ($\frac{1}{342}$ mol) in a solution of 100.6×10^{-6} m^3 at a temperature of 36 °C (309 K). Taking

$R = 8.3$ J mol K^{-1} we have

$$\Pi = \frac{8.3 \times 309 \times (1/342)}{100.6 \times 10^{-6}} = 0.75 \times 10^5 \text{ N m}^{-2} \approx 0.75 \text{ atm.}$$

Pfeffer's experimental value is 0.746 atm. (Of course, van't Hoff used Pfeffer's data the other way around to calculate R.)

The osmotic pressure increases linearly with concentration but at higher concentrations the observed value is usually greater than that given by equation (10.38).

Osmosis plays a very significant part in many biological processes involving diffusion through membranes; an important factor is that relatively small concentrations give rather large osmotic pressures.

10.11 Surface tension

10.11.1 *Surface energy*

It is common experience that liquids tend to draw up into drops; small drops form perfect spheres. Since a sphere is the geometric form which has the smallest surface area for a given volume we conclude that the surface of the liquid has a higher energy than the bulk; the liquid is therefore always attempting to attain its lowest energy state by reducing its surface area. This energy excess is called the free surface energy γ and is measured in J m^{-2}. For most simple liquids γ has a value of the order of 30–100 mJ m^{-2}. This unit is identical with the older unit, still current amongst some physical chemists, of erg cm^{-2} or dyne cm^{-1}.

If a liquid has area A its free surface energy is γA. If we increase the surface by doing work on it the work done is

$$\frac{\mathrm{d}(\gamma A)}{\mathrm{d}A} = \gamma + A\frac{\partial \gamma}{\partial A}. \qquad (10.39)$$

With a liquid, however much we stretch the surface, the surface structure regains its initial equilibrium configuration in a very short interval of time. Thus the equilibrium structure remains unchanged so that $\partial\gamma/\partial A = 0$. This is in contrast to a solid.

Hence

$$\text{work done per unit area of extension} = \gamma. \qquad (10.40)$$

Recent experiments have shown that very thin films of water have the same surface energy as two surfaces separated by bulk water. The identity holds down to films as thin as 2 nm. This implies that the molecular forces responsible for the surface energy (see below) are very short range.

The free surface energy is equivalent to a line tension acting in all directions parallel to the surface. Consider a liquid surface (as shown in figure 10.11) of width L and length X. Suppose at the barrier AB we apply a force F parallel to the surface and normal to AB so that we extend the length of the surface by an amount x. The area increase is Lx and the work done in increasing the surface area is $Lx\gamma$. This must be equal to the external work done Fx.

Hence

$$Fx = Lx\gamma,$$

or

$$\frac{F}{L} = \gamma. \tag{10.41}$$

Thus the surface energy is equivalent to a surface tension or line tension $\gamma\,\mathrm{N\,m^{-1}}$.

10.11.2 *Molecular interpretation*

The free surface energy of a liquid lends itself to a very simple molecular interpretation. Molecules in the bulk are subjected to attraction by surrounding molecules; the field is symmetrical and has no net effect. At the surface, however, the surface molecules are pulled in towards the bulk of the liquid. Apart from a few vapour molecules there is no attraction in the opposite direction. Consequently if we wish to increase the area we have to pull molecules up to the surface from the bulk against this one-sided attraction. This accounts for the surface energy.

Alternatively we may say that in moving molecules from the bulk towards the surface we are continuously breaking and reforming bonds. At the top, however, roughly half the bonds are not reformed. Since bond formation implies a decrease in energy the incomplete reformation of bonds at the surface means that the surface has a higher energy than the

Figure 10.11. Sketch showing equivalence of surface energy per unit area and the tension per unit length.

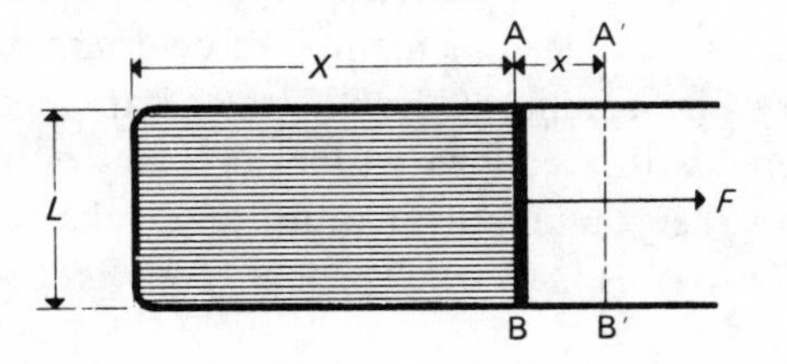

bulk. Practically the whole of this process is confined to the last few molecular layers and it is for this reason that extremely thin liquid films have the same surface energy as 'bulk' liquid (see above). Without going into further details it is evident that the surface energy is of the order of one-half the energy required to break all bonds, i.e. one-half the latent heat of vaporization of a molecular layer. This is essentially the same as the model we discussed earlier for solids and we emphasize again that the intermolecular forces responsible for the surface tension of a liquid are of the same nature as those responsible for the pressure defect of the corresponding gas.

If L is the latent heat of vaporization per mole and M the molecular weight and ρ the density, we have, as an approximate relation (cf. equation (6.9)),

$$\gamma = 0.3\,\frac{L}{N_{\mathrm{A}}}\left(\frac{N_{\mathrm{A}}\rho}{M}\right)^{\frac{2}{3}}. \tag{10.42}$$

For liquid argon this gives a value of 14 mJ m^{-2} for the surface tension compared with the observed value of 13; for liquid neon 4 compared with 5.5; for nitrogen 11 compared with 10.5; for oxygen 13 compared with 18; for benzene 110 compared with 40 and for mercury 630 compared with 600 mJ m^{-2}.

Thermodynamics provides another quantity, the total surface energy h. When a surface is increased in area, apart from the surface energy term γ, there is also a 'latent heat' term. Heat must be provided to keep the temperature constant. Its magnitude is $-T\,\partial\gamma/\partial T$. Hence the total surface energy is

$$h = \gamma - T\frac{\partial \gamma}{\partial T}. \tag{10.43}$$

Equation (10.42) does not distinguish between these two concepts; it is really a model for γ at 0 K so that agreement with the room temperature experimental values must not be taken too seriously. On the other hand the difference between h and γ is often significant. For example for water $\gamma = 72$ and $h = 118$ mJ m^{-2}. If fine droplets of water are allowed to coalesce so as to destroy their surface area the temperature rise is determined by h, not by γ. This provides a very ingenious method of determining areas of fine particles. They are first equilibrated with water vapour so that their surface is fully covered with a condensed film of water only a few angstroms thick. They are then immersed in water in a delicate calorimeter. If the heat given out is ΔQ the surface area is given by $\Delta Q/0.118$ m^2.

10.11.3 *Contact angle and contact equilibrium*

It is generally found that liquids with low surface tension readily wet most solids, giving a contact angle of zero; those with high surface tension often show a finite contact angle. In molecular terms we may say that if the cohesion between the molecules of the liquid is greater than the adhesion between the liquid and solid, the liquid will not 'want' to wet the solid, consequently it will show a finite contact angle.

Contact equilibrium is shown in figure 10.12. The free surface of the liquid has a surface energy γ_L; the solid-liquid interface energy γ_{SL} and the exposed portion of the solid adjacent to the liquid where vapour has been adsorbed a surface energy γ_{SV}. If we consider a virtual movement of the contact region so that an additional m^2 of solid is wetted we have

surface energy increase at solid–liquid interface $= \gamma_{SL}$,

surface energy decrease at solid–vapour interface $= \gamma_{SV}$,

surface energy increase of water surface $= \gamma_L \cos\theta$.

By the principle of virtual work

$$\gamma_{SV} = \gamma_{SL} + \gamma_L \cos\theta. \tag{10.44}$$

This equation, which was first derived by T. Young, without proof, in 1805 and by A. Dupré in 1869, is exactly what we would obtain if we took the horizontal components of the surface tension forces. This at once raises the question: what has happened to the vertical component $\gamma_L \sin\theta$? The answer is simple. The force is very small and its effect on a solid is negligible. If, however, the solid consists of a very thin flexible sheet, for example, of mica, the vertical forces may be sufficient to distort the surface visibly. One is then aware of the reality of the vertical component.

Equation (10.44) has been studied experimentally by Miss S. Kay, using the bifurcation of mica as a means of determining directly the surface energies of the solid. In the first experiment she determined γ for a mica

Figure 10.12. (*a*) Contact angle equilibrium. (*b*) Contact on a thick specimen. (*c*) Contact on a thin mica sheet where the vertical component of the tension distorts the shape of the sheet.

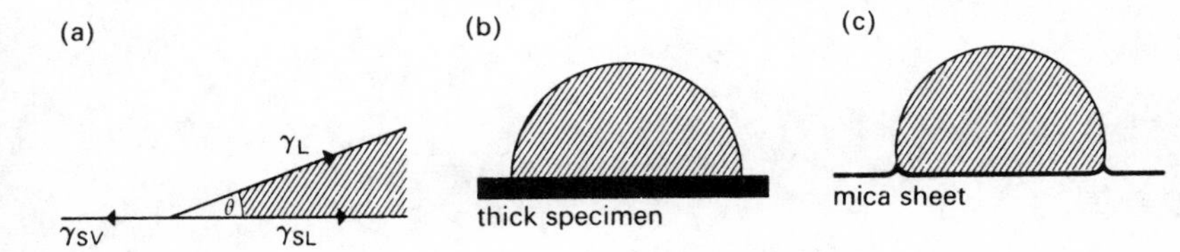

Table 10.2. *Surface energy of mica in presence of vapour and liquid*

	Surface energies (mJ m^{-2})			
Fluid	γ_{SV}	γ_{SL}	$\gamma_{SV}-\gamma_{SL}$	γ_L
Water	183	107	76	73
Hexane	271	255	16	18

surface exposed to water vapour; this gives γ_{SV}. In the second experiment she determined γ when the surfaces were immersed in water; this gives γ_{SL}. Since water completely wets mica ($\theta=0$) we should expect to find, from equation (10.44) that

$$\gamma_L=\gamma_{SV}-\gamma_{SL}. \tag{10.45}$$

Results obtained with water are given in Table 10.2. The table includes results for hexane which also wets mica. The agreement is very satisfactory.

10.11.4 *Pressure difference across a liquid–vapour interface*

We shall now discuss a few of the general properties of capillarity which are shown by liquids. Before doing so we derive a simple relation for the pressure difference across a curved liquid interface.

Let ABCD be a portion of a surface marking the boundary between air below and liquid above (see figure 10.13). Over a small region a curved

Figure 10.13. Pressure difference across a liquid–air interface.

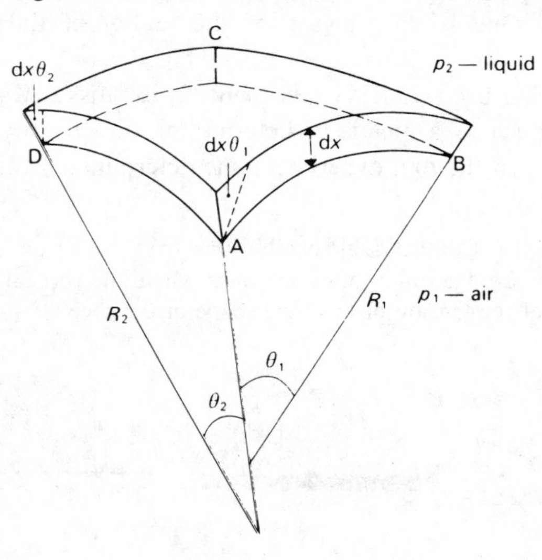

surface can always be described in terms of two radii of curvature in mutually orthogonal planes. These are known as the principal radii of curvature R_1 and R_2. Suppose we assume that the pressure on the air side is p_1 and on the liquid side is p_2. We now apply the principle of virtual work displacing the surface parallel to itself through a distance $\mathrm{d}x$. Then AB increases in length by $\mathrm{d}x\ \theta_1 = \mathrm{d}x(\mathrm{AB}/R_1)$, similarly AD increases in length by $\mathrm{d}x\ \theta_2 = \mathrm{d}x(\mathrm{AD}/R_2)$. The increase in surface area, neglecting terms in $(\mathrm{d}x)^2$, is then

$$\Delta\alpha = \left(\mathrm{AB} + \mathrm{AB}\frac{\mathrm{d}x}{R_1}\right)\left(\mathrm{AD} + \mathrm{AD}\frac{\mathrm{d}x}{R_2}\right) - \mathrm{AB} \times \mathrm{AD}$$

$$\Delta\alpha = \mathrm{AB} \times \mathrm{AD} \times \mathrm{d}x \times \left(\frac{1}{R_1} + \frac{1}{R_2}\right). \tag{10.46}$$

This involves an increase in surface energy of $\gamma\Delta\alpha$. The work done by the pressure difference $p_1 - p_2$ is

$$(p_1 - p_2)\mathrm{AB} \times \mathrm{AD} \times \mathrm{d}x = \gamma\,\Delta\alpha.$$

Therefore

$$p_1 - p_2 = \gamma\left(\frac{1}{R_1} + \frac{1}{R_2}\right). \tag{10.47}$$

The pressure on the concave side is greater than on the convex side of a liquid interface, by an amount depending on the surface tension and on the curvature. This relation becomes clearer in the following two very simple cases. For a spherical bubble of radius R within a liquid the surface area is $4\pi R^2$. If a small expansion occurs increasing r by $\mathrm{d}R$ the increase in area of surface is $8\pi R\,\mathrm{d}R$. The work that must be done by the pressure in the bubble is $4\pi R^2 p\,\mathrm{d}R$. Equating this to the increase in surface energy we have

$$4\pi R^2 p\,\mathrm{d}R = 8\pi R\,\mathrm{d}R\gamma, \qquad p = \frac{2\gamma}{R}. \tag{10.48}$$

This is exactly the result of equation (10.47) if $R_1 = R_2$. The second case is for a soap bubble. Here two surfaces are involved and the pressure inside the bubble is

$$p = \frac{4\gamma_s}{R}, \tag{10.49}$$

where γ_s is the surface tension of the soap solution.

10.11.5 *Capillary rise*

Suppose a uniform tube of small radius R is held vertically and lowered into a liquid of density ρ_L which wets it ($\theta = 0$). The liquid rises to a height h above the common level. This is sometimes explained in terms of the surface forces around the periphery of the meniscus. The meniscus has peripheral length $2\pi R$ and for zero contact angle the upward force is $2\pi R\gamma$. This is balanced by the downward weight of the liquid column $h\pi R^2\rho_L g$ (figure 10.14(*a*)). Thus

$$h\rho_L g = \frac{2\gamma}{R}. \tag{10.50}$$

Although equation (10.50) is corect, this is largely fortuitous. For example if the capillary was conical in shape (figure 10.14(*b*)) the capillary rise would still be given by equation (10.50) where R is the capillary radius at the point where the meniscus formed. Clearly $2\pi R\gamma$ could be considerably less than the weight of the column of liquid.

The proper method is to treat the problem in terms of the pressure difference across the meniscus. The meniscus for $\theta = 0$ is a hemisphere of radius $R_1 = R_2 = R$. According to equation (10.47) the pressure at B will be less than at A (see figure 10.14(*c*)) by the amount

$$\gamma\left(\frac{1}{R_1} + \frac{1}{R_2}\right) = \frac{2\gamma}{R}. \tag{10.51}$$

But the pressure at A is the same as that at A′ (it is atmospheric) and is the same as that at C except for a minute air pressure drop A′C. Thus the pressure in the liquid at B is less than the pressure in the liquid at C.

Figure 10.14. (*a*) Simple calculation of capillary rise in terms of surface tension forces. (*b*) Inapplicability of method (*a*) to a conical capillary. (*c*) Calculation of capillary rise in terms of pressure defect.

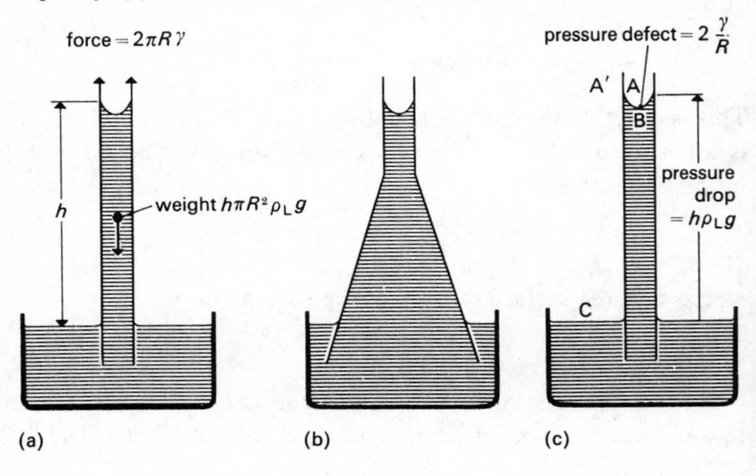

Consequently the liquid is drawn up the tube to a height h, where

$$h\rho_L g = \frac{2\gamma}{R}.$$

We should obtain the same result for a conical capillary in agreement with experimental observation.

This treatment also explains why the meniscus in a narrow capillary can be considered to be spherical in shape. If the capillary rise is large compared with the diameter of the tube, the height of rise is practically constant over the whole width of the capillary, i.e. the pressure defect $h\rho_L g$ is constant. This means (see equation (10.50)) that the curvature of the meniscus must be constant. For a circular capillary the two principal radii of curvature must, by symmetry, be equal. Consequently the meniscus will be a portion of a sphere of almost constant radius, the value of which depends on the contact angle. For zero contact the shape will be hemispherical, the radius of curvature being equal to that of the tube R. If the contact angle is θ, it will be equal to $R/\cos\theta$ (see figure 10.15(a)) and the pressure defect giving rise to the capillary ascent will be equal to $2\gamma\cos\theta/R$. As the capillary gets wider the height of the liquid column becomes small compared with the capillary (see figure 10.15(b)). There is a very large difference between the pressure defect at L and at M. Indeed at M where the height of rise is negligible the curvature is almost zero ($R\to\infty$) whilst it is finite at L and N. The detailed shape of the meniscus

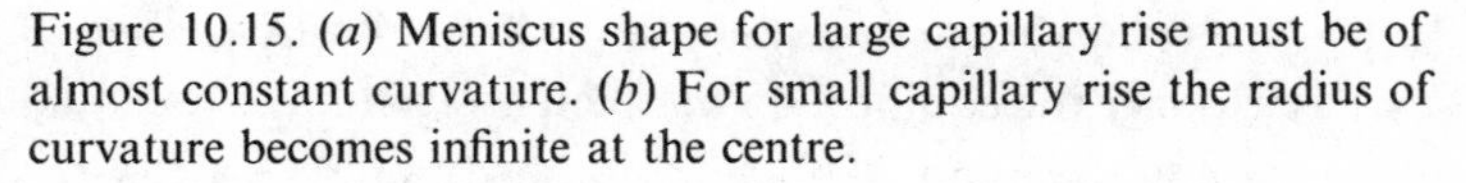

Figure 10.15. (a) Meniscus shape for large capillary rise must be of almost constant curvature. (b) For small capillary rise the radius of curvature becomes infinite at the centre.

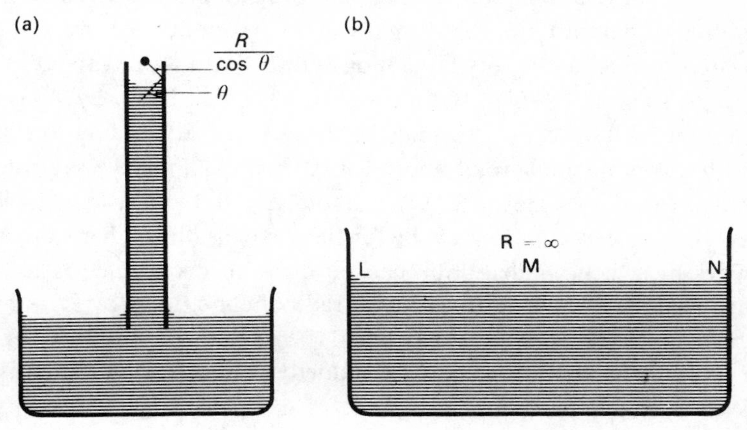

can be solved theoretically by equating the pressure defect

$$\gamma(1/R_1+1/R_2)$$

at any given point to the value of $h\rho_L g$ at that point, where h is the height of the liquid above the bulk level of the liquid at a large distance from the walls of the container.

10.11.6 *Vapour pressure over a liquid meniscus*

We see from figure 10.14(c) that the vapour pressure over the liquid at A is less than over the liquid at C by $h\rho_V g$ where ρ_V is the mean density of the vapour. We may write:

vapour pressure defect

$$=h\rho_V g=h\rho_L\times\frac{\rho_V}{\rho_L}\times g=\frac{2\gamma}{R}\times\frac{\rho_V}{\rho_L}. \tag{10.52}$$

This result is part of a more general statement: the vapour pressure over a concave meniscus surface is less than that over a flat liquid surface by the quantity given in equation (10.52). Similarly the vapour pressure over a convex meniscus is greater than over a flat surface by a similar amount. We may see this in a different way by considering the behaviour of a spherical liquid drop of radius R. The liquid within the drop is subjected to a surface tension pressure of amount $p=2\gamma/R$. As we saw in a previous section this increases the vapour pressure by the amount

$$p\frac{\rho_V}{\rho_L}=\frac{2\gamma}{R}\times\frac{\rho_V}{\rho_L}. \tag{10.53}$$

10.12 **Nucleation in condensation: the Wilson cloud chamber**

Consider the equilibrium between a vapour and its liquid and the problem of condensation. If there are already present a few droplets of radius of curvature r, the droplets exert a vapour pressure of $(2\gamma/r)\times(\rho_V/\rho_L)$. This can be very large. For example with water vapour at 293 K, ρ_V/ρ_L is about 2×10^{-5}. But if r is only 0.5 nm the factor $2\gamma/r$ has a value of about 3×10^8 N m^{-2} so that the droplets are attempting to evaporate with a vapour pressure of about 6×10^3 N m^{-2}. Since the saturated vapour pressure of water at 293 K is about 2×10^3 N m^{-2}, the droplets will evaporate; they cannot grow by further condensation. There are only two ways by which condensation can be made to occur. One is to provide artificial nuclei with relatively large radii of curvature: this is, in practice, generally achieved by dust particles. The second is to produce supersaturation of the vapour, usually by sudden cooling. This is the basis of the Wilson cloud chamber.

Consider a vessel closed with a movable piston in which the vapour is in equilibrium with the liquid. Charged particles are projected through the vapour. Because the electrostatic energy of a point charge is lowered by the condensation of a dielectric on it, some condensation may occur. If the piston is now suddenly raised there is a marked adiabatic temperature drop. Although there is some volume increase the main effect is that the existing vapour pressure is much higher than the saturated vapour pressure at the new reduced temperature. Condensation occurs readily and the path of the particles is revealed as a series of liquid droplets.

10.13 Superheating

A similar problem is involved in the heating of a liquid to produce boiling. If the initial bubbles have a radius of curvature r, the vapour pressure in the bubbles must exceed $2\gamma/r$ if the bubbles are to grow. For a bubble of radius $r = 0.5$ nm in a liquid such as water, this pressure is of the order of 3000 atmospheres. Thus unless the vapour pressure reaches this value the bubble will collapse. To obtain vapour pressures of this order the liquid must be very strongly superheated. With water, temperatures of the order of several 100 K above the normal boiling point are required. Experiments with water from which dissolved gases have been eliminated do, in fact, show that bubbles will not form until temperatures of this order are reached. Of course, once the bubble begins to form at this high vapour pressure, the excess pressure necessary to produce bubble growth begins to decrease as r increases, so that the bubble grows at a catastrophic rate – the liquid explodes.

Steady boiling at the true boiling temperature is achieved by providing suitable nuclei. These are often air pockets trapped at solid surfaces or gases dissolved in the liquid itself.

10.14 The energy for capillary rise

We raise here two points concerning capillary rise that are not often discussed in elementary textbooks. The first is, what is the source of the energy for capillary rise? Consider a tube of radius R. The liquid rises to a height h given by

$$h = \frac{2\gamma_L}{R\rho g}. \tag{10.54}$$

The potential energy ε_l gained by the liquid is equivalent to the raising of a mass of liquid $\pi R^2 h\rho$ through a height $\frac{1}{2}h$. Hence

$$\varepsilon_l = \tfrac{1}{2}\pi R^2 h^2 \rho g. \tag{10.55}$$

This energy comes from the wetting of the walls of the tube by the liquid. Descriptively we may see this in the following way. The surface energy of a solid arises from the asymmetric forces at the free surface. If we cover the surface with another material, for example, a liquid, we reduce this asymmetry and hence reduce the surface energy. The energy liberated in this process is available for pulling the liquid up the tube. We may make this quantitative as follows.

Consider a liquid with contact angle $\theta=0$. The equilibrium condition, as we saw earlier, is

$$\gamma_{SV}=\gamma_L+\gamma_{SL}. \tag{10.56}$$

If the liquid advances along the tube so as to cover 1 m^2 of the surface we destroy 1 m^2 of γ_{SV} and gain 1 m^2 of γ_{SL}. There is no change in the area of the liquid meniscus. The energy given up by the system is thus

$$\gamma_{SV}-\gamma_{SL}. \tag{10.57}$$

From equation (10.56) we see that this is equal to γ_L. Thus each m^2 of capillary surface wetted releases energy γ_L. The energy released for a rise of h is then

$$\varepsilon_2=\gamma_L 2\pi Rh. \tag{10.58}$$

Substituting from equation (10.54) for γ_L we have

$$\text{loss of surface energy } \varepsilon_2=\pi R^2h^2\rho g. \tag{10.59}$$

This is exactly twice the gain in potential energy given by equation (10.55). This implies that if the liquid were non-viscous it would rise to a height $2h$ and oscillate between 0 and $2h$ with its mean position at h. In practice viscosity consumes the excess energy very quickly.

The behaviour is like that of a vertical spiral spring onto the end of which a weight is suddenly attached. The weight falls instantaneously to double its equilibrium depth and oscillates between this and zero until the kinetic energy is consumed by friction and static equilibrium is achieved.

The second query is the following. Some solids are known to have much higher surface energies than others. For example if they are first thoroughly cleaned and then immersed in a liquid the amount of heat released may vary very widely; nevertheless if a liquid wets these surfaces ($\theta=0$), the capillary forces do not depend on the material of which the capillary is made; they depend only on the surface energy of the liquid. Why is this? The answer is implicit in equation (10.56). The adsorbed vapour on the solid will always adjust its thickness so that the quantity

$$\gamma_{SV}-\gamma_{SL}$$

is equal to γ_L (for $\theta=0$). The energy gain is thus dependent only on γ_L.

We may describe this differently. The vapour adsorbed on the solid presents a surface to the advancing liquid which closely resembles the liquid itself. On a high-energy surface the adsorbed film will tend to be thicker than on a low-energy surface. In both cases (for $\theta = 0$) the liquid is virtually wetting itself so that only γ_L is involved in the process. In this connection we may note that the range of forces is small and that the adsorbed film needs to be only 1–2 nm thick in order to mask almost completely the influence of the underlying substrate.

10.15 **Liquid crystals**

Before we leave this chapter on liquids we briefly consider a state of matter which is intermediate between a truly amorphous isotropic liquid and a crystalline solid. These materials are known as liquid crystals. They are mainly found in certain organic molecules which have a high degree of anisotropy; for example, long rod-like or disc-like molecules that usually but not always include a polar group. In most of the examples described here we shall deal with thermotropic liquid crystals, i.e. those which are formed by heating the solid.

10.15.1 *Nematic liquid crystals*

Consider for example the rod-like molecules

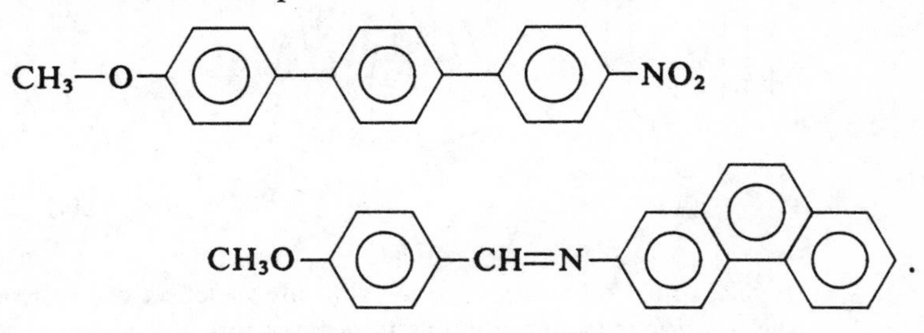

They contain benzene rings and double bonds and are relatively rigid. In the solid state the molecules are held together by van der Waals and dipole forces and are packed parallel to one another in a well defined structure (figure 10.16(*a*)). As the temperature is raised the molecules acquire more thermal energy until at T_c (see figure 10.17) they are able to free themselves from their neighbours. They still retain a fair measure of orientation but their long axes are randomly tilted away from a truly parallel arrangement (see figure 10.16(*b*)). This constitutes the nematic liquid crystal phase, so called because when viewed between glass plates in the microscope they appear to consist of threads (nematic = thread-like) as a result of defects. In bulk form the material appears turbid because of domains in the arrangement of the molecules. In this state the material will flow without the application of a finite yield stress. As the temperature is raised above T_c the additional thermal energy enables the molecules to deviate farther from the ordered alignment. When the average angle of tilt exceeds about 40° another marked change occurs. Thermal energy 'overcomes' the self-aligning tendency of the molecules, they acquire a random isotropic orientation and become a true liquid (figure 10.16(*c*)). They are optically clear. If at this temperature the liquid is subjected to hydrostatic pressure, the reduction in volume reduces the freedom of disorientation and the material becomes nematic once more. (At even higher pressures it becomes solid.) Certain nematic liquid crystals containing mobile ions are exploited in display devices. A thin layer of the liquid crystal is retained between two glass plates, the inner surfaces

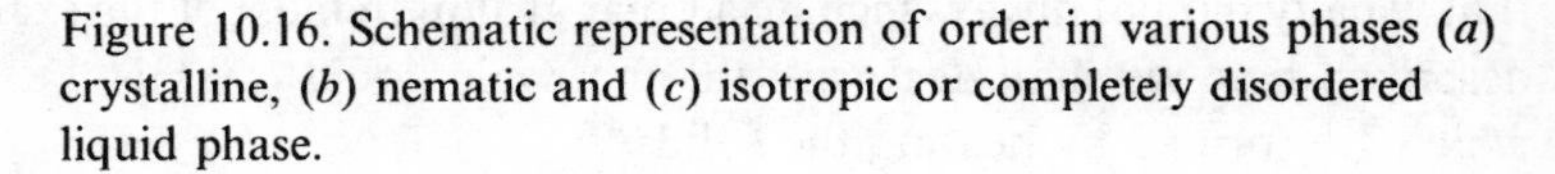
Figure 10.16. Schematic representation of order in various phases (*a*) crystalline, (*b*) nematic and (*c*) isotropic or completely disordered liquid phase.

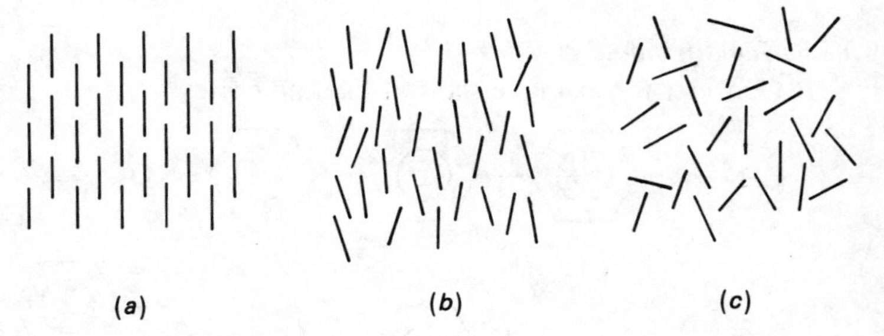

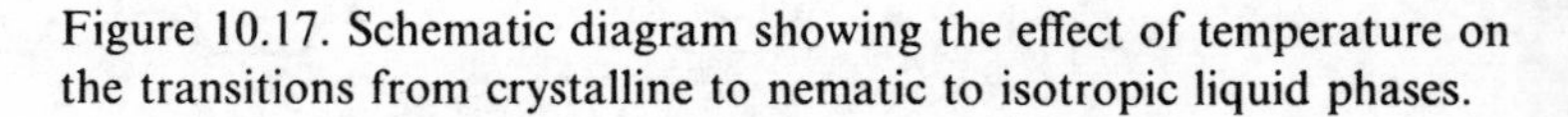
Figure 10.17. Schematic diagram showing the effect of temperature on the transitions from crystalline to nematic to isotropic liquid phases.

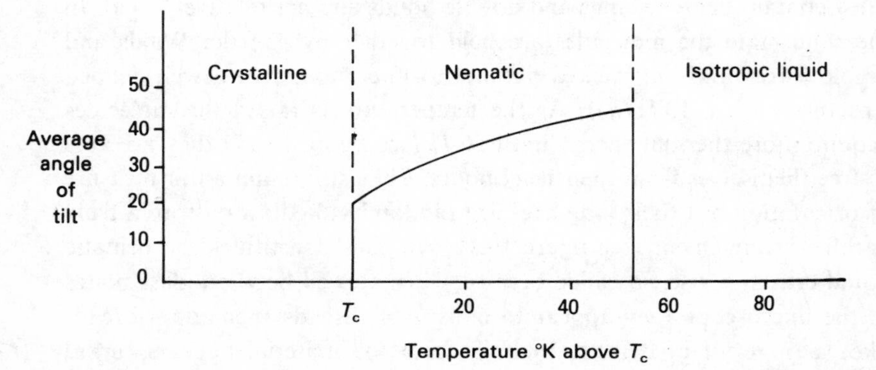

of which are covered with a thin transparent conducting material. When an electric field is applied across the plates turbulence is set up in the layer producing intense scattering of light, i.e. the material appears black.

Certain types of nematic molecules can be polymerized and in the drawn state form fibres of high tensile modulus and strength (see Chapter 13, p. 354).

10.15.2 *Cholesteric liquid crystals*

In nematic liquid crystals the molecules have a high degree of long-range orientational alignment with a mean tilt that is never greater than about 40° but they show no long-range translational or packing order. Nematic materials in which the molecules are optically active acquire a significantly different structure. If contained within parallel plates the molecules behave as though arranged in parallel sheets in each of which the molecules are aligned as in the pure-nematic phase. However, because of the additional interaction provided by the optically active components the molecular orientation in each sheet is twisted relative to its neighbours and the material acquires a helical structure as shown in figure 10.18. We could say that a true nematic liquid crystal has a helical structure of infinite pitch (the distance over which the orientation of the molecules turns through 360°). It is interesting to note that if a very small quantity of a cholesteric substance is added to a nematic the mixture will adopt a typical helical configuration. This emphasizes a point made at various places in this book that very small energy changes can produce strikingly different structural forms. The cholesteric phase is so called because it was first observed with derivatives of the cholesterol molecule.

Figure 10.18. The cholesteric liquid crystal. The molecules are not arranged strictly in parallel sheets but it is convenient to show the twist in orientation of the molecules in this schematic way.

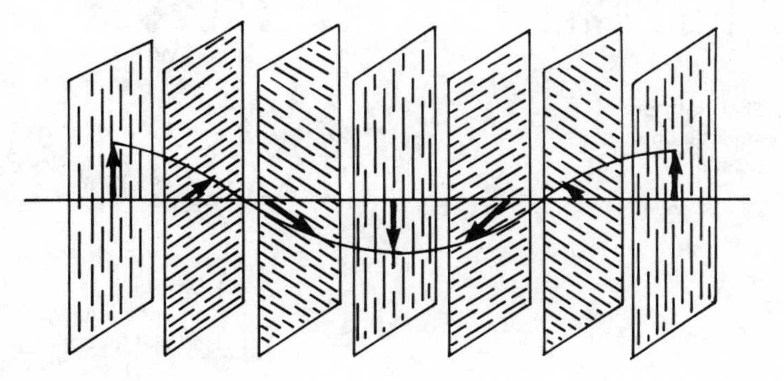

It is sometimes called the chiral-nematic phase, the word chiral being a Greek word meaning twisted. There are two main differences between the cholesteric and the pure nematic phases. The first is that in certain flow configurations the cholesteric liquid shows an enormous increase in apparent viscosity at very low shear rates as though its structure produces a blockage and prevents easy flow. Its behaviour indeed resembles that of a Bingham fluid (see figure 11.5). By contrast the pure nematic phase is almost Newtonian in its flow properties. Secondly the material is optically active, and the dominant wavelength of scattered light is determined by the pitch of the helix. As a result of the conflict between order and thermal energy the pitch decreases as the temperature rises. Thus if the material is illuminated with white light the colour of the scattered light changes with temperature. For example with a mixture of 20 per cent cholesterol carbonate, 80 per cent cholesterol nonanoate the dominant wavelength of scattering at 25 °C is 0.55 μm (5500 Å ≈ bright yellow) and at 45 °C is 0.39 μm (3900 Å ≈ deep violet). Thus this material can be used for visual display of surface temperatures.

Figure 10.19. Smectic liquid crystals (*a*) arrangement of molecules in a thermotropic liquid crystal just above the melting point, (*b*) the lyotropic smectic form which occurs with metallic soaps in the presence of water.

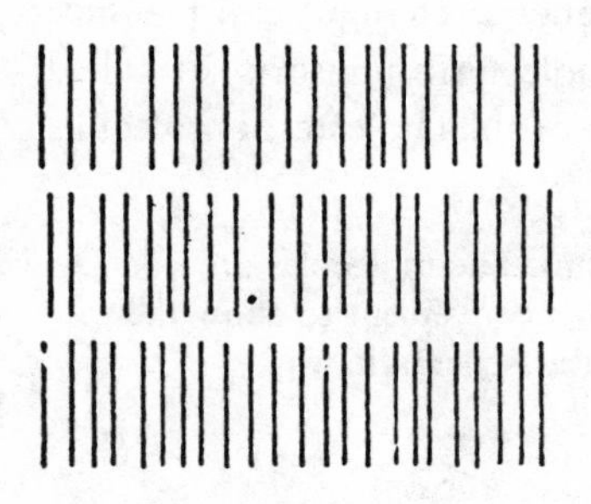

(*a*) Thermotropic smectic liquid crystal

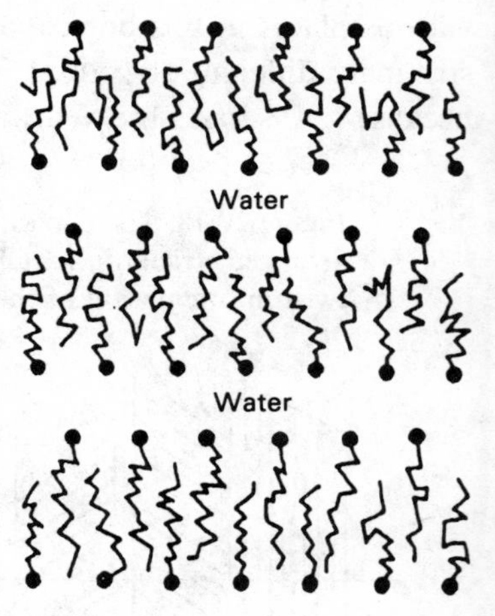

(*b*) Lyotropic smectic liquid crystal (soap)

10.15.3 *Smectic liquid crystals*

Smectic liquid crystals are so named because they are often observed in certain soaps (Greek smegma). When these melt they form a liquid crystal in which the chains are orientated more or less parallel to one another but not regularly spaced laterally (figure 10.19(*a*)). A new and characteristic feature of the smectic phase is that the molecules have a stratified or layered structure. The thickness of the layer is roughly equal to the length of the molecule, or of the dimer. The lateral forces between the molecules *within* a layer are strong, whereas the interaction *between* the layers is relatively weak. The material is 'slippery'. The smectic form with which we are most familiar in everyday life is that which occurs with soaps in the presence of water: They are known as lyotropic liquid crystals (lyo = liquid). A typical arrangement for a sodium soap is shown in figure 10.19(*b*) where it is seen that the individual layer is somewhat longer than the molecular length while the layers are separated by a thin film of water. The layers slide easily over one another and this property gives the material its typically soapy feel. In some cases where the pH is appropriate the separation may be maintained by electrical charges in the thin water film. This effect is discussed in Chapter 12 in relation to the stabilization of colloids by electrically charged double layers.

As a matter of general interest it may be noted that lyotropic liquid crystals, both smectic and nematic, are found in body fluids such as mucus and saliva, and for modestly loaded surfaces they act as low-friction lubricants. There is some evidence that synovial fluid which lubricates anatomical joints, may also fall into this category.

11

Liquids: their flow properties

In this chapter we shall discuss the flow properties of liquids, and derive or state some of the more standard equations of flow. However, our main attention will be directed to a molecular model of viscous flow; this follows closely the theory proposed by Eyring.

11.1 Flow in ideal liquids: Bernoulli's equation

As with gases we can assume the existence of ideal liquids in which internal forces play a trivial part, so that they have negligible surface tension or viscosity. Their flow properties are determined solely by their density. Of course, no liquids can have zero internal forces, they would not be liquid if this were so, but they can often behave as ideal liquids in flow if the inertial forces dominate.

Consider the flow of such an idealized liquid and let us follow the path of any particle in it. If it moves in a continuous steady state we may draw a line such that the tangent at any point gives the direction of flow of the particle. Such lines are called streamlines. They are smooth continuous lines throughout the liquid and can never intersect – no fluid particles can flow across from one streamline to another. Let AB represent an imaginary tube in the liquid bounded by streamlines (figure 11.1(a)). At A the liquid is at height h_1 above a reference level, the pressure acting at that point is p_1, the cross-sectional area of the tube is α_1, and the liquid flow velocity is v_1; at B the corresponding quantities are h_2, p_2, α_2, v_2. Consider the energy balance during a short time interval dt. The pressure p_1 at A drives a volume of liquid $\alpha_1 v_1$ dt through the tube: the work done is

$$w_1 = p_1 \alpha_1 v_1 \, \mathrm{d}t. \tag{11.1}$$

The mass of liquid involved is $\rho\alpha_1 v_1\, dt$ so that the kinetic energy it brings with it is

$$w_2 = \tfrac{1}{2}(\rho\alpha_1 v_1\, dt)v_1^2, \tag{11.2}$$

and its potential energy is

$$w_3 = (\rho\alpha_1 v_1\, dt)gh_1. \tag{11.3}$$

Since the motion is assumed steady and the liquid is ideal the same amount of energy leaves the tube at B. Hence

$$(w_1 + w_2 + w_3)_A = (w_1 + w_2 + w_3)_B. \tag{11.4}$$

Because the same mass of liquid leaves at B as enters at A,

$$\alpha_1 v_1\, dt = \alpha_2 v_2\, dt. \tag{11.5}$$

Equation (11.4) becomes

$$\frac{p_1}{\rho} + \tfrac{1}{2}v_1^2 + h_1 g = \frac{p_2}{\rho} + \tfrac{1}{2}v_2^2 + h_2 g. \tag{11.6}$$

This is Bernoulli's equation of flow for an ideal liquid: the three terms on each side of the equation correspond respectively to the pressure energy to move the liquid, the kinetic energy and the potential energy.

A simple example is illustrated in figure 11.1(*b*). A container is filled with liquid to a height h_1. At the bottom of the container is a small hole. What is the velocity v_2 of the outflowing liquid? If we construct a tube of flow as shown, the conditions at the top of the tube are p_1 = atmospheric = A, height = h_1, velocity, $v_1 \approx 0$. At the bottom of the tube $p_2 = A$, height =

Figure 11.1. (*a*) Streamlines of flow of a liquid in a gravitational field. (*b*) Flow of liquid from an orifice at a depth h below the free level ($p_1 = p_2$ = atmospheric pressure A).

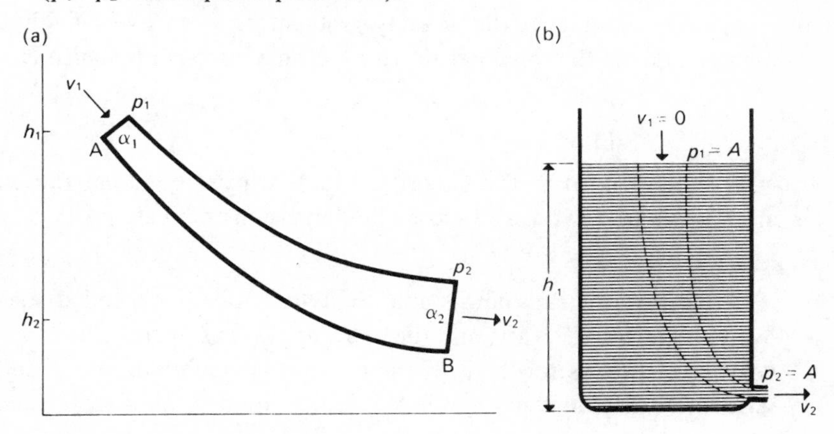

0, velocity $= v_2$. Starting with equation (11.6),

$$\frac{p_1}{\rho}+\tfrac{1}{2}v_1^2+h_1g=\frac{p_2}{\rho}+\tfrac{1}{2}v_2^2+h_2g,$$

$$\frac{A}{\rho}+0+h_1g=\frac{A}{\rho}+\tfrac{1}{2}v_2^2+0, \tag{11.7}$$

$$v_2^2=2gh_1.$$

The flow of an ideal liquid does not involve any particular molecular model; density is the only property concerned, and we shall not discuss it further. We may, however, remark that in general an increase in flow velocity is accompanied by a drop in pressure. This accounts for the suction experienced by a ship approaching a quay – the water flows faster at the regions of nearest approach. It also accounts for the downward curvature of the flight of a tennis ball that has been given top spin. Again, because the top surface of an aerofoil is longer, the air velocity in flight is higher and the pressure lower than over the bottom surface; this produces aerodynamic lift. At a more humdrum level, the spilling of liquid down the outside of a bottle (or down the spout of a teapot) as liquid is poured out occurs because liquid at the lip flows faster than farther away, the liquid pressure is less and the air pushes the liquid against the lip and on to the outer surface. This can be avoided by using a flared lip with a rather sharp edge so that liquid has to turn through nearly 180° in order to flow on the outside (J. B. Keller, *J. Appl. Phys.* **28**, 859–64 (1957)).

11.2 Flow in real liquids, viscosity

11.2.1 *Steady and unsteady flow, turbulence*

Real liquids experience a viscous resistance to flow. We have already defined viscosity in discussing the behaviour of gases but we may recapitulate it here. If the velocity gradient between two neighbouring planes is $\mathrm{d}u/\mathrm{d}z$ the force per m^2 to overcome viscous resistance is

$$f=\eta\frac{\mathrm{d}u}{\mathrm{d}z}, \tag{11.8}$$

where η is defined as the viscosity. Fluids which obey this relationship are known as Newtonian fluids. The dimensions of η are

$$[\eta]=ML^{-1}T^{-1}. \tag{11.9}$$

If the flow is steady and stable the work done is expended solely in overcoming the viscosity, and the work appears as heat. If, however, the velocity of flow is too high or there are other unfavourable features, a series of vortices may develop in the liquid. Some of the work is expended in providing kinetic energy for the vortices. Vortices often assemble on a solid boundary and this is known as the boundary layer. The condition for turbulent motion was first established by O. Reynolds. If a liquid of density ρ, viscosity η flows along a channel of lateral dimension a, there is a critical velocity at which orderly streamlined flow changes to turbulent motion. By dimensional analysis it is easy to show that this occurs when

$$v_c = \frac{k\eta}{\rho a}. \tag{11.10}$$

The quantity k is called Reynolds' number. It is dimensionless and the transition from streamline to turbulent motion occurs when its value is around 1000 to 2000.

For $\dfrac{v\rho a}{\eta} < k$ the flow is streamlined.

For $\dfrac{v\rho a}{\eta} > k$ the flow is turbulent.

Turbulence is essentially a *condition* of instability and the force involved in flow is not necessarily a clear indication of whether turbulence has begun or not. The force is determined primarily by viscous or inertial factors. An interesting simple example of these two conditions is furnished by considering the steady movement of a solid sphere of radius a through a fluid. The resistance to motion can be derived by dimensional arguments. One solution for the resistive force is

$$F = Aa\eta v. \tag{11.11}$$

This is Stokes' law; analysis shows that A has the value 6π.

Another solution is

$$F = Bv^2\rho a^2. \tag{11.12}$$

This does not involve viscosity but kinetic energy, the force being determined by the momentum of the oncoming fluid.

11.2.2 *Viscous flow through a tube*

Consider a cylindrical tube of length l, radius r. Liquid enters one end at pressure p_1 and leaves the other end at pressure p_2. If the flow is uniform the streamlines are all parallel to the tube axis, there is no slip of liquid at the solid boundary and the velocity profile is parabolic. The rate of flow of liquid per second is

$$Q = \frac{\pi r^4}{8\eta}\left(\frac{p_2 - p_1}{l}\right). \tag{11.13}$$

Poiseuille first used this equation to determine the viscosity of blood in horses' arteries.

11.2.3 *Viscosity of suspensions*

If small particles are suspended in a liquid of viscosity η_0, the viscosity η of the suspension is given by:

$$\eta = \eta_0(1 + 2.5\phi) \tag{11.14}$$

where ϕ is the volume fraction of the particles in the liquid. This result, first deduced by Einstein in his doctoral thesis in 1905, assumes that each particle contributes as a separate entity to the increase, i.e. there is no interaction between the liquid flow paths around the particles. It is valid only for low values of ϕ, of the order of 1 per cent.

Einstein applied equation (11.14) to the viscosity of dilute solutions of sucrose where he assumed that the sugar molecule behaved like a particle. By combining published viscosity data with other data on diffusion in sucrose solutions Einstein was able to deduce the size of the sucrose 'particle' and Avogadro's number. A few weeks later he submitted his paper on Brownian motion (see Chapter 4, section 4.7.4). Both papers underline the universality of Avogadro's number in very diverse systems. This was the same year in which, apart from these contributions to classical physics, he produced his special theory of relativity and his paper on light quanta.†

When the concentration of particles is greater than 1 or 2 per cent there is interaction between the liquid flow paths and between the particles themselves. Continuous formation and rupture of links between the particles takes place. The links are due to surface forces, and their rupture involves a considerable dissipation of energy. This is reflected in the increased viscosity. Slurries are an extreme form of suspension and finite stress is often required to initiate flow: they resemble Bingham fluids described in figure 11.5.

11.2.4 *Rotation of a shaft in a bearing: hydrodynamic lubrication*

Consider a shaft of radius a rotating with angular velocity ω in a bearing of radius $a+c$ (figure 11.2(a)). If the shaft carries a light load W and the space between shaft and bearing is filled with oil, they will remain concentric. The velocity gradient across the film is $\omega a/c$. The

† A fascinating account of the state of classical physics and chemistry at this period and of Einstein's contributions is given in Chapter 5 of "'*Subtle is the Lord*'. *The Science and life of Albert Einstein*", A. Pais, Oxford University Press, 1982.

viscous force f per m^2 is this quantity multiplied by η. If the length of the shaft is l, the resultant tangential force that must be exerted at the edge of the shaft to maintain steady rotation is

$$F = f \times 2\pi a l = \frac{\omega a}{c}\,\eta 2\pi a l = \frac{Na}{c}\,\eta a l \times 4\pi^2, \tag{11.15}$$

where $N = \omega/2\pi$ is the number of revolutions per second.

We define the coefficient of friction μ as the ratio of F to W. Then

$$\mu = \frac{Na}{c}\,\eta\,\frac{al}{W} \times 4\pi^2. \tag{11.16}$$

Since the nominal pressure P on the bearing is $W/2al$ we may write

$$\mu = \frac{4\pi^2 N\eta}{P}\left(\frac{a}{2c}\right). \tag{11.17}$$

For a bearing and shaft of fixed dimensions the quantity $a/2c$ is a constant so that μ is linearly proportional to $N\eta/P$ (see figure 11.2(*b*)).

Figure 11.2. (*a*) Rotation of a shaft of radius a in a bearing of radius $a+c$ (Petroff). If the liquid has viscosity η and the angular velocity is ω the viscous force F that has to be overcome is proportional to $\eta\omega$. (*b*) If the normal load is W the coefficient of friction is proportional to $\eta\omega/F$ or to $N\eta/P$, where N is the number of revolutions per second and P the normal pressure. At high pressures or low speeds the hydrodynamic film is penetrated and the friction rises. (*c*) In practice, a convergent oil wedge must be formed so that sufficient pressure can be developed in the liquid to support the applied load; as a result the shaft is not quite concentric with the bearing (Reynolds).

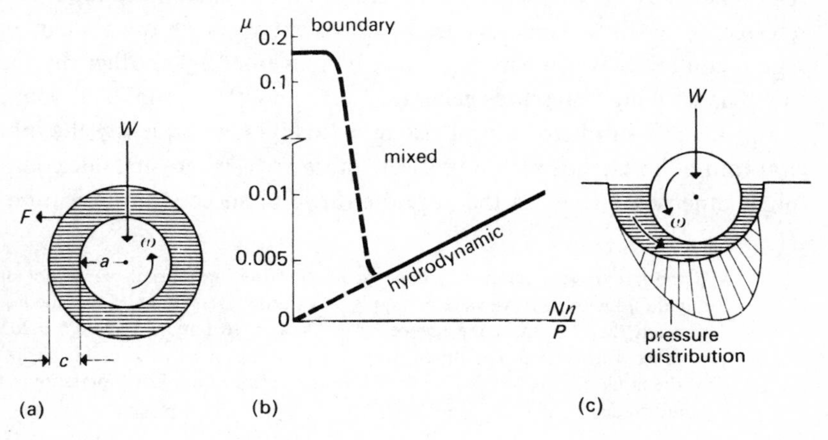

This relation was first derived by A. A. Petroff in 1883 and describes adequately the behaviour of a lightly loaded bearing. However, it is evident that for loads which are not negligible, the journal cannot run in the central position. If the oil film is to support an appreciable load the journal must rotate eccentrically relative to the bearing, so that the lubricant is squeezed through the converging gap between the surfaces. This convergence in a viscous liquid produces a hydrodynamic pressure in the oil wedge as O. Reynolds showed in 1886 and under suitable conditions it is sufficient to support the applied load.†

For a journal in a 'half' bearing the pressure distribution is as shown in figure 11.2(*c*) and similar results are obtained for a 'full' bearing. It is seen that the (asymmetric) pressure distribution gives a resultant upthrust which would not occur if the journal rotated centrally in the bearing. Nevertheless the viscous resistance is not very different from that given by the simple Petroff theory and, for a well-designed bearing, friction coefficients of the order of 0.001 are obtained.

A far more important difference between the Petroff and Reynolds treatment concerns the distance of nearest approach of journal to bearing. As the load increases the oil convergence must increase and this leads to a decrease in the separation. At some critical stage the separation necessary for ideal hydrodynamic lubrication may be less than the height of the surface roughnesses. The hydrodynamic film is penetrated and contact occurs between the metals themselves or, at best, between metals covered by a thin adsorbed lubricant layer only one or two molecules thick. This is referred to as boundary lubrication and involves higher friction ($\mu = 0.1$) and possible wear. By contrast hydrodynamic lubrication gives a very low coefficient of friction ($\mu \sim 0.001$) and, in principle, no wear at all. The region between the classical hydrodynamic and boundary regime is often referred to as the mixed regime. More recent work shows that in this region contact between asperities may be prevented by another mechanism. The very high pressures generated at the points of imminent contact can produce a prodigious local rise in viscosity. Consequently the lubricant cannot be extruded from precisely those areas where metallic contact might otherwise occur. In this way the range of successful lubrication is

† If the liquid were completely non-viscous the rotating journal would not drag any liquid with it. If we were to supply an external pressure and force the liquid through the gap, the convergence, as we saw in section 11.1, would actually produce a pressure *drop* in the convergent gap. It is primarily the viscous property of the liquid that is responsible for the development of a positive pressure in the liquid wedge.

extended into a region of far more severe conditions. This type of lubrication, where high elastic stresses have a favourable effect on the viscosity of the lubricant, is known as elasto-hydrodynamic lubrication. When the elasto-hydrodynamic film itself is finally penetrated or destroyed we are back again at classical boundary lubrication where only molecular adsorbed layers remain.

11.2.5 *Simple ideas on the viscosity of liquids*

The viscosity of gases is explained in terms of molecular momentum-transfer from one plane to a neighbouring plane. The mechanical analogy is to consider two trains travelling parallel to one another at slightly different speeds. People are continuously jumping from one train to the other, so that those leaving the slower train, slow up the faster; whilst those jumping from the faster train speed up the slower. There is no net change in the numbers of people in any one train but there is clearly a momentum transfer tending to equalize the velocities of the two trains. Work must be done on the trains to maintain their difference in velocity. It is this process on a molecular scale which is responsible for the viscosity of gases.

In liquids the momentum transfer process also takes place but it is a trivial part of the viscous resistance. The train model may still be used but now jumping passengers find themselves caught by their coat-tails. They must grab hold of the other train and pull very hard before they can detach themselves from their own train. This will have a marked effect in tending to equalize the train velocities. In molecular terms one may say that in a liquid the molecules are so close together that considerable energy must be expended in dragging one molecular layer over its neighbour. No rigorous theory exists but a simple molecular model may be given. We first give it descriptively and in the next section deal with it analytically.

Consider an instantaneous picture of the liquid. There is short-range order and we may represent a small part of two neighbouring molecular planes by an array similar to that of a solid (figure 11.3). To drag molecule A2 from its position to the neighbouring equilibrium site A2′ involves overcoming the attraction of neighbours. There is thus a potential hill to be climbed, and energy is needed for this. (Note that in falling from the summit X to site A2′ the energy that is restored appears as heat.) Occasionally molecules will have enough thermal energy to do this but there is just as much chance of the molecule jumping to site A1 as to A2′. There is thus no net flow in any direction. If now we apply a shear stress to the right of the molecules in plane A mechanical work is done on the molecules. The thermal energy necessary to move the molecule to the right is reduced whilst that required to move it to the left is increased (figure 11.3(*b*)). Thus thermally activated flow to the right is favoured. Another way of looking at the process is to say that before molecule A2 can move to site A2′ it must make available a vacancy in the liquid to receive it. The energy to do this depends on the amount of opening up of the liquid that is necessary and on the latent heat of vaporization. Indeed the energy needed to make a vacancy available is roughly equal to $\frac{1}{3}-\frac{1}{2}$ of the latent heat of vaporization of the vacancy.

11.2.6 *The Eyring theory of liquid viscosity*

We consider a small cluster of molecules as shown in figure 11.3. In order to transpose molecule A2 to position A2′ against the attractive forces of neighbours we have to surmount a potential energy barrier ε. The rate at which this is possible as a result of thermal fluctuations is

$$C\exp\left(-\frac{\varepsilon}{kT}\right), \tag{11.18}$$

where C is a frequency term. Eyring suggests that this is the vibrational

Figure 11.3. The Eyring molecular model of liquid viscosity (*a*) to transpose a molecule from position A2 to A2′ intermolecular forces must be overcome. This involves a potential energy barrier ε, (*b*) if a shear stress is applied to the right the potential energy curve is distorted, transitions to the right are facilitated, to the left hindered, (*c*) a shear stress f acting on a molecule occupying a nominal area α exerts a shear force $f\alpha$.

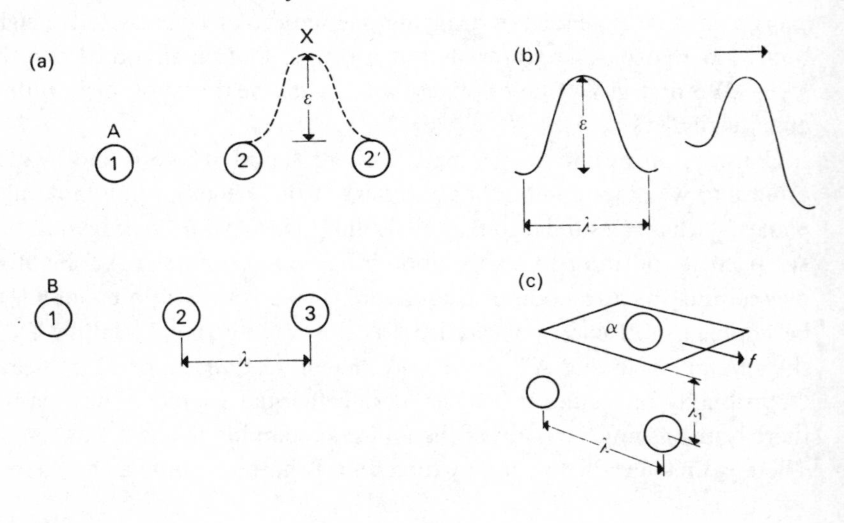

frequency of the molecule and is given approximately by kT/h, where h is Planck's constant. Hence the frequency with which a molecule can jump to a neighbouring site as a result of thermal activation is

$$\mathcal{N}_0 = \frac{kT}{h} \exp \frac{-\varepsilon}{kT}. \tag{11.19}$$

When no stress is applied to the liquid there is an equal rate of jumping to left and right so that no net flow occurs. Suppose now a shear stress f is applied (see figure 11.3(c)). If the average area occupied by a molecule is α the force on the molecule is $f\alpha$. The only mechanical work we can perform is expended in carrying the molecule to the top of the potential barrier (on the other side of the barrier peak the molecule is assumed to give up all its energy as heat). The work done is

$$\frac{f\alpha\lambda}{2}, \tag{11.20}$$

where λ is the distance between neighbouring molecules in the plane.

It follows that the thermal energy required for a jump to the right is reduced from ε to $\varepsilon - \frac{1}{2}f\alpha\lambda$, i.e. the rate of jumping is increased. Similarly the thermal energy required for a jump to the left is increased to $\varepsilon + \frac{1}{2}f\alpha\lambda$ so that the jumping rate is reduced. The net rate of jumping to the right is now,

$$\mathcal{N} = \frac{kT}{h}\left[\exp\left(-\frac{\varepsilon - \frac{1}{2}f\alpha\lambda}{kT}\right) - \exp\left(-\frac{\varepsilon + \frac{1}{2}f\alpha\lambda}{kT}\right)\right], \tag{11.21}$$

$$= \frac{kT}{h}\exp\left(-\frac{\varepsilon}{kT}\right)\left[\exp\left(\frac{f\alpha\lambda}{2kT}\right) - \exp\left(-\frac{f\alpha\lambda}{2kT}\right)\right]. \tag{11.22}$$

For the exponentials in the square bracket, if the shear stresses are not too high, i.e. if $f\alpha\lambda \ll kT$, we may take the first terms of the expansion. This gives

$$\mathcal{N} = \frac{kT}{h}\exp\left(-\frac{\varepsilon}{kT}\right)\left(2\frac{f\alpha\lambda}{2kT}\right) = \frac{f\alpha\lambda}{h}\exp\left(-\frac{\varepsilon}{kT}\right). \tag{11.23}$$

We must now link $\mathcal{N}$ with the macroscopic flow process. The velocity gradient across the two molecular layers separated by a distance λ_1 is

$$\text{velocity gradient} = \frac{\text{velocity difference}}{\lambda_1}$$

$$= \frac{(\text{distance per jump}) \times (\text{number of jumps per second})}{\lambda_1}$$

$$= \frac{\lambda\mathcal{N}}{\lambda_1}. \tag{11.24}$$

But by definition the viscosity η is given by

$$\eta = \frac{\text{force per unit area}}{\text{velocity gradient}} = f \times \frac{\lambda_1}{\lambda \mathcal{N}} \tag{11.25}$$

Inserting the value of $\mathcal{N}$ from equation (11.23) we obtain

$$\eta = \frac{h\lambda_1}{\alpha\lambda^2} \exp\left(\frac{\varepsilon}{kT}\right). \tag{11.26}$$

Putting $\lambda_1 \approx \lambda$ and treating $\alpha\lambda_1$ as the effective volume occupied by a molecule (v_m), equation (11.26) becomes

$$\eta = \frac{h}{v_m} \exp\left(\frac{\varepsilon}{kT}\right)$$

or

$$\eta = \frac{hN_A}{V_m} \exp\left(\frac{E}{RT}\right), \tag{11.27}$$

where N_A is Avogadro's number, V_m the volume per mole of the liquid and E the molar activation energy for surmounting the energy barrier. This relation is in full accord with the very marked decrease in viscosity observed with increasing temperature.

The barrier energy E may also be regarded as the energy to create a hole in the liquid big enough to receive a molecule. This ought, therefore, to be comparable with the latent heat of vaporization (L); however, because there is already some free volume in the liquid, the work to open up a molecular hole is less than this. For a very large range of liquids

$$E \approx (0.3 \text{ to } 0.4)L.$$

Consequently for a very wide range of liquids we may write

$$\eta = \frac{hN_A}{V_m} \exp\left(\frac{0.4L}{RT}\right). \tag{11.28}$$

However, for liquid metals it is found that E is very much less than $0.4L$, presumably because the metal ions are much smaller than the metal atoms.

11.2.7 *Effect of external pressure on viscosity*

To examine the effect of pressure on liquid viscosity it is simplest to regard the energy barrier E as the energy required to open up a hole for the transposed molecule. In the absence of applied pressure we may write equation (11.27) in the form

$$\eta_0 = D \exp\left(\frac{E}{RT}\right). \tag{11.29}$$

If now an external pressure P is applied the work to form a hole is increased by PV_h where V_h is the volume of the hole. The thermal energy for activated flow is then increased from E to $E + PV_h$. The viscosity then becomes

$$\eta = D \exp\left(\frac{E + PV_h}{RT}\right)$$

$$= \eta_0 \exp\left(\frac{PV_h}{RT}\right). \tag{11.30}$$

This type of relation is found to hold over a fairly wide range of pressures. If $\ln \eta$ is plotted against P a straight line of slope V_h/RT is obtained. It turns out that V_h is about 15 per cent V_m for simple liquids and about 5 per cent V_m for liquid metals. Attempts to explain away these discrepancies are not convincing. One must accept this as part of the inadequacy of the model. In broad outline it is surprisingly successful.

The effect of high pressures in increasing viscosity can be very significant in lubrication. It is probable that many sliding mechanisms operate successfully because local high pressures, which would normally be expected to squeeze out the lubricant and cause seizure, so increase the viscosity of the oil that it remains trapped between the surfaces. For example with heavily loaded gear teeth where elastic deformation in the contact zone may produce pressures of the order of 10 000 atmospheres or 100 kgf mm^{-2}, the viscosity of a typical lubricant may be increased a million-fold. In this way nature is much kinder to the engineer than he might have supposed. This type of elasto-hydrodynamic lubrication, (mentioned previously on p. 303) is an area of great importance in current lubrication practice.

11.2.8 *Effect of pressure and high shear rates on viscosity*

We may take these conclusions one stage further. We first observe that in equation (11.24) the velocity gradient or shear rate $\dot{\gamma}$ is $\dot{\gamma} = \lambda\mathcal{N}/\lambda_1 \approx \mathcal{N}$ if $\lambda_1 \approx \lambda$.

We now use the symbols often used in the literature. In equation (11.22) for algebraic convenience we take $a\lambda/2$ instead of $a\lambda$ to be the effective volume and for the molar situation write it as Ω. This is referred to as the stress activated volume: then equation (11.22) becomes

$$\mathcal{N}=\frac{kT}{h}\exp-\frac{E}{RT}\left(\exp\frac{\tau\Omega}{RT}-\exp-\frac{\tau\Omega}{RT}\right), \tag{11.31}$$

the term in the parentheses being a sinh function. Now allowing for external pressure P we replace E by $E+P\phi$ where ϕ is the effective molar volume of the hole. This is referred to as the pressure activation volume. We then have

$$\mathcal{N}=\dot{\gamma}=\frac{kT}{h}\exp\left(-\frac{E+P\phi}{RT}\right)\sinh\frac{\tau\Omega}{RT}. \tag{11.32}$$

This equation was first derived by Eyring† in 1936. A typical curve of τ versus $\dot{\gamma}$ is shown in figure 11.4. At low values of τ and $\dot{\gamma}$ we see that τ is proportional to $\dot{\gamma}$, the slope being the viscosity η. This is the Newtonian regime. The slope gradually decreases so that the effective viscosity is less than it would be if the behaviour was Newtonian over the whole τ–$\dot{\gamma}$

Figure 11.4. Schematic representation of equation (11.32) showing variation of shear stress τ with shear rate $\dot{\gamma}$. At high pressures and high rates of shear τ is almost independent of $\dot{\gamma}$. In this plateau regime $\tau \approx \tau_0 + \alpha p$. The broken line shows typical experimental results: the fall in τ is attributed to appreciable viscous heating at high rates of shear. Newtonian flow occurs when τ is less than RT/Ω.

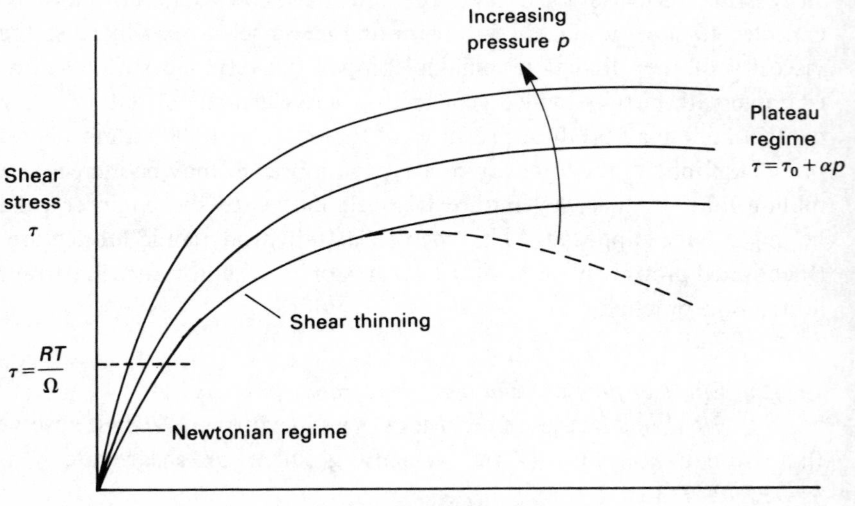

† H. Eyring, *J. Chem. Phys.* **4**, 283 (1936).

range. This phenomenon is sometimes referred to as shear thinning and sets in when the argument of the sinh function is about unity, i.e. when $\tau = RT/\Omega$: it has a value of order 10^6 N m^{-2} for many lubricants.† Finally, τ reaches an almost constant value independent of $\dot{\gamma}$. In this regime the behaviour resembles the shearing of a plastic solid.

We may estimate the value of τ at this stage by noting that in equation (11.31) the term $\exp(-\tau\Omega/RT)$ tends to zero as τ increases and the behaviour is dominated by the term $\exp(\tau\Omega/RT)$. Equation (11.32) then becomes

$$\dot{\gamma} = \frac{kT}{h}\exp[-(E+\rho\phi)+\tau\Omega]/RT. \quad (11.33)$$

Taking logarithms and assuming that the temperature is constant

$$\tau\Omega - E - \rho\phi = \ln\dot{\gamma} + \text{const.}$$

Hence

$$\tau = \frac{1}{\Omega}[E+\rho\phi+\ln\dot{\gamma}+\text{const}]. \quad (11.34)$$

Since $\ln\dot{\gamma}$ varies very slowly with $\dot{\gamma}$ we may treat it as a constant and group it together with the other constants such as E and Ω. We call the constant $\Omega\tau_0$. Then

$$\tau = \tau_0 + p\phi/\Omega = \tau_0 + \alpha P, \quad (11.35)$$

where α for a wide range of lubricants has a value between 0.08 and 0.1. Furthermore, if equation (11.33) is studied in greater detail it is found that temperature has little effect on the final result of equation (11.35). This relation thus resembles the shear strength of polymers and suggests that at this stage the liquid has been solidified by pressure. Direct measurements do in fact indicate that such a phase change can occur at pressures of the order of 10^9 Pa (10^4 atmospheres). There is some evidence that the phase change is relatively sluggish. Thus the use of equation (11.35) to explain the behaviour of lubricated systems operating at very high contact pressures, such as ball bearings, must be applied with some caution since the lubricant film is in the contact region for a very short time.

The Eyring model was originally developed for simple spherical molecules and we might not expect it to apply to larger molecules which are relatively complex. The fact that it so well describes the behaviour of long straggly molecules (50 or more carbons with sidegroups) over such a wide range of pressures and shear-rates is truly remarkable. What is even more

† C. R. Evans and K. L. Johnson, *Proc. Instn. Mech. Eng.* **2000C,** 313–24 (1986). W. Hirst and J. W. Richmond, *ibid,* **202C,** 129–44 (1988).

surprising is that the values of the activation volumes deduced (Ω and ϕ) are physically reasonable. Whatever the criticisms levelled at some of its theoretical assumptions, the Eyring equation provides a very useful basis for describing the real behaviour of lubricants.

11.2.9 *Summary*

Although the model of liquid viscosity which we have discussed above is greatly simplified it brings out three important points. Liquid viscosity arises from intermolecular forces and the ability of thermal fluctuations aided by an applied shear stress to produce flow in the direction of shear. Consequently the viscosity decreases if the temperature is increased. Secondly the viscosity increases if the pressure on the liquid is increased. This is because the molecules are pushed closer together, or, in terms of the alternative way of describing the molecular jumps, more work must be done to open up a vacant site.

A third conclusion is that viscosity, being determined by intermolecular forces, is closely related to surface tension γ and both are related to the latent heat of vaporization L. For many simple organic liquids there is a fair correlation between $\ln \eta$ (see equation 11.28) and γ/Z, where Z is a factor allowing for the size of the molecule.

It may be mentioned that these models are most suitable for unassociated liquids with van der Waals bonding between the molecules. Highly polar molecules do not fit in so well.

The basic differences between the mechanisms responsible for viscosity in gases and in liquids are clearly brought out in Table 11.1.

11.3 **Rigidity of liquids**

The molecules in a liquid jump about with a frequency of 10^{10}–$10^{12}\ \mathrm{s}^{-1}$. If the rate of shear is sufficiently great there may not be time for the molecules to advance to a neighbouring site. In this case the liquid will not show viscous flow, but will show a finite elastic rigidity. For simple liquids the rates of shear for this to occur are enormous but for

Table 11.1. *Viscosity of gases and liquids*

Fluid	Effect of temperature T	Effect of pressure P
Gases	η increases as $T^{\frac{1}{2}}$	None (but see Fig. 3.12)
Liquids	η decreases as $\ln \eta = A + \frac{B}{T}$	η increases as $\ln \eta = A + kP$

liquids with large bulky molecules it may occur at moderate rates of shear. There is a similar behaviour in polymers such as polyethylene, nylon and perspex. The flow of polymeric chains which occurs at slow rates of loading resembles the viscous flow of liquids; this may be impossible at high rates of strain. The material will then deform elastically. A very striking example of this type of behaviour is silicone putty. This material will bounce with a very high resilience, since there is not enough time for the molecules to flow. Again if the material is in the form of a rod and is pulled rapidly it will first stretch and then snap in a brittle fashion. On the other hand at very low rates of deformation, it will flow in a viscous manner. For a more general discussion see Chapter 13, p. 346.

11.4 Non-Newtonian flow

Many liquids encountered in daily life are non-Newtonian. This is particularly true of liquids with large bulky molecules or those which have some additional structure, for example slurries. A finite stress is often needed to break down the structure and initiate flow (see figure 11.5(*a*)). Such materials are known as Bingham fluids. In many cases, once flow starts the molecules tend to orient along the direction of shear. The interaction of the molecules is reduced and the viscosity diminishes as the rate of shear increases (see figure 11.5(*b*)). This is known as thixotropy and often occurs with paints. Brushing reduces the viscosity and the paint flows easily. On the other hand, since the effect is only slowly reversible, the paint will not drip easily when left undisturbed.

Figure 11.5. Schematic diagram of stress against shear rate for (*a*) a Bingham fluid, (*b*) a Bingham fluid showing thixotropy.

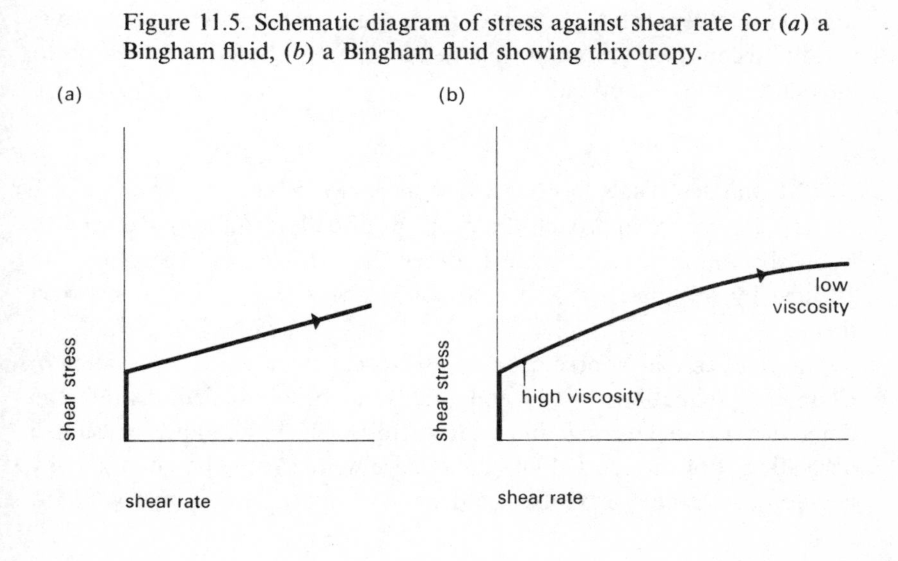

12

The colloidal state of matter

In this chapter we describe briefly some of the main features of the colloidal state. The colloidal state is very widespread in biological systems and in many practical situations. Basically a colloid consists of two distinct phases, a continuous phase (the dispersion medium) and a particulate phase, where the particles generally have dimensions ranging between 20 and 2000 Å (2 and 200 nm). The two phases can be liquid-in-liquid (milk), solid-in-liquid (paint), liquid-in-gas (aerosol) and other combinations. The basic problem with colloidal systems is their stability. Clearly the particles must not be too large otherwise gravity will produce ready sedimentation. However, if the density of the two phases is similar this tendency will be reduced. The other important factor is the attractive force between the particles: if this is too large the particles will cling together and separate as a cluster from the dispersion. To prevent this, various techniques are employed to reduce the attractive forces or to introduce repulsive forces.

12.1 van der Waals forces between macroscopic bodies

Before we approach the problem of colloid stability we need to know the van der Waals forces between the particles and how these are affected by the presence of a continuous phase (e.g. a liquid) between them.

The van der Waals force between macroscopic materials was quoted in Chapter 1 (equations (1.21) and (1.22)) in order to demonstrate the difference between normal and retarded forces. Here we indicate a simple derivation. For unretarded interactions we write for the potential energy between two atoms or molecules distance r apart (see equations (1.18) and (1.19)),

$$u = -C/r^6. \qquad (12.1)$$

This relation applies to dipole–dipole interactions (the Keesom energy), to dipole–induced dipole interactions (the Debye energy) as well as to dispersion forces. We assume that the interaction of a single molecule with a bulk solid made up of similar molecules is the sum of all its interactions with all the molecules in the body, i.e. we assume simple additivity and ignore the problem of how the induced fields are affected by intervening molecules.

12.1.1 *Molecule–surface potential interaction*

Consider a single molecule at a distance h above a flat semi-infinite solid (figure 12.1(a)). We construct a circular ring (parallel to the surface) of radius a, width da and thickness dx. The number dn of molecules in the ring will be $2\pi\rho a\,\mathrm{d}a\,\mathrm{d}x$ where ρ is the number density of molecules in the solid. The potential energy du of the molecule and the ring is:

$$\mathrm{d}u = \frac{-C}{r^6} = \frac{-C\,2\pi\rho a\,\mathrm{d}a\,\mathrm{d}x}{(x^2+a^2)^3}. \qquad (12.2)$$

Keeping x constant we integrate for a ranging from 0 to infinity

$$U = -C\,2\pi\rho\,\mathrm{d}x \int \frac{a\,\mathrm{d}a}{(x^2+a^2)^3} = \frac{2\pi\rho\,\mathrm{d}x}{4}\,\frac{1}{x^4}. \qquad (12.3)$$

Figure 12.1. (a) A single molecule at a distance h from a semi-infinite solid. (b) Two parallel surfaces distance H apart: calculation of attractive force per unit area.

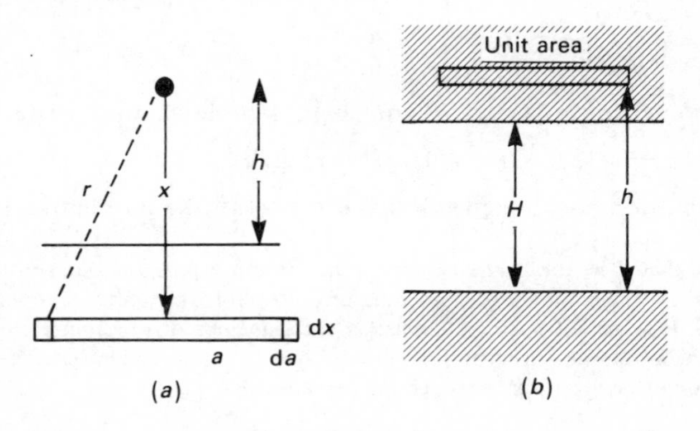

We now integrate over x where x ranges from h to ∞ and obtain

$$u=\frac{-C\pi\rho}{6h^3}. \tag{12.4}$$

12.1.2 *Two parallel surfaces (semi-infinite solids)*

A molecule at a distance h from a flat surface has potential energy given by equation (12.4). The potential energy of two semi-infinite parallel surfaces is infinite; we must restrict our calculations to unit areas. We construct a flat disc of unit area and thickness dh parallel to the surface (figure 12.1(b)). It contains $\rho\,\mathrm{d}h$ molecules. The potential energy is

$$\Delta u=\frac{-C\pi\rho}{6h^3}\rho\,\mathrm{d}h. \tag{12.5}$$

Integrating from $h=H$ to $h=\infty$ we obtain

$$U=\frac{-C\pi\rho^2}{12H^2}. \tag{12.6}$$

The force between the solids is $F=-\partial U/\partial H$. Thus

$$F/\text{unit area}=\frac{C\pi\rho^2}{6H^3}=\frac{C\pi^2\rho^2}{6\pi H^3}=\frac{A}{6\pi H^3} \tag{12.7}$$†

where A is called the Hamaker constant. If the bodies are of different materials the Hamaker constant A_{12} is $C_{12}\pi^2\rho_1\rho_2$.

12.1.3 *Sphere and flat surface*

Consider a sphere of radius R: its nearest distance from the flat semi-infinite solid is H (figure 12.2). At a height h from the solid surface we construct a disk of radius a, thickness dh within the sphere: it contains $\rho\pi a^2\,\mathrm{d}h$ molecules each of which has potential energy given by equation (12.4). Thus for the disk

$$\Delta u=\frac{-C\pi\rho}{6h^3}\rho\pi a^2\,\mathrm{d}h. \tag{12.8}$$

From the geometric properties of chords in a circle we may write

$$a^2=(h-H)[2R-(h-H)]\approx(h-H)2R. \tag{12.9}$$

The approximation is valid since the major part of the interaction occurs

† Note that if we adopt the Lennard-Jones repulsive potential, i.e. constant/r^{12}, the above integration procedure gives a repulsive force per unit area proportional to r^9. It is a moot point as to whether such integration is meaningful since the repulsion vanishes for separations greater than two atomic diameters. The repulsive power law is merely an algebraic convenience.

in the region where $h-x$ is small compared with R. Then

$$\Delta u = \frac{-C\pi^2\rho^2 2R}{6}\left(\frac{h-H}{h^3}\right)dh$$

$$u = \frac{-AR}{3}\int_H^{\infty}\left(\frac{1}{h^2}-\frac{H}{h^3}\right)\mathrm{d}h$$

$$= -\frac{AR}{6H} \tag{12.10}$$

and the force

$$= \frac{AR}{6H^2}. \tag{12.11}$$

12.1.4 *Two spheres*

It turns out that for the system we are considering we may apply the following† geometric short-cut to derive the potential between two spheres of radius R. We subtract the curvature from one sphere converting it into a plane surface and add the curvature onto the other sphere converting it into a sphere of radius $R/2$. We then have a sphere of radius $R/2$ and flat configuration for which

$$u = \frac{-AR}{12H} \quad \text{and} \quad f = \frac{AR}{12H^2}. \tag{12.12}$$

Figure 12.2. Sphere of radius R at distance H from a semi-infinite solid.

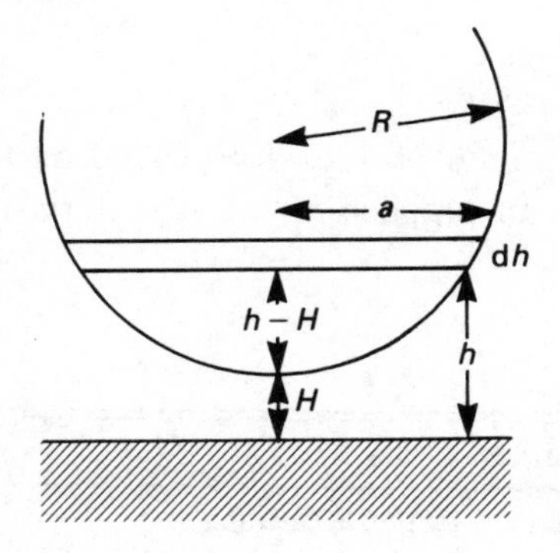

† This geometric approach cannot be applied if the index n in the potential law between atoms $-C/r^n$ is less than 3. It is consequently inapplicable to gravity where $n=1$. (See J. N. Israelachvili, *Intermolecular and Surface Forces*, Academic Press, 1985.)

This is the attractive force between two colloidal particles of radius R in the absence of any intervening medium. It may be checked by more complete calculations along the lines given above for other geometries.

Similar derivations may be made for retarded forces. We note that retardation only applies to dispersion forces. Dispersion forces are dominant in most colloidal systems unless the particles include polar components. Typical results are given in Table 12.1 for the force between two spheres of radius R distance H apart and for two semi-infinite solids.

The value of C for unretarded forces is given by equation (1.19) if we remove x^6 from the denominator. We can now make an estimate of the Hamaker constant A. We have

$$A = -\tfrac{3}{4}h\nu\left(\frac{\alpha}{4\pi\varepsilon_0}\right)^2 \pi^2\rho^2. \tag{12.13}$$

In a later chapter we shall show that the polarizability α of an atom (or a molecule of roughly spherical shape) of radius r is given by

$$\alpha = 4\pi\varepsilon_0 r^3. \tag{12.14}$$

Hence

$$A = -\tfrac{3}{4}h\nu r^6\pi^2\rho^2 \tag{12.15}$$

for a cubic array of atoms the number density $\rho = (1/2r)^3$.

Hence

$$A = -\tfrac{3}{4}h\nu\pi^2/64 \approx h\nu/8. \tag{12.16}$$

The quantity $h\nu$ is comparable with the first ionization potential and is of order 5–10 eV, say 7 eV $= 7 \times 1.6 \times 10^{-19}$ J.

Hence

$$A \approx 1.4 \times 10^{-19}\ \text{J}. \tag{12.17}$$

We see that for this rather simple model A does not depend on the size or nature of the atom (molecule). In practice for most materials in the

Table 12.1

Force	Potential	Dispersion forces: Force between spheres	Force/area flat on flat	Hamaker constant
Unretarded	$-C/x^6$	$AR/12H^2$	$A/6\pi H^3$	$A = C\pi^2\rho^2$
Retarded	$-C_r/x^7$	$\pi BR/3H^3$	B/H^4	$B = 0.1C_r\pi\rho^2$

For two different materials $A_{12} = C_{12}\pi^2\rho_1\rho_2$

condensed phase A lies between 0.4 and 4×10^{-19} J. Similar calculations give an average value of $B\approx10^{-28}$ J m^{-1} for retarded forces.

These calculations are for surfaces separated by air or vacuum where the transition to retarded forces begins at separations of order 10–20 nm: but in the presence of an intermediate medium it begins for smaller separations because the velocity of the electric field generated by the electron fluctuations is less. Consequently in colloidal systems which are near coagulation the retarded force equations are more usually applicable.

12.1.5 *The influence of the intervening continuous phase*

There is no simple way of modifying the fluctuating dipole model for pairwise interactions described in Chapter 1 to cope with the presence of an intervening medium between macroscopic bodies. A different approach is required. This treats the interactions between bodies in terms of the dielectric properties of the materials involved. This is not so surprising in view of the fact that (as we shall see in Chapter 13) the dielectric constant or relative permittivity of a material can be expressed in terms of its polarizability. If body 1 and body 2 are separated by a medium 3 for which the dielectric constants are respectively ε_1, ε_2, and ε_3 the theory shows that the Hamaker constant A is proportional to

$$\left(\frac{\varepsilon_1-\varepsilon_3}{\varepsilon_1+\varepsilon_3}\right)\times\left(\frac{\varepsilon_2-\varepsilon_3}{\varepsilon_2+\varepsilon_3}\right). \tag{12.18}$$

The full relation includes additional terms which vary in a similar way which we may ignore. There are a number of interesting and unexpected results. If $\varepsilon_1>\varepsilon_3>\varepsilon_2$ the value of A is negative, that is, there is a repulsive force. It is difficult to explain this effect in terms of pairwise interactions. One might say that ε_3 likes itself more than its neighbours.

In colloidal systems bodies 1 and 2 are identical and we have

$$A=c\left(\frac{\varepsilon_1-\varepsilon_3}{\varepsilon_1+\varepsilon_3}\right)^2 \tag{12.19}$$

so that A is positive whether ε_3 is greater or less than ε_1; that is, the van der Waals force is always attractive. However, the intervening medium always reduces the forces as compared with the force observed if the medium is air or vacuum ($\varepsilon_3=1$). Another result from equation (12.19) is that for two similar media interacting across another medium the Hamaker constant is unchanged if the media are interchanged; that is, if ε_1 and ε_3 are interchanged. This leads to the surprising result that if no other forces operate, a liquid film in air will always tend to thin. It is as though the two surfaces of the liquid film are attracted to one another and squeeze out the liquid between them.

12.2 The principles of stabilization

We consider here as a model system the solid-in-liquid type of colloid. There are two broad classes, lyophilic and lyophobic. The lyophilic (liquid-loving) colloids are those in which there is a strong affinity between the liquid and the particle. The liquid is strongly adsorbed on to the surface of the particle so that the interface between the liquid-covered particle and the liquid is very similar to the interface between the liquid and itself. This is intrinsically a relatively stable system particularly if some suitable ions are also available for adsorption on the surface of the particles (see below). In general lyophilic colloids are more stable the smaller the particle size.

In lyophobic (liquid-hating) colloids the liquid does not show affinity for the solid particle. If the forces between the particles are due to van der Waals interactions, it is easy to see that there will be a strong tendency for them to stick together whenever they come into contact, i.e. the colloid will be unstable and coagulation will occur. The reason is as follows. As we showed above, the potential energy of two particles of radius R distance H apart is $u = -AR/12H$. Then assuming that this relation is valid down to atomic contact we may calculate the value of u putting $H \approx 3 \times 10^{-10}$ m. For typical particles in an aqueous medium A usually has a value of the order of 10^{-20} J. Thus if R is, say 100 nm, u has the value 3×10^{-19} J. But the thermal energy of the particles is of order kT. At room temperature (300 K) this has the value 4×10^{-21} J. This is far smaller than the van der Waals potential energy u for particles in atomic contact. Thus once the particles touch they will never be able to separate: the thermal energy will be insufficient to overcome the attractive energy. (The situation resembles the behaviour of a gas below its critical temperature.) Consequently the particles will coagulate unless some method is devised for overcoming the attraction by introducing repulsive forces between the particles. This may be achieved in two ways:

(i) by adding to the liquid phase ions which produce electrostatic repulsion between the particles,

(ii) by adsorbing long chain polymers onto the surface of the particles: they produce steric or entropic repulsion. Some polymers, e.g. polyelectrolytes, combine both electrical and steric effects.

12.3 Stabilization by the diffuse electrically charged double layer

In this type of stabilization, ions are added to the liquid phase: they adsorb on to the surface of the particles and produce a diffuse double layer of charge between the particles. The general principle is as follows: Suppose the particle itself carries no net surface charge, but positive ions are preferentially adsorbed on to its surface. In the complete absence of thermal motion an equal number of negative ions (counterions) would adsorb on to the positive charge and neutralize it. Because of thermal agitation such a compact double layer does not form. Instead a diffuse double layer is formed as shown in figure 12.3. There is a fairly rapid change in concentration of negative and positive ions as we move away from the surface (see figure 12.4(*a*)). As a result the potential falls off as shown in figure 12.4(*b*), according to a relation of the form $\psi = \psi_0 \exp(-\kappa x)$. At a distance $x = 1/\kappa$ the potential has fallen to $1/e$ of its value at the surface and beyond this the change in potential is small. Thus $1/\kappa$ may be considered as the thickness of the double layer and is known as the Debye length. The higher the concentration of ions the more the 'centre of gravity' of the charge moves towards the surfaces and the smaller the thickness of the double layer. For example, for a monovalent electrolyte the thickness is about 10 Å (1 nm) for a 10^{-1} M solution and about 100 Å (10 nm) for a 10^{-3} M solution, since it turns out that κ varies as (concentration)$^{-\frac{1}{2}}$.

Figure 12.3. Distribution of positive and negative ions in a liquid adjacent to a solid structure: the electrical double layer.

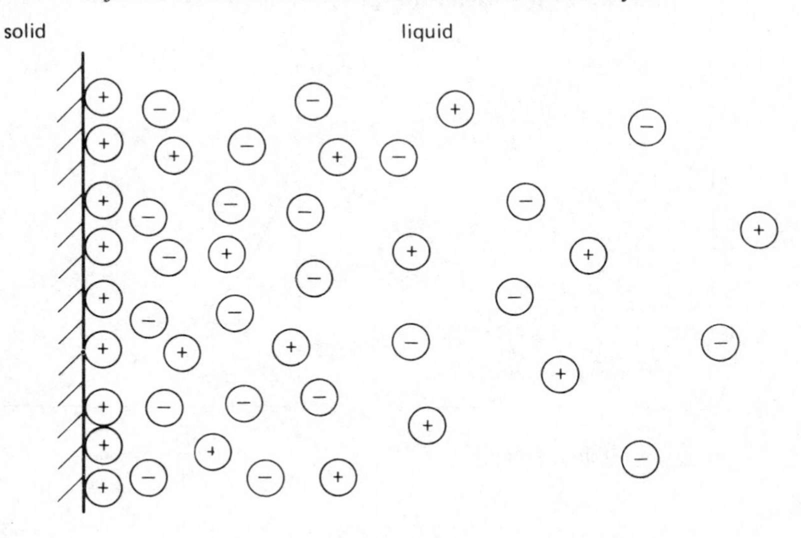

We consider the simplest case of two parallel surfaces of the same material at a distance H apart with electrolyte between them. The potential within the double layer consists of two symmetrical curves resembling figure 12.4(b) and the net effect is roughly additive (figure 12.5). It is seen that as H decreases the overlap of the double layer causes the electrical potential of the system to rise. This increase in potential energy implies a repulsive force. We shall not give the derivation but quote the repulsive energy V per unit area of surface for $H > 1/\kappa$:

$$V = \frac{C}{n^{\frac{1}{2}}}\exp(-2\kappa H) \tag{12.20}$$

where C contains details concerning the nature of the electrolyte, the potential ψ_0 and the temperature: n is the number of anions or cations per unit volume and is therefore proportional to the ionic concentration. But the main factor determining the behaviour is the exponential term. The potential increases exponentially with decreasing separation. The repulsive potential is also larger for small κ, i.e. for smaller concentrations of electrolyte.

A similar type of repulsive potential applies to spherical surfaces. Consequently as two colloidal particles approach one another the double layer repulsion opposes the attraction arising from van der Waals interactions.

Figure 12.4. (a) Concentration of ions in the double layer as a function of distance from the solid surface. (b) Potential ψ within the double layer as a function of distance x from the solid surface. At $x = 1/\kappa$ (the Debye length) the potential has fallen to $1/e$ of its value at the surface.

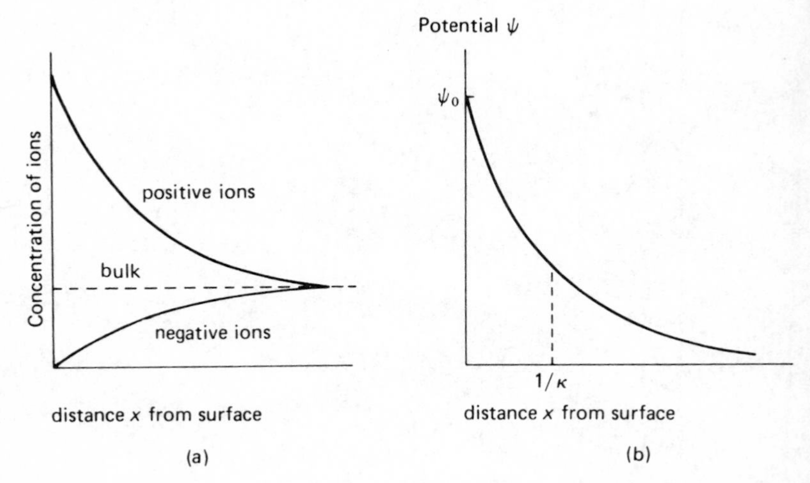

tions. The overall effect is most easily described in terms of potential rather than forces, the repulsive potential being positive and the attractive potential negative. Figure 12.6 shows typical potential energies for the van der Waals attraction and the double layer repulsion. In this figure the resultant potential shows a primary deep minimum at M_1 corresponding approximately to contact between the particles and a secondary shallower minimum at M_2. If the particles are initially far apart and approach one another to a separation M_2, and if their thermal energy (kT) is small compared with the depth of this minimum, the particles will not be able to escape from one another. The particles will be loosely held together at fairly large separations. This type of instability is known as flocculation. The colloid can be restabilized by heating (increasing kT) or by changing

Figure 12.5. Potential of two parallel surfaces with an electrolyte between them. (*a*) large separation, (*b*) small separation: the electric potential of the system is increased implying a repulsive force.

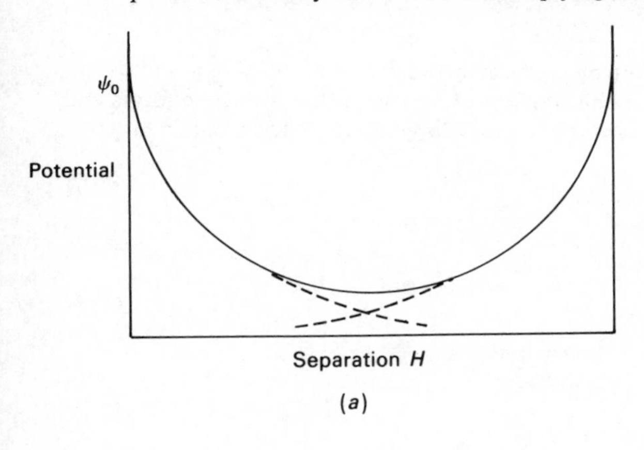

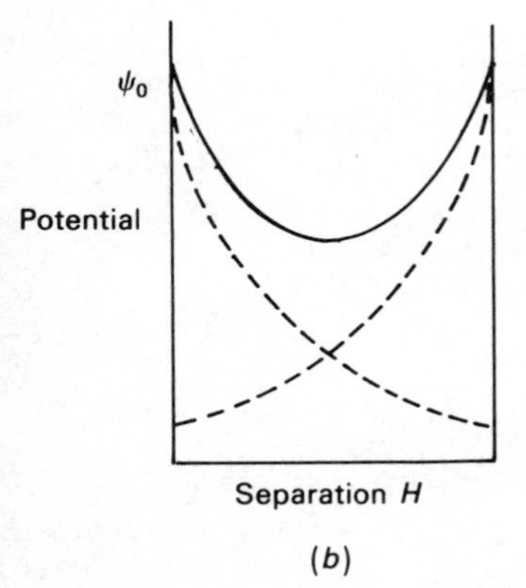

the ionic strength to increase the repulsive force. Figure 12.7 summarizes four main types of potential energy curve. In curve A the repulsive forces are weak and the curve is not greatly different from the van der Waals curve itself. These particles will attract one another and reach equilibrium at the primary minimum M_1. The particles are then virtually in contact and whole clusters will come out of suspension. This is known as coagulation, in contrast to flocculation where the particles are loosely held together at fairly large separations (see figure 12.6).

In curve B where the repulsive forces have been increased there is a very shallow secondary minimum at M_2, but it is so shallow compared with kT that the particles are easily able to escape from one another. There is little danger of flocculation. On the other hand the maximum P_1 represents an energy barrier comparable with kT. Thus particles coming from a large separation may well be able to surmount this barrier and fall to the primary minimum M_1. Since this is relatively deep compared with kT the particles will not have enough energy to escape from one

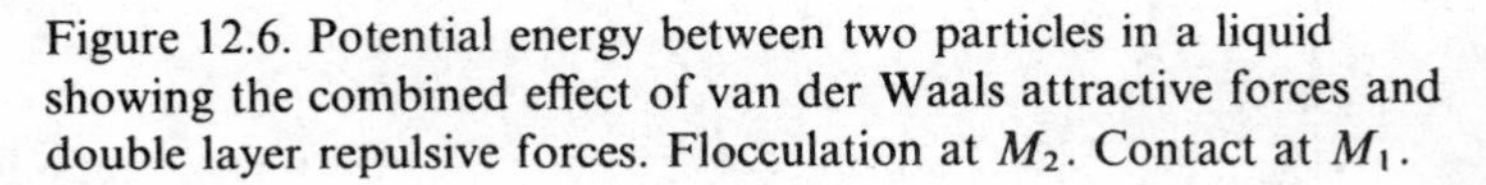
Figure 12.6. Potential energy between two particles in a liquid showing the combined effect of van der Waals attractive forces and double layer repulsive forces. Flocculation at M_2. Contact at M_1.

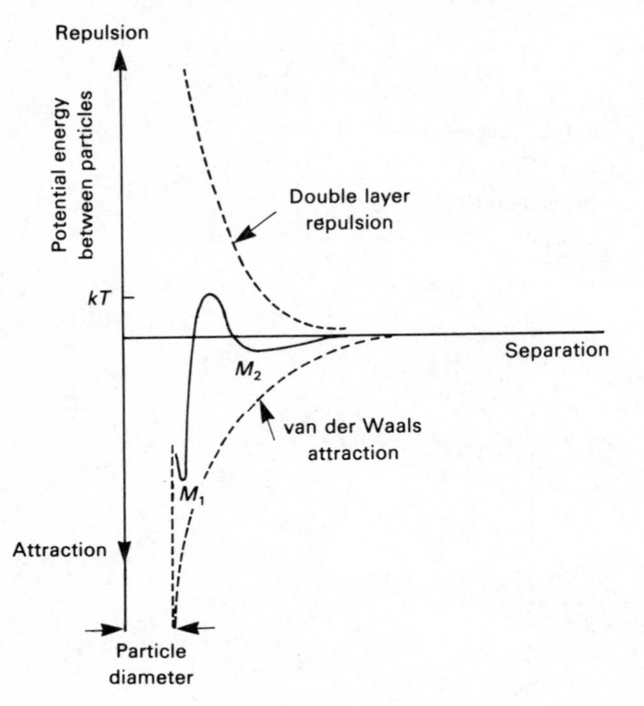

another. In that case irreversible coagulation will have occurred. It is interesting to note that paint is a classical example of a flocculated colloid of this type: if left undisturbed for a prolonged period it may become permanently coagulated.

In curve C the repulsive forces are so large that there is no secondary minimum. Consequently there is no tendency even for transient floccu-

Figure 12.7. Four examples of the resultant potential between two particles. Curves A, B, C and D represent the effect of increasing the repulsive forces. A: immediate contact at M_1 (coagulation); B: temporary flocculation at M_2 and possibility of coagulation; C: no flocculation or coagulation; D: spontaneous dispersion. A crucial parameter is the thermal energy kT.

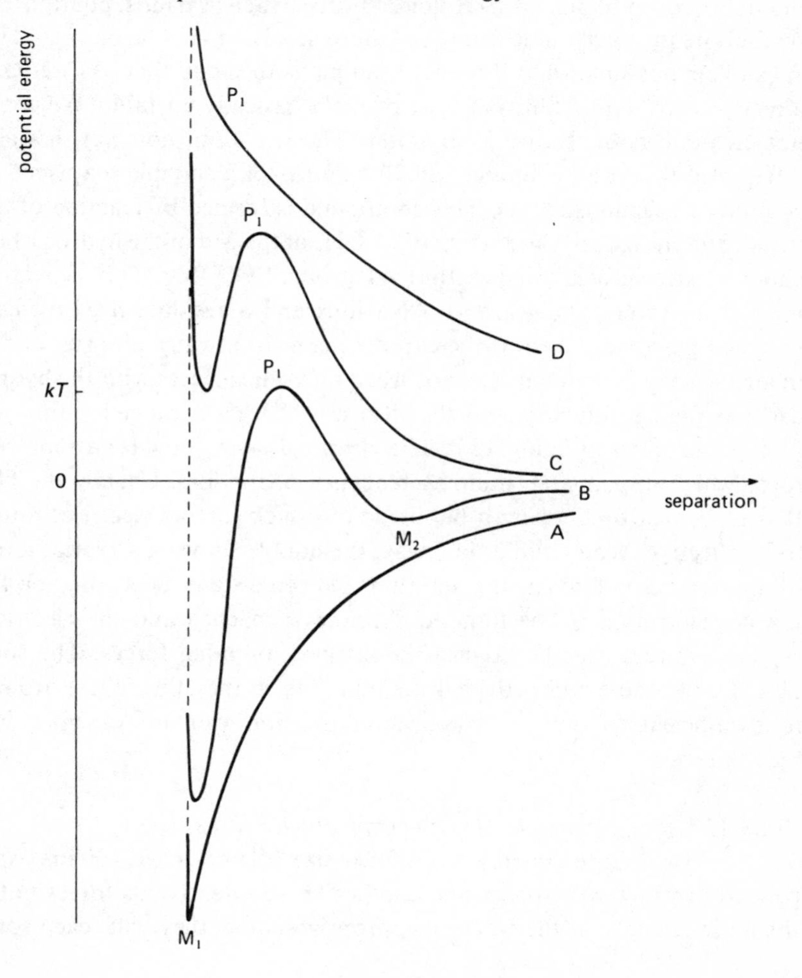

lation to occur. Further, the maximum P_1 is so much greater than kT that there is little chance of particles ever being able to surmount this barrier and fall into the primary minimum. Consequently coagulation is almost impossible.

Finally, in curve D the attractive forces are so weak at all separations compared with the repulsive forces that there is not even a primary minimum. The particles always repel one another: this corresponds to an extremely stable colloid. If particles are added to the liquid, spontaneous dispersion occurs.

12.3.1 *Stabilization of soap films*

The electrically charged double layer plays an important part in the stabilization of numerous colloidal systems such as paints, pharmaceuticals, food products: indeed most of those involving an aqueous medium. A familiar but somewhat different example is the soap film. As we have already noted, liquid films on their own are basically unstable. We could not create a viable thin film of water. The situation, however, is quite different if the water contains a small quantity of a suitable soap such as sodium or calcium stearate. This compound is formed by reaction of the metal with the carboxylate group (CO–OH) at the end of the hydrocarbon chain in stearic acid to give the end group CO–ONa. This is readily ionized in water to give positive Na ions and a residual negative carboxylate group attached to the carbon chain. In a water film suspended in air the soap is preferentially absorbed on both surfaces with the hydrocarbons facing outwards and the electrical charges arranged within the film as indicated in figure 12.8. The electrical layer provides a repulsive force which opposes the thinning tendency of the film. Usually the film that is formed by blowing a bubble is too thick for the electrical forces to be effective. Water will drain away, the bubble showing a typical series of interference colours as the film thins. At some stage, depending on the ion concentration in the film, equilibrium is reached and the electrical repulsive forces exactly balance the intrinsic thinning forces. The soap film is now stable and provided the humidity in the atmosphere around it is sufficient to prevent evaporation the film will, in principle, last indefinitely.

12.3.2 *A simple example of self organization: self assembly*

Polystyrene spheres of uniform size (diameter *ca.* 100 nm) suspended in water will attract one another by van der Waals forces but if the ionic strength in the water is appropriately low they will reach some equilibrium separation of the order of 300–400 nm. This equilibrium is at the minimum of the potential corresponding to flocculation. The particles will gradually arrange themselves in a close-packed array, typically a body-centred cubic (b.c.c.) lattice. Particles at the edge of the assembly may escape into the liquid while others will leave the bulk liquid and join the assembly. The free particles resemble the vapour phase around a crystal. These arrays can be viewed with a microscope and are exactly like an atomic b.c.c. crystal except that the 'atoms' are a few hundred times larger in diameter and the separation far larger. Similar effects may be obtained with fine silicon particles. Although these assemblies may be poured from one vessel to another (where they will reconstitute their crystalline arrangement) they are not, as Professor F. C. Frank has pointed out, liquid crystals but true plastic crystals with a very small but finite yield stress.

Because of the regular spacing between the particles in the assembly they behave as a three-dimensional diffraction grating and in white light give coloured reflections, the colour depending on the angles of illumination and of viewing. The flashing iridescent colour-changes observed might well be described as opalescence since research† shows that the brilliant opal gem stone consists of close packed spherical particles of silica of uniform diameter. (The diameter varies from 150 to 400 nm in various opals and this accounts for their different dominant colours.) It

Figure 12.8. A water film with a monolayer of sodium stearate adsorbed at each surface. The electrical charge distribution within the film produces a repulsion which prevents the film from thinning.

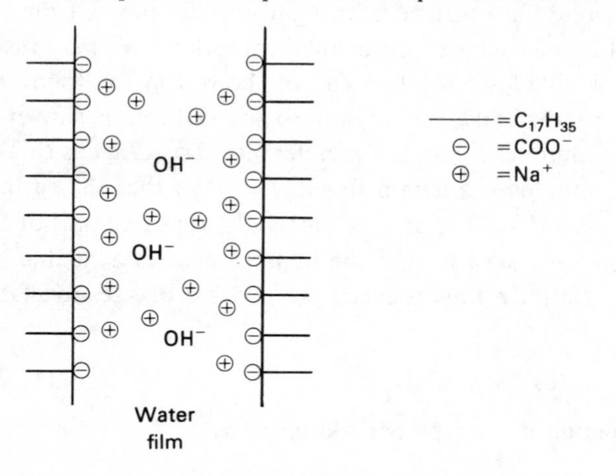

† J. V. Sanders, *Nature*, **204**, 1151–3 (1964).

is the voids between the spheres or, more correctly, the contrast in refractive index between the voids and spheres that provides the diffraction lattice.

The array we have just described resembles, though it is not identical with, the process of self-assembly. This term is used to describe the behaviour of certain long chain molecules, particularly if they have polar end groups, which come together and form local structural units (micelles, bilayers) that remain stably dispersed in solution. The lateral attraction between the chains gives lateral stability whilst the end groups interact with one another or with the aqueous phase and are stabilized either by electrical charges or by hydrogen bonds. They are of great importance in cell structures, especially in cell membranes.

12.4 Stabilization due to adsorbed polymeric films: entropic repulsion

A completely different type of stabilization can be achieved particularly in non-aqueous media by adsorbing suitable polymers on to the surface of the particles. Entropic effects can produce a repulsive force which may balance the attractive van der Waals forces and prevent the particles from coagulating.

Consider the highly simplified model shown in figure 12.9 for two parallel surfaces immersed in a liquid. Suppose a rigid rod-shaped molecule of length l, cross sectional area a, is adsorbed on to one of the surfaces. If the surfaces are far apart the rod can sweep out a whole hemisphere of area $2\pi l^2$. We divide the surface of the hemisphere into contiguous regions each of an area a. Then the rod can occupy $2\pi l^2/a$ sites. If now the second bare surface is brought to a distance x from the first surface ($x<l$) it will exclude part of the hemisphere. We may easily calculate the area available to the free end of the rod in the following way. If the radius vector makes an angle θ to the surface, as shown in figure 12.9(*b*), the annulus r has a circumference $2\pi r = 2\pi l \cos\theta$. The curved rim of the annulus has length $\mathrm{d}s = \mathrm{d}x/\cos\theta$ so that the surface area of the annulus is $2\pi l\,\mathrm{d}x$. The area of the whole truncated part of the hemisphere is, therefore, $2\pi lx$ so that the number of sites available for the rod is $2\pi lx/a$. We have thus reduced the number of sites available from

$$W_0 = 2\pi l^2/a \quad \text{to} \quad W_\mathrm{d} = 2\pi lx/a. \tag{12.21}$$

This causes a reduction in entropy per rod given by

$$\Delta S = k \ln W_\mathrm{d}/W_0 = k \ln(x/l). \tag{12.22}$$

The corresponding change in free energy per rod is

$$\Delta G = -T\Delta S = +kT \ln l/x \tag{12.23}$$

hence ΔG increases as x decreases which implies a repulsive force f, per rod

$$F = \frac{d\Delta G}{dx} = kT/x. \tag{12.24}$$

Suppose there are $(2l)^{-2}$ rods per unit area the repulsive forces per unit area will be

$$\frac{kT}{4l^2x}$$

as compared with the attractive force per unit area $A/6\pi x^3$. These are equal when

$$\frac{kT}{4l^2x} = \frac{A}{6\pi x^3}. \tag{12.25}$$

Using $kT = 4.2 \times 10^{-21}$ J; $A = 10^{-20}$ J we obtain

$$l \approx 1.4x. \tag{12.26}$$

Figure 12.9. Simple model illustrating entropic repulsion. (*a*) A rigid molecule of length l, cross sectional area a is adsorbed on one surface. It can sweep out an area $2\pi l^2$. (*b*) If a second surface approaches to a distance x the area available to the molecule is reduced to $2\pi lx$. (*c*) Closer packing of the molecules.

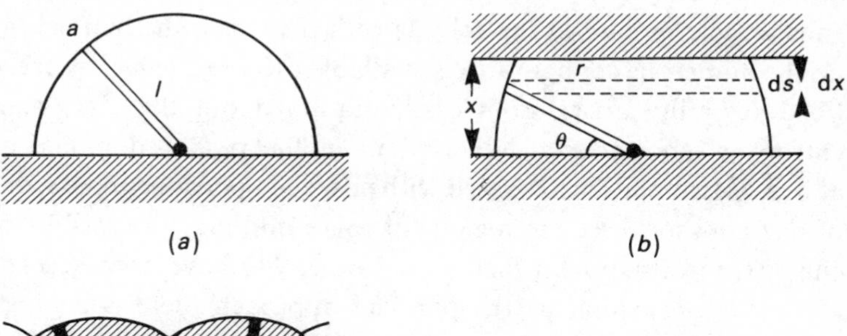

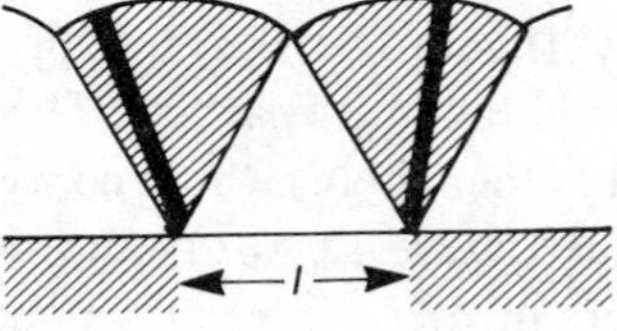

Thus for the attractive and repulsive forces between the surfaces to balance at a separation of 20 Å (2 nm) we should need cylindrical rods of length *ca.* 30 Å. Clearly since the attractive forces increase as x^{-3} and the repulsive forces as x^{-1} if the surfaces approach closer than 20 Å the attractive forces will win. This model system illustrates the principle of steric stabilization but cannot in fact achieve it. The entropy change is not large enough.

We could of course increase the concentration of rods on the surface. For example if the rods were distance l apart instead of $2l$ we should have four times as many rods. A much more important effect, however, is the limited space available for the free ends. If we assume that the zones cannot overlap, the free ends have available an area about one-eighth of that described above (see figure 12.8(*c*)). The entropy is therefore smaller: as a result any approach of the second surface will produce only a small reduction in entropy and it is the change in the entropy which is important. In the limit if the rods were close-packed their free ends would have virtually no free space. Their entropy would be zero and would be unaffected by the approach of the second surface. Furthermore the rods would now constitute a mass of high density and this would exert strong attractive van der Waals forces on the second surfaces. We should be back where we started.

The way to overcome this is to choose materials which have low density (to give negligible attractive forces) but which experience large entropy changes when the second surface approaches. This is achieved by using long-chain polymers which are reasonably soluble in the liquid and which adsorb on to the surface, attaching themselves at many points. The free ends waggle around in the solvent with their own thermal energy taking up the numerous configurations available to a flexible hydrocarbon chain. (The flexibility arises from easy rotation about the C–C bond.) The entropy is enormously greater than in the rigid rod model described above. If two particles covered with adsorbed polymer come within range of one another there are regions of space no longer accessible to a given chain. Conformations which would otherwise have been accessible to it are excluded so that as the particles approach there is an appreciable reduction in configurational entropy (figure 12.10(*a*)). This may be described as a need for the chain to avoid the other particle. There is also an entropic need for the chains to avoid other chains whose concentration is greatly increased in the contact zone (figure 12.10(*b*)) and this again implies a reduction in configurational entropy. This corresponds to an osmotic pressure and in the literature it is sometimes described in these terms.† The resultant overall reduction in entropy implies an increase in free energy and associated with this a repulsive force between the particles. These forces increase very rapidly with decreasing separation and are able to overcome the attractive forces even at very small separations so that absolute stability of the colloid can be achieved. A typical result (taken from J. Klein, *Physics World*, June 1989, pp. 35–8) is shown in figure 12.11 for the interaction between mica surfaces in the presence of a cyclopentane solution of polystyrene with each polymer molecule containing 2×10^4 monomers. The effective distance that the molecule extends from the surface is, at high surface coverage, of order 20 nm. It is seen that repulsion sets in at a separation of about 50 nm and completely overcomes the van der Waals attractive forces.

Of course, entropy is also lost when polymer segments adsorb on to the surface of the particle. In principle this might be expected to inhibit or prevent adsorption. In practice a relatively large number of monomers may contact the surface at any instant, and if these are attracted the energy of adsorption may easily compensate for the entropy loss. If the net attraction is strong enough the adsorption becomes effectively irreversible.

Figure 12.10. (*a*) Sketch of adsorption of a long chain polymer (from polymer solution) on the surface of particle 1. (If a large number of monomers are involved in the adsorption it is strong.) When particle 2 approaches there is an appreciable reduction in configurational entropy implying a repulsive force. (*b*) Sketch illustrating high concentration of chains in the contact zone compared with concentration in the surroundings. This leads to a repulsive force resembling osmotic pressure.

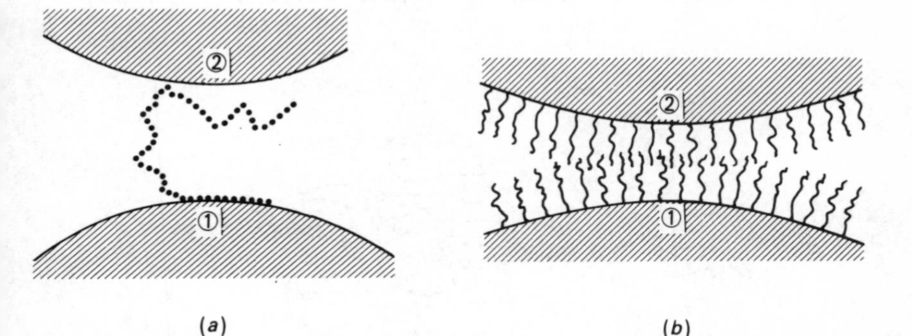

† The molar concentration of solvent is less in the contact zone than in the solution so that, as described in Chapter 10, section 10.10.3, solvent from outside will attempt to enter the zone and so produce an osmotic pressure.

Another way of ensuring firm attachment to the particle is to use a polymer which does not readily adsorb but which has an 'anchor' group at one end (figure 12.12). In some cases the surface polymers can be chemically grafted to the surface. In these ways it is possible to combine strong attachment to the surface and marked steric repulsion.

Figure 12.11. Interaction energy between mica surfaces in the presence of a solution of polystyrene in cyclopentane. Each polymer chain contains 2×10^4 monomers, radius of gyration about 20 nm. (*a*) In the absence of polymer there is simple van der Waals attraction. (*b*) At moderate surface coverage the entropic repulsion begins to counteract the attractive forces at separations less than 20 nm. (*c*) At high surface coverage the interaction is repulsive at all separations.

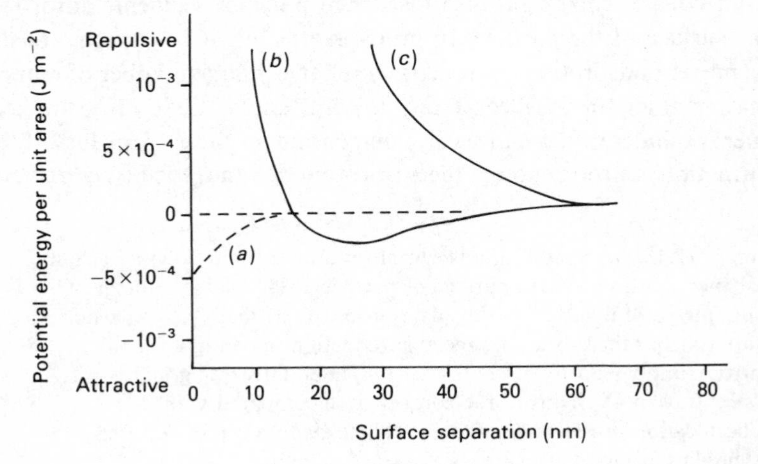

Figure 12.12. Configuration of a polymer with an anchor group at one end: (*a*) low surface concentration, (*b*) high surface concentration.

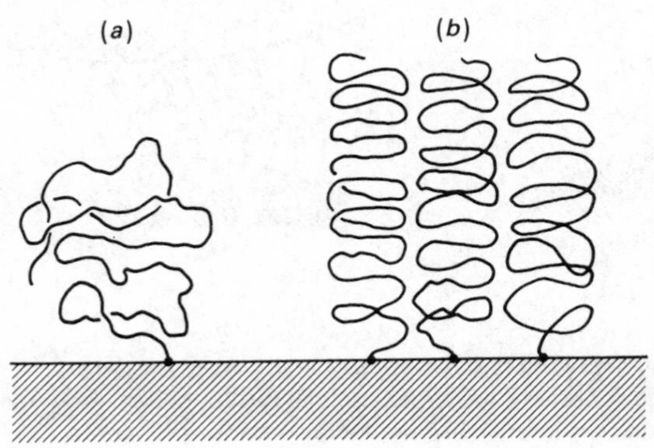

Polymeric surface films producing a net repulsion by entropic and osmotic processes can provide a very successful method for stabilizing dispersed particles and in preventing coagulation.

As a postscript to this chapter we may note that over 4500 years ago the Egyptians applied this principle empirically, in the production of 'instant' ink for writing on papyrus. The ink was prepared by mixing lamp black (fine particles of soot) with a solution of a natural steric stabilizer, such as the resin from the acacia tree (gum arabic), casein from milk, or egg albumen. These materials provide water soluble polymeric chains which adsorb on the carbon particles. The carbon black was then moulded into a pencil-like shape and dried. When the ink was required part of the pencil was broken off and dipped into water. The sterically stabilized carbon black particles redispersed spontaneously.

Steric stabilization is today exploited in a wide range of industrial products such as paints, glues, food emulsions, detergents and lubricants, and plays a part in many biological systems.

13

Some physical properties of polymers

Most of the polymers that we meet in everyday life are carbon based and they owe their existence to the capacity of the carbon atom to hybridize its s and p orbitals to give four valency bonds at fairly well defined angles. Silicon and boron are also tetravalent and can form polymers but we shall deal mainly with polymers based on carbon.

Consider for example the molecule ethylene

$$\begin{matrix} H & & H \\ C & = & C \\ H & & H \end{matrix}$$

In the presence of a suitable catalyst the double bond is broken and the monomer

$$\begin{matrix} & H & & H & \\ - & C & - & C & - \\ & H & & H & \end{matrix}$$

is able to link up chemically with other monomers to form a polymer with many thousands of carbon atoms in the chain.

There are two main classes of polymers. Those which are so strongly cross-linked that they are no longer rubbery: they are known as thermosetting polymers, are hard and brittle and do not melt. We shall deal with these only briefly at the end of this chapter. The others are thermoplastic polymers: they do not have cross-links and melt when heated. We shall deal mainly with these. Five of the most common polymers of this type are

Chemical name	Formula	Common name
Polyethylene (PE)	$\left[\begin{matrix} H & & H \\ C & - & C \\ H & & H \end{matrix}\right]_n$	Polythene

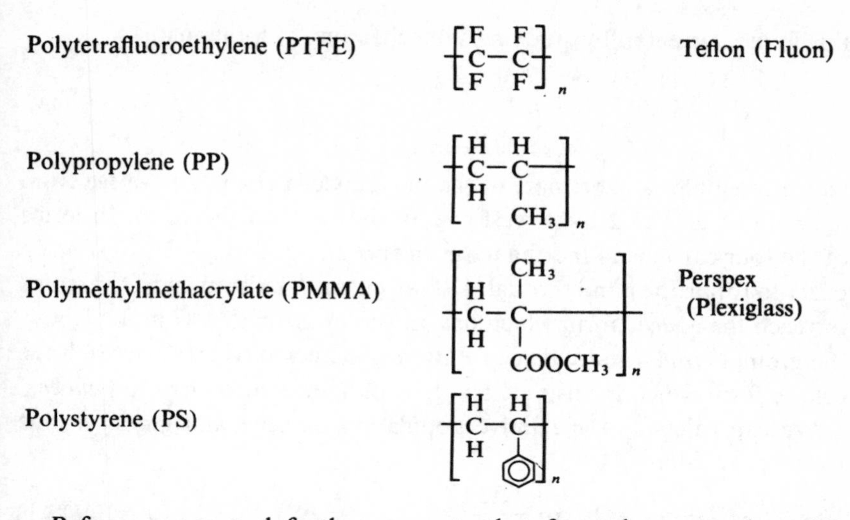

Before we proceed further we note that for polymers such as PP, PMMA, PS, there is choice as to which way round the monomers are incorporated into the chain (tacticity or ordering). This affects the flexibility of the chains and the packing. In general the isotactic form is least closely packed. We give here these forms for PP:

Isotactic
(equal ordering)

```
  H  H     H  H     H  H
 -C--C-----C--C-----C--C-
  H  |     H  |     H  |
    CH3      CH3      CH3
```

Atactic
(no ordering)

```
                     CH3
  H  H   H  H   H    |
 -C--C-  C--C-  C--C-
  H  |   H  |   H  H
    CH3    CH3
```

Syndiotactic
(alternate ordering)

```
            CH3
  H  H   H  |    H  H
 -C--C-  C--C-   C--C-
  H  |   H  H    H  |
    CH3             CH3
```

In what follows we shall deal mainly with PE which is representative of polymers which are able to crystallize – at least in part – and with atactic PS which is typical of polymers which are always amorphous.

13.1 **Conformations of the polyethylene chain**

In Chapter 7 in our discussion of rubber elasticity, we showed how the configuration of a long chain organic molecule may be derived assuming there is freedom of rotation about the C–C bond. We consider this in greater detail by studying the behaviour of n-butane

```
  H   H   H   H
HC—C—C—CH
  H   H   H   H
```

which resembles a short part of the polyethylene chain. As we see from figures 13.1 and 13.2 the lowest energy state is the *trans* form. In terms of the four carbons in the chain the valence bonds form a flat zig-zag. In order to rotate the bond through 120° an energy barrier must be overcome to reach the *gauche* form which has an energy $\Delta E = 3.34$ kJ mol^{-1} above the ground (*trans*) state. If the rotation is increased to 180° we reach the eclipse form which is unstable and the configuration returns to *gauche*.

We can calculate the relative population of *trans* and *gauche* forms from the relation

$$\frac{n_{gauche}}{n_{trans}} = \frac{n_g}{n_t} = 2\exp\left(-\frac{\Delta E}{RT}\right). \tag{13.1}$$

The factor 2 is introduced because there are two *gauche* forms for every *trans*. For $\Delta E = 3.34$ kJ mol^{-1} we obtain the results shown in Table 13.1.

Thus above room temperature (i.e. above 300 K) the *gauche* form is easily accessible and we consider rotation about the C–C bond to be unrestricted. The same situation applies to the long isolated polyethylene molecule, though in molten PE the activation energies will be larger.

Figure 13.1. The n-butane molecule showing the three main arrangements of the carbon bonds: (*a*) trans (planar), (*b*) gauche, (*c*) eclipse. The dihedral angle is measured from the trans position. The lower figures show that the CH_3 groups are farthest away in the trans, and overlap in the eclipse forms.

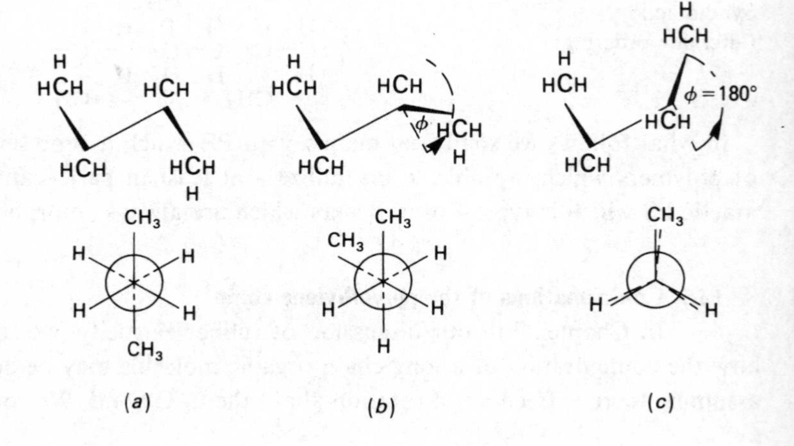

Table 13.1. *Trans* (t) *and gauche* (g) *forms in n-butane*

Temperature (K)	450	400	300	200	100
Equilibrium population (n_g/n_t)	0.81	0.74	0.52	0.26	0.036
Time of *trans–gauche* transition (s)	3×10^{-10}	6×10^{-10}	2×10^{-9}	3×10^{-8}	10^{-3}

These calculations apply to the equilibrium population of *trans* and *gauche* forms. The *rate* of transition, however, depends on the ability of the system to surmount the barrier ΔE_{tr} which has a value of about 14 kJ mol^{-1}. The transition time is given by

$$\tau = \tau_0 \exp \Delta E_{tr}/RT \tag{13.2}$$

where for butane and for polyethylene τ_0 is of order 10^{-11} s. The relevant values of τ are given in Table 13.1 and it is seen that at 200 K the transition is 100 times slower than at 450 K. At 100 K it is over a million times slower. Thus the response of the chain depends on the temperature. If the deformation rate is greater than τ^{-1} rotation is restricted, the chain behaves rigidly; if the deformation rates are well below τ^{-1} the chain is

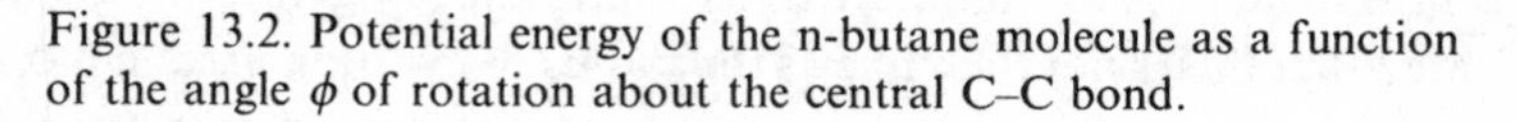

Figure 13.2. Potential energy of the n-butane molecule as a function of the angle ϕ of rotation about the central C–C bond.

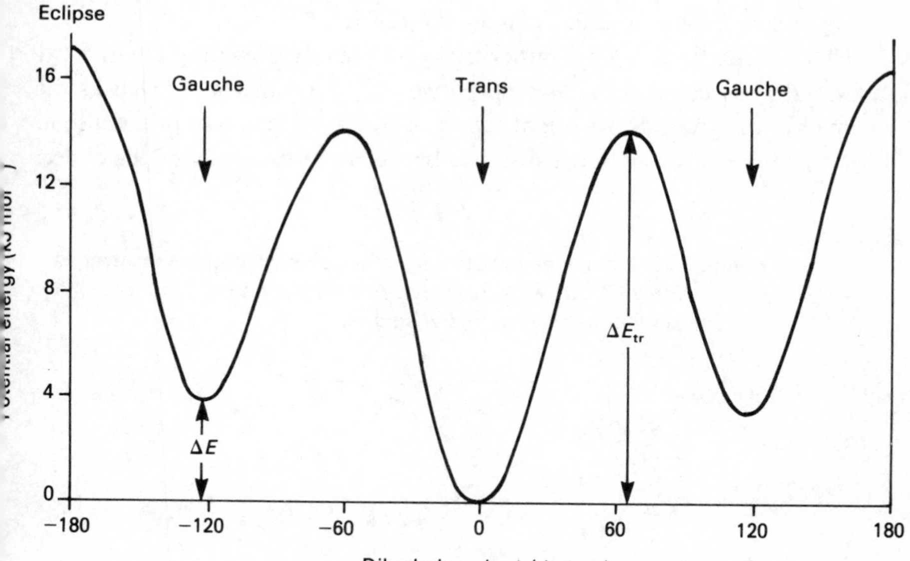

flexible. Thus the chain response depends on both deformation rate and temperature. A high deformation rate is equivalent to a low temperature. We shall discuss examples of this when we deal with the polymer in bulk, especially in sections 13.7 and 13.8.

13.2 The effective size of a free polyethylene chain: radius of gyration

On grounds of symmetry we should expect a free polymer chain, or a chain in dilute solution, to occupy overall a roughly spherical shape. It will of course continuously change its shape and configuration under the action of random thermal or Brownian motion so that its actual dimensions can only be averaged over the numerous configurations that it can assume. One way of determining this average size is by studying the scattering of light or some other radiation impinging on the polymer. If the chain contains N units each of mass m the angle of scattering is determined by the wavelength of the radiation and an effective radius R_G where

$$R_G^2 = \sum_1^N mR_i^2/Nm \tag{13.3}$$

where R_i is the distance from the centre of mass to the ith unit (figure 13.3(a)). Thus R_G is the average distance of the whole molecule from the centre of mass. It is referred to as the radius of gyration. It should be noted that this is not the same as the radius of gyration used in mechanics: the latter is a 2-D concept about an axis, whereas in scattering R_G is a 3-D concept about a point, namely the centre of mass.

The calculation of R_G is rather tricky and involves treating the interval $(R_i - R_j)^2$ as equal to $\frac{2}{3}\lambda^2|n_i - n_j|$. However, on geometric as well as on dimensional grounds, we might expect R_G to be a fraction of or a multiple of r_m where r_m is the linear distance between the two ends of the coiled

Figure 13.3. Radius of gyration R_G of a polymer molecule as defined in equation 13.3. (a) R_i is measured from the centre of mass O. (b) The summation for vectors R_i and R_j.

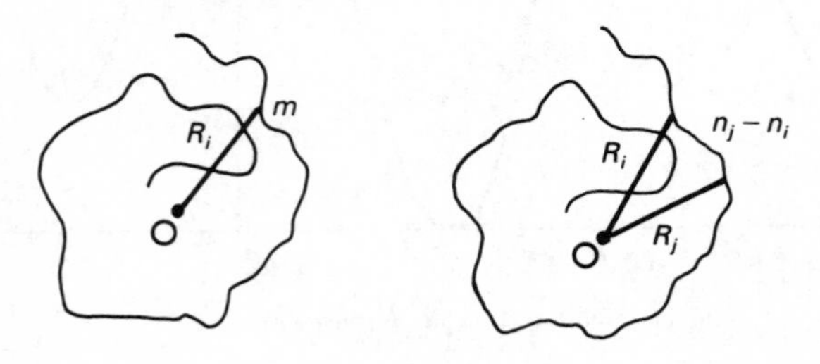

polymer. In fact the analysis gives

$$R_G^2 = \tfrac{1}{6}r_m^2 = \tfrac{1}{9}N\lambda^2. \tag{13.4}$$

If we take $(R_G^2)^{\frac{1}{2}}$ to be the effective radius of the polymer coil we may calculate the amount of empty space between the chains. The apparent volume V_a of the sphere is

$$V_a = \tfrac{4}{3}\pi(R_G^2)^{\frac{3}{2}} \approx \tfrac{1}{6}N^{\frac{3}{2}}\lambda^3. \tag{13.5}$$

The volume V_c occupied by the molecular material itself is

$$V_c = Nv \tag{13.6}$$

where v is the volume of each unit of the chain. Then the fraction ψ of the sphere actually occupied by molecular matter is

$$\psi = \frac{V_c}{V_a} \approx \frac{6v}{\lambda^3 N^{\frac{1}{2}}}. \tag{13.7}$$

For PE the estimated volume of v, the H–C–H group, based on X-ray data of a crystal of a long-chain paraffin, is 25×10^{-30} m^3 and $\lambda = 1.54\times10^{-10}$ m. For PE of molecular mass 1.4×10^6 we have $N=10^5$ and $\psi=0.13$. Thus 87 per cent of the volume occupied by the molecular coil is empty space. In many polymers the unit involved in free rotations consists of several monomers. Although this increases v the increase in λ^3 is much greater. Thus the fraction of empty space is even greater.

13.3 Dilute polymer solutions: chain configuration: the Zimm theory

We have seen that the radius of gyration is proportional to $N^{\frac{1}{2}}$. In the simple random walk from which this result is derived it is assumed that any 'step' can be occupied any number of times. In reality a position occupied by a segment cannot be occupied by other segments. There is thus a certain amount of excluded volume and this leads to a slight change in the relation for R: it becomes more nearly proportional to $N^{\frac{3}{5}}$. To some extent this effect is counterbalanced by the van der Waals attraction between chain segments. However, the most marked effect on the configuration of the chain is that produced by solvents. In a good solvent it is energetically unfavourable to displace solvent and bring segments together: the chain swells. Conversely in a poor solvent it is favourable for the segments to come together: the chain contracts. It is possible to choose a solvent, at a particular temperature θ, to produce a radius of gyration given by the simple random walk theory (13.4). This is known as a theta solvent.

13.3.1 *The viscosity of dilute polymer solutions: the Zimm Theory*

The viscosity η of a dilute solution in a solvent of viscosity η_0 is given by the Einstein equation

$$\eta = \eta_0(1 + 2.5\phi)$$

where ϕ is the volume fraction of the polymer. Thus

$$\frac{\eta - \eta_0}{\eta_0} = 2.5\phi. \tag{13.8}$$

We have N links each of length λ, mass m. The molecular mass is $M = Nm$. At the θ point the radius of gyration of the chain from equation (13.4) is

$$R_G = \tfrac{1}{3}N^{\frac{1}{2}}\lambda \tag{13.9}$$

so that the effective volume V_a of the spherical coil is

$$V_a = \alpha N^{\frac{3}{2}}\lambda^3 \tag{13.10}$$

where α has a value of approximately $\frac{1}{6}$ (see equation (13.5)).

If there are n polymer chains per unit volume and the concentration is so small that they do not overlap

$$\phi = nV_a = \alpha n N^{\frac{3}{2}}\lambda^3 \tag{13.11}$$

while the mass concentration c per unit volume is

$$c = n\frac{M}{N_A} \quad \text{or} \quad n = \frac{cN_A}{M} \tag{13.12}$$

when N_A is Avogadro's number.

Substituting from equation (13.12) into equation (13.11) we have

$$\phi = \alpha\frac{cN_A}{M} \times N^{\frac{3}{2}}\lambda^3 = \alpha\frac{cN_A}{M}\left(\frac{M}{m}\right)^{\frac{3}{2}}\lambda^3,$$

that is

$$\phi = \alpha\left(\frac{N_A\lambda^3}{m^{\frac{3}{2}}}\right)M^{\frac{1}{2}}c. \tag{13.13}$$

Thus if the polymer coils move through the liquid as solid spheres the viscosity change should be given by

$$\frac{\eta - \eta_0}{\eta_0} = \text{constant} \times M^{\frac{1}{2}}c. \tag{13.14}$$

Experiment shows that this relation is indeed observed. Thus, surprisingly, although the polymeric coils are quite open to the solvent, flow does not occur through the polymer chains themselves. The coils carry with them the solvent within them. This result is associated with the Zimm or hydrodynamic theory.

Experiments with polymers in solution provide little or no evidence for the flow of solvent *through* the whole chain, except perhaps for very good solvents which swell the coil and even tend to straighten the chains. However, such a situation does apply when the solvent is the polymer itself, i.e. when we deal with the molten polymer.

13.4 Conformation of the molten polymer: neutron scattering from deuterated polymers

Before proceeding further we may ask what form the polymer adopts in the molten state. It cannot be in the form of an assembly of individual coils for then its density would be very low and would depend on the chain length as shown by equation (13.7) and the example quoted. In fact the density is about the same for a polymer containing 500 carbons as for one containing 100 000 carbons: and it is of the same order as the density of a crystalline hydrocarbon. The chain segments must be very close to one another. However, it proved impossible to determine the conformation of the polymer chains because of their complicated interpenetration until it was discovered that neutron diffraction from deuterated polymers could solve the problem in the following way. A few deuterated polymers otherwise identical to the parent molecule are introduced into the molten polymer. It is then found that neutron scattering from the deuterated polymer is sufficiently different from that of the host polymer for its configuration to be identified. The results show that it has essentially the same radius of gyration as that of the classical free random coiled configuration. The 'empty' spaces in the coil are filled with other polymer chains with similar random coiled configurations. (This was mentioned in our discussion of the elastic modulus of bulk rubber in Chapter 7.) The reason for this surprising result is that the free random coil experiences close contact between segments of itself: in the molten state the polymer cannot distinguish between its own segments and those of neighbouring chains.

13.5 Viscosity of the molten polymer: the Rouse theory

In the molten state the chains flow through and around one another: they are also stretched, deformed and retracted. All these processes are involved in the viscous resistance of the polymer to flow. Rouse suggested that the chain could be treated as a series of beads linked by springs. By assuming that the friction on each bead is independent of its neighbours he deduced that the bulk viscosity of the molten polymer is proportional to its molecular mass M, i.e.

$$\eta = cM. \tag{13.15}$$

For chains containing fewer than a few hundred rotating links (the links may be C–C bonds or monomer–monomer bonds), this relation is well obeyed experimentally as shown in figure 13.4. We emphasize again that in this range the chains flow freely through one another.

13.6 Entanglements and reptation in the molten polymer

A marked change in the flow properties occurs when the number of freely rotating links exceeds several hundred. The viscosity increases rapidly (see figure 13.4) and can be described by a relation of the form

$$\eta = gM^{3.4} \tag{13.16}$$

where g is a constant which varies from one family of polymers to another. The behaviour is descriptively attributed to a log-jam resembling the situation where trees float down a river and, at some critical stage, jam and resist further flow. In molecular terms the molecules are so heavily constrained by their neighbours that they can no longer flow easily through or past one another. They can only move by weaving their way

Figure 13.4. Viscosity of polyisobutylene as a function of the number N of monomers in the chain. There is a fairly sharp change in behaviour at $N \sim 500$. Temperature 217 °C (480 K). (T. G. Fox & P. J. Flory, *J. Phys. Chem.* **55**, 230 (1951).)

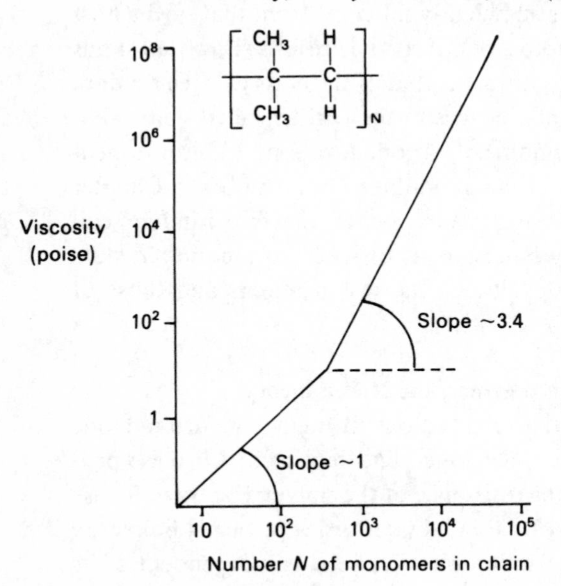

through available space between other molecules. At any instant each molecule may be regarded as existing in a contorted tube through which it can wriggle or worm its way forward. This process, for which there are highly sophisticated treatments, is known as reptation (figure 13.5). By contrast, for polymers containing less than a few hundred links there are so many free ends per unit volume that tube-like constraints are virtually non-existent.

We may describe the flow of the molten polymer in the following way. When the polymer is in repose in its virtual tube the polymer chain, by virtue of the Brownian motion of all its segments, acquires an average equilibrium configuration in which it maximizes its entropy. This has been described above. When a shear stress is applied to produce flow we may regard the shear as being equivalent to a tension at 45° to the shear in the direction of flow and a compression at right angles to the tension. The molecule in the tube is distorted and in response to Brownian motion produces a pressure against the walls of the tube and a retractive reaction along the tube. Work must be expended in overcoming these reactions and this constitutes part of the viscous work. In addition segments of the polymer may come into close contact with other segments and work must be expended in breaking these attachments. In the polymer literature these attachments are generally referred to as entanglements but it is still uncertain whether loops or knots are actually formed between one polymeric chain and its neighbour, though with PE it is possible that crystallites which are continuously forming and dissolving may trap parts of adjoining polymer chains. The main point about these attachments or

Figure 13.5. (*a*) The reptation of one long-chain molecule through its neighbours. (*b*) The virtual tube through which the molecule moves: its width is taken as the sideways excursion of the molecule in traversing its own length and is of order 30–50 Å (from J. Klein, *Nature*, vol. **271**, 1978, p. 143). (*c*) An applied shear stress during flow is equivalent to a compression and extension at 45° to the shear direction.

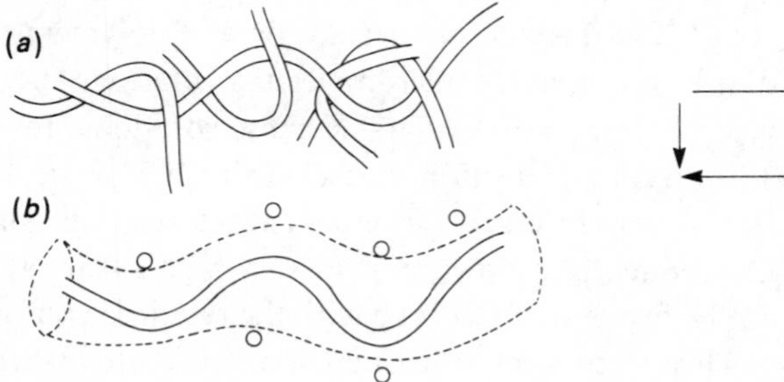

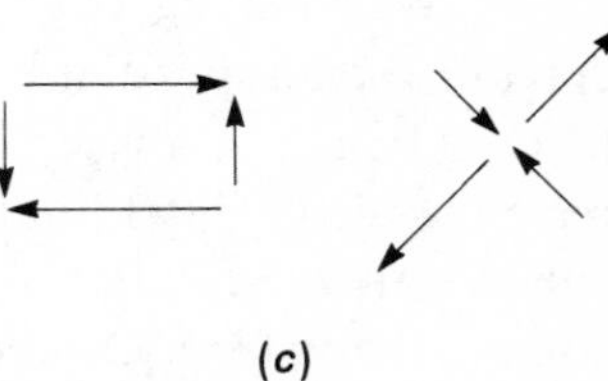

entanglements is that they are transient. They have a limited lifetime and if the flow is sufficiently slow they do not impede the reptation of the chain in its tube though they may contribute to the viscous work expended.

13.7 Diffusion in the molten polymers

According to the theory of reptation which we cannot develop here, the coefficient of diffusion D in the molten polymer is proportional to the inverse square of the molecular weight

$$D = cM^{-2}. \tag{13.17}$$

This has been confirmed experimentally. Because diffusion involves an activated process it depends on temperature. For $T > T_g$

$$D = D_0 \exp(-\Delta E/RT) \tag{13.18}$$

where ΔE, the activation energy, has a value of order 30 kJ mol^{-1}.

If the polymer is subjected to hydrostatic pressure p the diffusion coefficient is reduced and may be expressed in the form

$$D = D_0 \exp(-pv/RT) \tag{13.19}$$

where v is the same activation volume that is involved in the diffusion process. It turns out that v is almost independent of the size of the diffusing molecule and has a value between 50 and 100×30^{-30} m^3. This is roughly the volume involved in a *trans–gauche* or crankshaft molecular movement. It would seem that the micromolecular mechanism of reptation involves the cooperative rotation of a small number of adjacent C–C bonds, as illustrated in figure 13.6, which move the chain along by a distance of approximately one C–C spacing. This resembles a wrinkle or ruck or dislocation moving along the chain. The energy involved in moving this volume against the internal pressure within the molten polymer is responsible for about half the observed activation energy ΔE in equation (13.18).

13.8 Effect of shear rate on the flow of molten polymers

We have suggested that the macromolecular flow of polymers by reptation is made possible by coordinated micromolecular rotations about the C–C bond. At a typical temperature of 400 K these rotations, for a free molecule (Table 13.1) have a transition time of order 10^{-9} s. In the molten state the rotation times are much longer but the transitions still require a finite time. Consequently if flow involves very high shear rates such transitions may not be fast enough to cope with the required conformations of the chain. This often occurs in industrial processes where polymer is extruded through a narrow orifice. At some critical shear rate the polymer ceases to be fluid and the flow becomes unstable: the polymer becomes rubbery (see below) and the material which spurts from the orifice breaks up into pellets. As soon as they are free the pellets, of course, revert to the liquid state. To avoid spurting the rate of extrusion must be reduced or the temperature increased. The shapes of the nozzle and of the orifice are also important.

13.9 The four main states of the polymer

We may now briefly describe the main changes that occur in the behaviour of a linear polymer as the temperature is reduced.

(a) *Liquid state: viscous flow*

At temperatures well above the melting point the polymer at low rates of deformation is a viscous liquid and many of its characteristics have already been discussed.

(b) *Viscoelastic melt: rubbery state*

As the temperature of the molten polymer is reduced, the lifetime of local entanglements increases. At some specified rate of deformation

Figure 13.6. Diagram indicating how a few co-operative rotations about C–C bonds can lead to a displacement of about one C–C spacing.

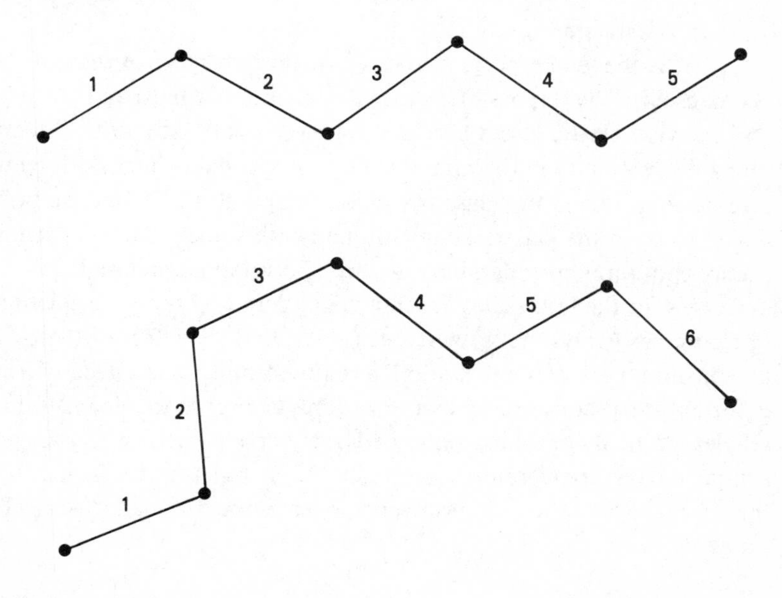

(usually greater than 10 cycles s^{-1}) there may not be sufficient time for the chains to free themselves from them: these entanglements then act as local cross-links between the chains. The material then resembles a classical elastomer and its behaviour is dominated by entropic factors. The main difference is that the cross-links are not permanent but transient. The elastic modulus is generally about 10^5 N m^{-2} and this would imply that the average effective length of segments between the 'entanglements' is of order $N=250$ (see Chapter 7, section 7.5).

(c) *Viscoelastic solid state*

The rubbery state persists over a range of a few tens of degrees. Below this temperature rotation of the C–C bonds becomes more sluggish: the attachments occur more and more numerously so that the length of the segments between the virtual cross-links decreases. The deformation of the chain segments is no longer determined by purely entropic factors: rotations about the C–C bonds must now be partly enforced by the external applied stress. In terms of equation (7.22) the internal energy involved in distorting the chain becomes increasingly important. The modulus rises rapidly by two or three orders of magnitude and the movement of chain segments past one another involves an appreciable dissipation of energy. This is the viscoelastic solid state.

(d) *Glassy state*

As the temperature is lowered further, rotation of the C–C bonds becomes virtually impossible: the material becomes hard and often brittle and the viscoelastic losses become relatively small. At a critical temperature a discontinuity in the rate of change of specific volume with temperature occurs. This is the glass transition temperature T_g and the polymer is said to be in the glasssy state although with some polymers such as PE it may contain a considerable amount of crystalline material.

Curves for the four states are shown in figure 13.7 for a typical amorphous polymer (polystyrene) where it is seen that the effect of temperature at constant rate of deformation is equivalent to the effect of rate of deformation at constant temperature. This is discussed below. With polyethylene, which can be partially or largely crystalline in the glassy state, similar curves are obtained (see figure 13.8) but the rise in modulus in the viscoelastic–solid regime occurs over a much smaller temperature range.

Figure 13.7. Four states of polystyrene. (*a*) Constant frequency, ~10 rad s^{-1} variable temperature. (*b*) Constant temperature (470 K), variable frequency. Data provided by Dr M. R. Mackley.

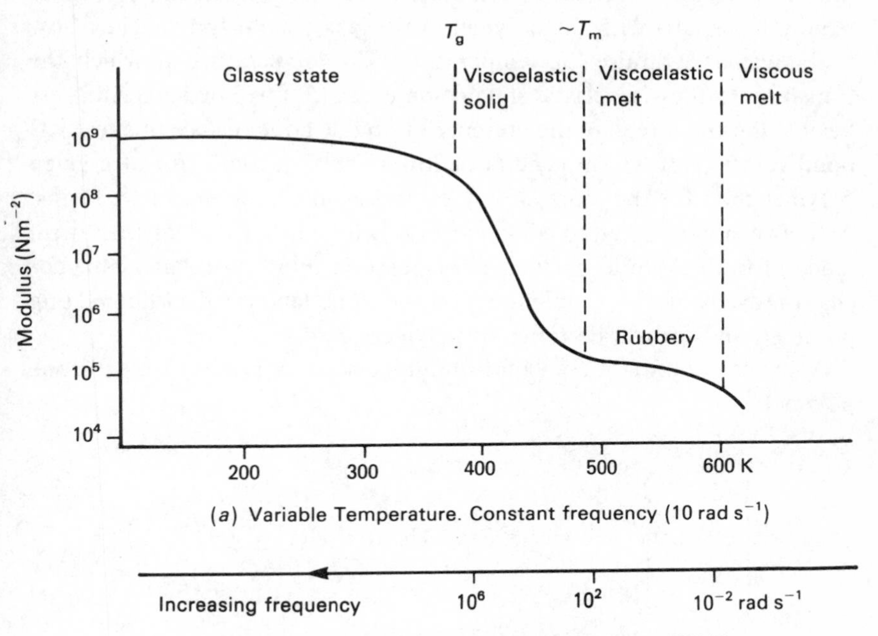

Figure 13.8. Four states of polyethylene as a function of temperature at constant frequency ~10 rad s^{-1}. The polymer is monodisperse linear PE, i.e. of uniform molecular mass 130 000. Data provided by Dr M. R. Mackley.

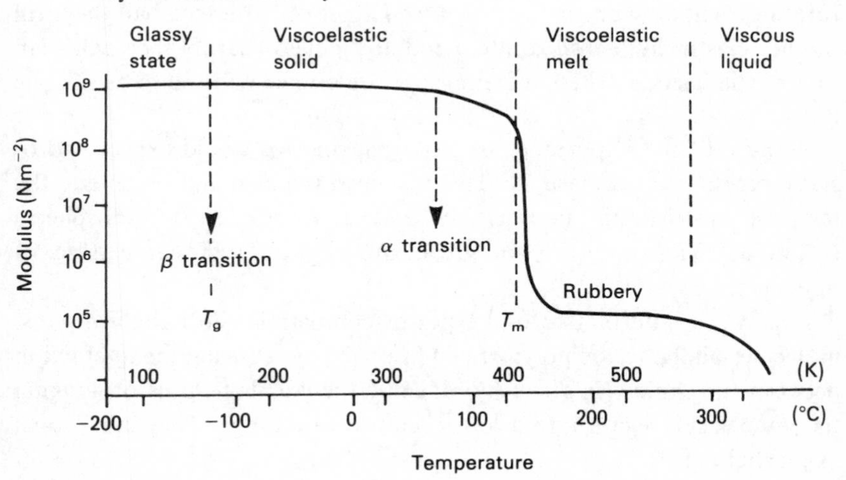

13.10 Factors affecting T_g

For a wide range of polymers the main factors affecting T_g are the flexibility of the chain and steric packing. This is understood more readily if we start with the polymer in the glassy state and consider how it changes as we raise the temperature. The temperature at which the transition to the viscoelastic state commences (T_g) will be higher the less flexible the chain (e.g. in the presence of double bonds) since this restricts bond rotation. It will in general be lower for isotactic forms of a given polymer than for the more closely packed syndiotactic or atactic forms. As between various types of polymers it will be lower for polymers with 'uncomfortable' steric packing since there is more space available for chain movement. As we might expect the same factors affect the melting point T_m and for a wide range of polymers $T_g = (0.6\text{–}0.8)T_m$.

A typical example of uncomfortable packing is provided by silicone polymers

$$\left[-O-\underset{R}{\overset{R}{\underset{|}{\overset{|}{Si}}}}- \right]_n \qquad \text{where } R = CH_3 \text{ or } C_2H_5.$$

The dimethyl silicones have a T_g well below 0 °C and are molten at room temperature. They can exhibit the whole range of viscous flow, rubbery viscoelastic melt, viscoelastic solid and brittle solid simply by changing the rate of deformation as indicated in figure 13.7. Thus at room temperature they will flow under their own weight; if rolled into a ball they will bounce elastically; if rolled into a rod and pulled sharply they will snap in a brittle fashion. These materials are sometimes referred to as silicone putty.

Because of the importance of steric packing we would expect hydrostatic pressure to increase T_g. This is indeed the case and in general the temperature rise lies between 10 and 20 K per 1000 atmospheres (0.1 GPa). This is clearly of great importance in the processing of polymers.

Finally we note that certain types of compatible short-chain organic molecules added to the polymer can lower T_g by reducing the interaction between the chains (in effect by reducing the packing), thus prolonging the viscoelastic region into a lower temperature range. They are known as plasticizers.

13.11 Time–temperature superposition: an example from stress-relaxation

The deformation behaviour of a polymer depends on temperature and rate of deformation. These two effects may be superposed by means of the Williams–Landel–Ferry (WLF) transform which we illustrate here in terms of stress-relaxation. If a polymer sample in the form of a rod is extended by a fixed amount the stress will gradually fall off with time in a roughly exponential manner. The material behaves as though its modulus is decreasing with time. Typical results for polyisobutylene

$$\left[\begin{array}{cc} CH_3 & H \\ | & | \\ -C & -C- \\ | & | \\ CH_3 & H \end{array}\right]_n$$

carried out at various temperatures are given in the left hand portion of figure 13.9. These curves may be displaced on the $\log_{10}$ (time) axis to give a composite curve which covers the whole range of temperature and time. The shift factor according to the WLF transform is

$$\log_{10} \alpha_T = \frac{A(T-T_0)}{(B+T-T_0)} \tag{13.20}$$

where A and B are fixed constants which apply to a wide range of polymers and T_0 is a reference temperature that is between 50 and 100 K greater than T_g. For the data of figure 13.9 the value of T_0 is T_g+45 K. At the reference temperature of 298 K, $\log_{10} \alpha_T$ is zero so that the relaxation-time curve at this temperature is transposed to the right covering exactly the same log (time) interval of approximately 10^2–10^{-2} h (hours). At a temperature of say 233 K, the insert shows that $\log \alpha_T$ is 4 so that the relaxation-time curve at this temperature is moved back by 4 orders and covers the range 10^{-2}–10^{-6}. In this way the whole of the right-hand side of figure 13.9 may be constructed. The result closely resembles the curve shown in figure 13.7 for the deformation–temperature behaviour of polystyrene.

The shift factors may also be related to the ratio of the viscosities of the polymer at temperatures T and T_0 (ignoring small changes in density with temperature), namely

$$\alpha_T = \eta/\eta_0 . \tag{13.21}$$

If the viscosities involve an activation energy Q we may write (see equation 11.18)

$$\eta = C\exp\frac{Q}{RT}, \qquad \eta_0 = C\exp\frac{Q}{RT_0}. \tag{13.22}$$

Hence

$$\log_{10}\alpha_T = \log_{10}\frac{\eta}{\eta_0} = -\frac{Q(T-T_0)}{2.3TT_0}. \tag{13.23}$$

This resembles the WLF relation. However, the denominator in the WLF relation suggests that to match equation (13.23) Q must vary such that

$$Q = Q_0\frac{TT_0}{B+T-T_0}. \tag{13.24}$$

This implies that the activation energy Q is not a constant but increases rather rapidly as the temperature T decreases. This would be consistent

Figure 13.9. Time–temperature superposition (the WLF transform) applied to stress-relaxation data for polyisobutylene. From J. M. G. Cowie: *Polymers: Chemistry of Physics of Modern Materials,* Blackie, 1973.

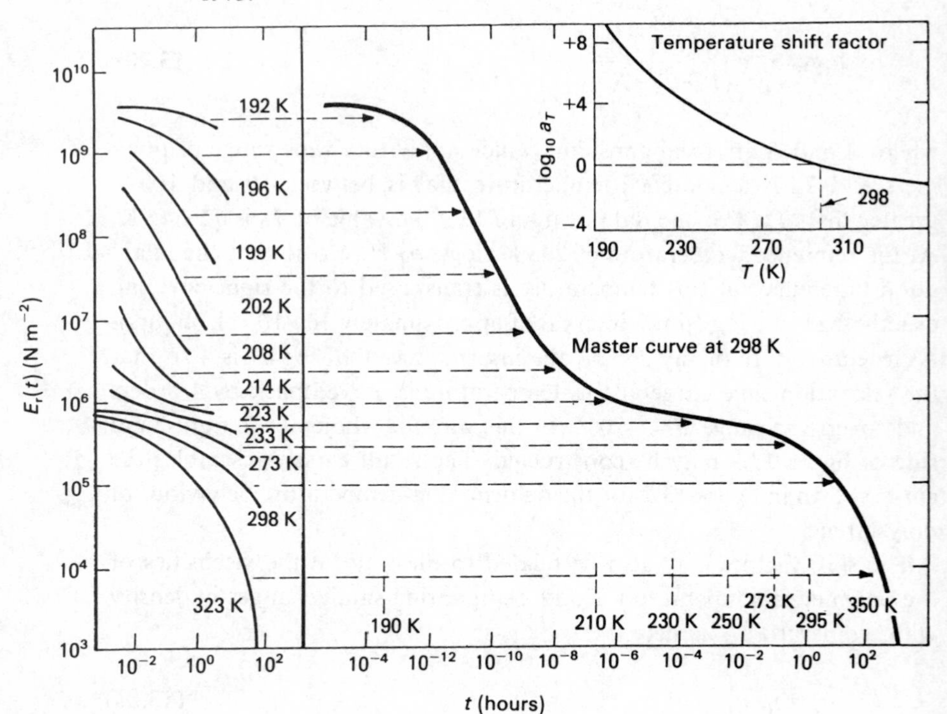

with the view that there is an additional term, maybe a pressure term, in the activation process such that as the polymer approaches T_g the constraints on bond rotation increase Q in a manner resembling equation (13.24).

The WLF transform provides a very convenient, if somewhat empirical way of correlating deformation rate and temperature effects. One would expect it to apply only at temperatures above T_g. In fact it extends successfully to temperatures well below this. One must therefore assume that the WLF procedure extrapolates into a supercooled non-glassy state.

13.12 The linear viscoelastic solid

It is often convenient to describe the rubbery–viscoelastic–glassy transitions in terms of a spring-and-dashpot model. The simplest of these is the arrangement known as the 'linear viscoelastic solid'. It consists of a weak spring of modulus E_a in parallel with a strong spring E_m attached to a dashpot of effective viscosity η (figure 13.10). Suppose an alternating stress is applied at different frequencies. At very low frequencies (i.e. quasi-static loading–unloading cycles), the viscous resistance of the dashpot will be negligibly small and no force will be transmitted to spring E_m.

Figure 13.10. Model illustrating behaviour of a linear viscoelastic solid under a variable stress σ. E_a is a weak spring, E_m a strong spring and the piston arrangement is a dashpot which provides viscous resistance. If the model is to exhibit stress-relaxation (see figure 13.9) a strong dashpot must be added as indicated within the broken lines.

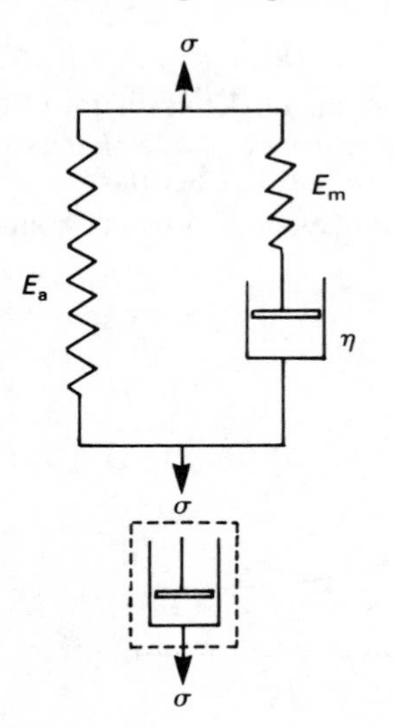

All the deformation occurs in spring E_a. The system behaves as a solid of low modulus and negligible loss, i.e. it resembles a pure rubber. Conversely at very high frequencies the viscous resistance of the dashpot will be so high that it will be virtually rigid and will not move. The stress will be carried by spring E_m in parallel with spring E_a. The system will behave like a solid of high modulus $(E_a + E_m)$ and of very low loss, i.e. it resembles the glassy state. At some intermediate frequency the loss (in the dashpot) will be a maximum and the behaviour corresponds to the viscoelastic state.

If the model is treated analytically it is found that the maximum dissipation in the dashpot occurs in the region where the effective modulus is changing most rapidly. These results are drawn schematically in figure (13.11). They show that the behaviour simulates many of the properties of an elastomer. The viscous dashpot is a reasonable substitute for the dissipative processes which occur when molecular chains 'rub' over one another or have to disengage themselves from mutual entanglements. However, the representation of the elastic properties by two springs of different but constant modulus is quite artificial. The polymer itself behaves as a spring of variable modulus, the variation depending on the frequency of deformation.

Note that if the system is extended by a fixed amount the stress in E_m will gradually fall to zero but in E_a it will remain at its original value. To achieve complete stress-relaxation and simulate the behaviour shown in figure 13.9, the model needs an additional powerful dashpot as indicated in the dotted frame of figure 3.10.

Figure 13.11. Characteristic behaviour of the model in Figure 13.10 showing the variation in modulus with frequency. This simulates the behaviour of an amorphous polymer. The insert shows the corresponding energy loss: the energy is dissipated in the dashpot.

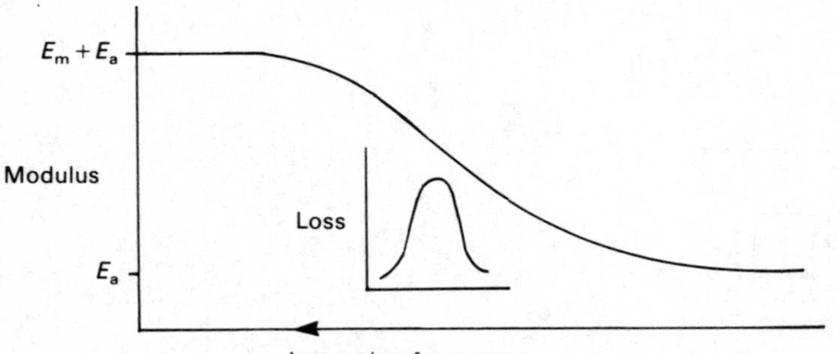

13.13 The morphology of the glassy state

13.13.1 *Amorphous polymers*

This may be summarized very briefly. The solid polymer has virtually the same morphology as the liquid state, with the molecules 'frozen' into the same configuration of tangles and intermingled chains.

13.13.2 *The crystalline state*

Many linear polymers such as PE, PTFE form crystals with a folded-chain structure. The crystallites consist of lamellae 10–20 nm thick with the chains perpendicular to the plane of the lamellae. By contrast, the molecular chain may be more than 1000 nm long (figure 13.12). Consequently the molecular chain within the crystallite lamella must be folded many times along its length. Further, the total volume of an individual lamella is greater than the volume of an individual molecule so that it must contain many molecules or portions of many molecules.

Figure 13.12. Sketch representing the morphology of a typical crystallite lamella in polyethylene indicating its connectedness with the polymer chains.

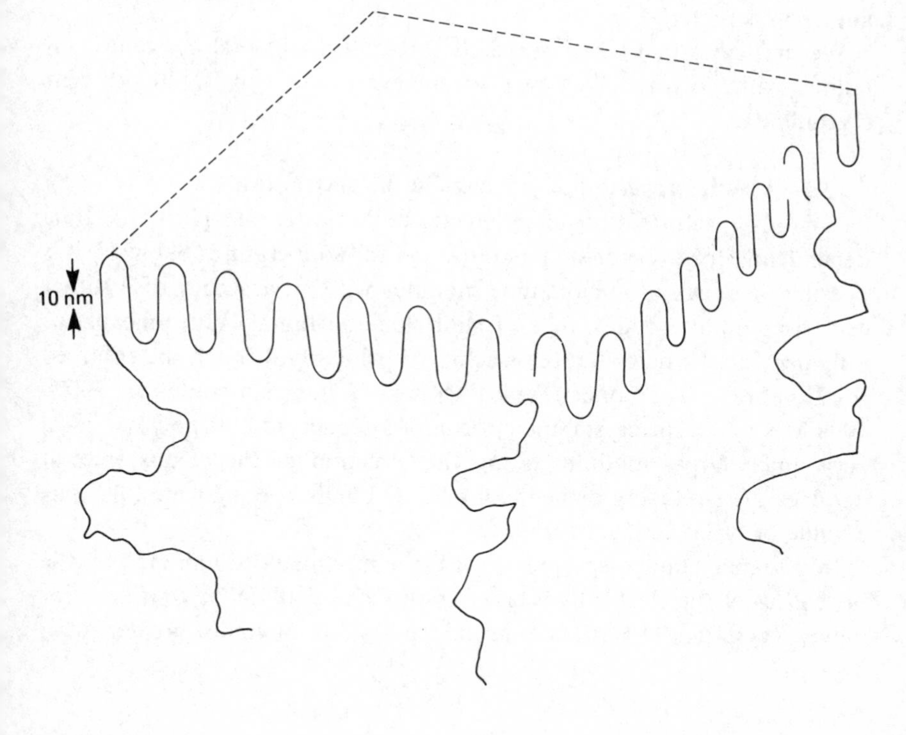

Neutron diffraction shows that the radius of gyration of the individual molecule in the solid state both for amorphous and crystalline polymers is almost the same as in the liquid state, implying that the rate of growth of the crystallites is far more rapid than the relaxation time of the whole molecule. This in turn implies that only a small part of the polymer is to be found in any individual lamella. Since the folds involve extra (bending) energy the lamellar structure is not the thermodynamically lowest energy state: it exists because of the kinetics of growth. The crystallinity depends on molecular mass and cooling rate.

Whereas the bulk structure of amorphous polymers resembles a vast agglomeration of closely packed randomly coiled spaghetti, the bulk structure of semicrystalline polymers must include crystalline lamellae, straggly amorphous regions and segments tying the lamellae together. With linear PE (which contains no sidegroups) polarized light studies show that the crystalline parts are in the form of radial leaves as sketched in figure 13.13. Apparently nucleation starts at some favourably situated region: fibrils and ribbons containing the crystallites grow out from the centre and impinge on one another and on similar growths from other nuclei so that their boundaries are polyhedral. These entities are known as spherulites. The gaps between the crystalline leaves are occupied by amorphous material.

We may note that whereas metals can be virtually 100 per cent crystalline, with polymers it is rare to achieve more than 30–40 per cent crystallinity.

13.14 Elastic properties of polymers in the glassy state

The deformation of polymers, both elastic and plastic, is time dependent. However for small strains and for temperatures below T_g it is possible to speak of a short-time modulus with an accuracy of ± 10 per cent over quite a wide range of short loading times. With amorphous polymers, elastic deformation is achieved primarily through the enforced rotation of the C–C bonds. Typical values of Young's modulus are 2000–4000 MN m^{-2}. The crystalline portion of a semicrystalline polymer has a very much larger modulus along the direction of the chains since it involves the stretching or bending of C–C bonds. Young's modulus has a value of order 2×10^5 MN m^{-2}.

In a semicrystalline specimen of PE the modulus is dominated by the properties of the elastically weaker amorphous material so that even for highly crystalline PE it is unusual to find a modulus greater than 5000 MN m^{-2}.

Table 13.2. *Typical values of Young's modulus E (GN m^{-2}) and Poisson's ratio ν at* 300 *K*

Material	T_g K	E (GN m^{-2})	ν
Steel	—	200	0.28
Inorganic glass	1000	80	0.23
Polystyrene	373	3	0.33
PMMA	318	4	0.33
Nylon 6-6	320	2	0.31
PE	188*	3	0.38
PE highly oriented along axis		~100	
Rubber		(0.1–1)10^{-3}	0.49

* PE like many other polymers behaves like a glassy polymer at temperatures above its true T_g. This is because there are other transitions which stiffen the chains before T_g is reached.

Figure 13.13. Spherulites in polyethylene. Crystalline leaves, consisting of an assembly of lamellae, grow from a nucleation centre. The material between is amorphous.

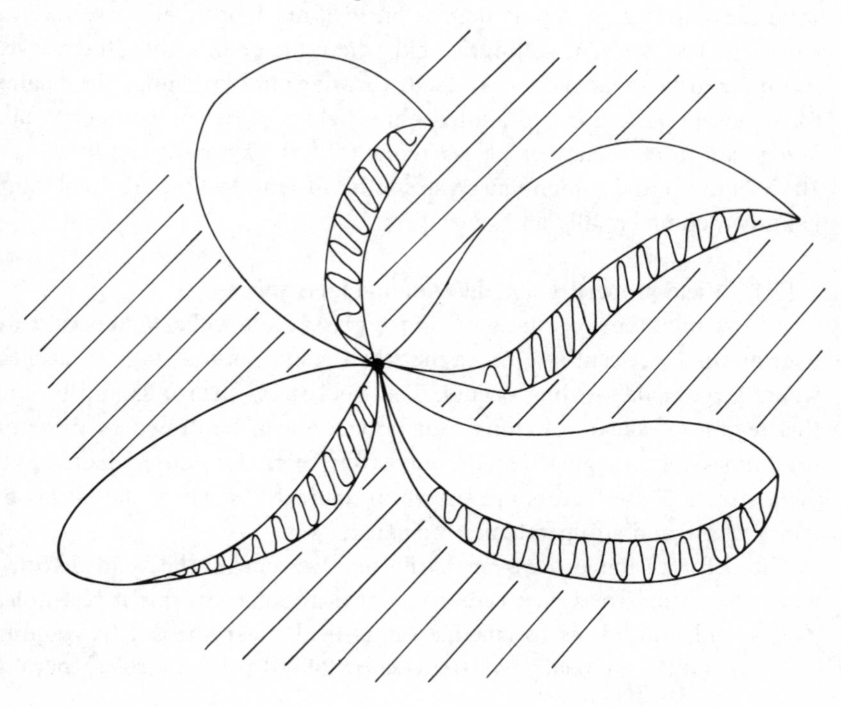

The modulus does not depend greatly on temperature or loading time at temperatures well below T_g: but at temperatures close to T_g it falls off appreciably. Some typical values at room temperature are given in Table 13.2.

The high modulus which characterizes the crystalline lamellae in PE may be approached in bulk specimens but by a different route. It is possible to draw the polymer from a suitable solvent so as to orient the chains parallel to the drawing direction. The elastic modulus in this direction is of order 10^5 MN m^{-2}. However, in the transverse direction the chains are held together only by van der Waals forces: the modulus is much lower and the cohesion relatively weak. As a result these drawn materials are exceptionally strong in tension but fibrillate rather easily.

This characteristic is also observed in certain polymers based on nematic liquid crystals. For example p-phenylene terephthalamide

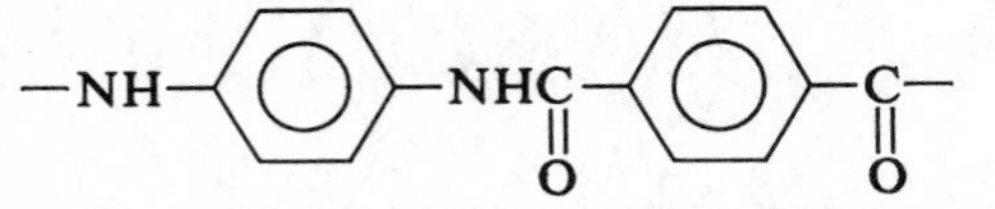

may be polymerized with some difficulty to produce polymers containing several hundred monomers. The polymeric chains are relatively stiff. The solid decomposes or depolymerizes on melting. It may however be dissolved in 100 per cent sulphuric acid where the chains align themselves spontaneously in nematic form. On drawing and spinning, the chains form closely packed interdigitated fibres held together by van der Waals and polar forces. They exhibit a remarkably high tensile modulus (*ca.* 10^5 MN m^{-2}) and a high tensile strength but tend to fibrillate. The fibre is known commercially as Kevlar.

13.15 **Yield properties and deformation mechanisms**

Under tension a polymer in the glassy state will extend elastically over about 1 per cent strain. As the stress is increased a stage is reached where large-scale yielding occurs. The yield stress then falls and it is in this regime of shear softening that polymers can be drawn to three or four times their original length: thereafter there is often an increase in yield stress. These factors are shown in figure 13.14 where the effects of temperature and strain rate are indicated.

The deformation process primarily involves shear: the solid deforms when the shear stress τ exceeds some critical value. In this it resembles metals and, indeed, as for metals the critical shear stress τ_c is roughly equal to half the uniaxial yield stress (see p. 208). Furthermore τ_c generally lies between one-twentieth and one-thirtieth of the elastic shear modulus (see p. 213) suggesting that τ_c is the 'ideal' theoretical shear strength of the polymer and that local slip occurs along the chains as it does in metal crystals over the atomic planes. There are, however, several differences. First, τ_c depends far more on shear rate and temperature than is the case with metals. Secondly, τ_c is markedly dependent on hydrostatic pressure p. In general one can write

$$(\tau_c)_p = \tau_c(1 + \alpha p) \tag{13.25}$$

where a is of order 0.1. In this it closely resembles the equation derived in Chapter 11 for the Eyring model of liquids in the high pressure range. For this reason the yield properties of polymers can often be expressed in terms of various thermally activated processes involving shear-stress and pressure activation energies. The main difficulty is to identify the particular process responsible for a specific activation energy.

A uniaxial tensile stress always involves a negative hydrostatic component: a compressive stress a positive hydrostatic component (p. 206). Consequently the uniaxial yield stress in compression is always larger than in uniaxial tension.

Figure 13.14. Typical stress–strain curve in tension of a polymer as a function of strain rate and temperature. The central portion corresponds to the region in which the polymer draws at a constant stress over a four-fold extension of its length.

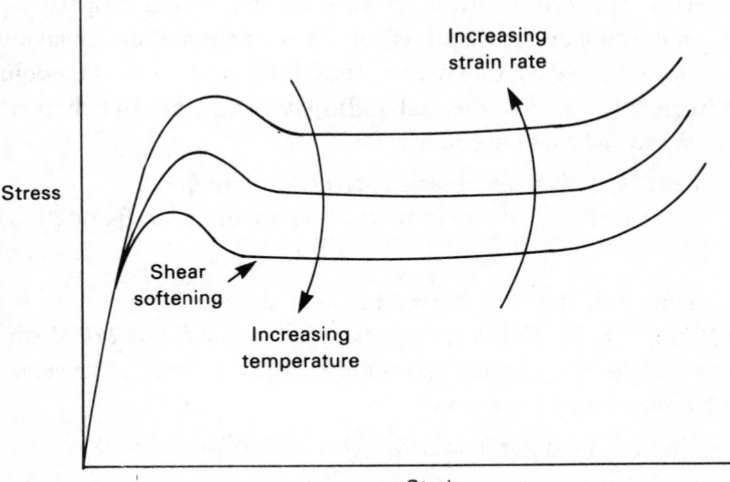

13.16 Morphological changes in shear and rupture

We first consider plastic flow under conditions where the specimen retains its integrity. A typical process is that of drawing. With truly amorphous polymers the chains are gradually aligned along the draw axis so that there is appreciable orientation compared with the original random arrangement. The negative hydrostatic pressure under tension tends to open up the polymer and so facilitates this process. However, the original chain entanglements remain substantially as they were. Thus if a specimen is drawn to four times its original length and then heated to a little above T_g it will return very nearly to its original shape and length.

With crystalline polymers the amorphous material takes up the first part of the drawing process. At a later stage the lamellae are pulled into a more favourable orientation or distorted in shear: in the final stages molecular chains are pulled out of the crystalline portions. Even under these conditions entanglements will remain and tend to pull the specimen back towards its original shape if the temperature is raised to T_g.

If the extension is more than four-fold the large increase in tensile stress will ultimately lead to rupture of the specimen. Part of this will be associated with chains slipping over and away from one another. However, it is most unlikely that the chain disentanglements will occur over the whole of the rupture area; consequently some chain scission is bound to occur. It is particularly marked in brittle fracture.

13.17 Thermal conductivity, diffusivity and adiabatic deformation

Before we turn to brittle fracture we deal with a property of a polymer that can have a major effect on its deformation behaviour, namely its low thermal conductivity K_s (primarily due to its low modulus) and consequently its low thermal diffusivity ($K_s/C\rho$). In Chapter 9 (p. 246) we showed that we could write

$$K_s = \tfrac{1}{3}\,(\text{specific heat capacity per unit volume}) \times (\text{velocity of sound waves}) \times (\text{phonon mean free path } \lambda). \tag{13.26}$$

For PE as for most plastics the specific heat at room temperature is of order 10^3 J kg^{-1} K^{-1}. This corresponds to between kT and $2kT$ thermal energy per CH_2 group. The density is of order 10^6 J m^{-3} K^{-1}. The velocity of sound waves may be taken as

$$(\text{Young's modulus/density})^{\frac{1}{2}} = (3\times 10^9/10^3)^{\frac{1}{2}} = 1.7\times 10^3 \text{ m s}^{-1}. \tag{13.27}$$

We take the phonon free path λ to be of order 1 nm (10^{-9} m). Hence

$$K_s = \tfrac{1}{3} \times 10^6 (1.7 \times 10^3) \times 10^{-9} = 0.6 \text{ J m}^{-1} \text{ s}^{-1} \text{ K}^{-1}$$
$$= 0.6 \text{ W m}^{-1} \text{ K}^{-1}. \qquad (13.28)$$

This agrees with observed values of K_s to within a factor of 2 for most amorphous polymers. It may be up to ten times greater for crystalline polymers and for highly drawn polymers in the direction of drawing partly because the modulus is greater and partly because λ is larger. However, the main conclusion is that the thermal conductivity is 100–1000 times smaller than that of metals. For example, for copper at room temperature $K_s = 400$ W m^{-1} K^{-1}. Since the specific heat per unit mass × specific gravity ($C\rho$) does not vary by more than a factor of 10 for most solids we are left with the conclusion that the thermal diffusivity ($K_s/C\rho$) is many hundreds of times smaller than for metals. For this reason if the shear rate exceeds a certain level there may not be time for the heat generated in the shear zone to diffuse away. The shear zone experiences an appreciable temperature rise; the polymer softens precisely in this region and shear becomes a 'runaway' process. Adiabatic shear of this type also occurs with metals but at very much higher rates of deformation.

13.18 Brittle behaviour

If a polymer is extended very rapidly at room temperature (but not rapidly enough to produce appreciable adiabatic effects) or if it is extended at room temperatures well below T_g there may be insufficient time for segmental motion within the chain to keep pace with the imposed extension. The specimen will snap. This is a typical example of tensile fracture and it often occurs under conditions of impact where the loading is short and sharp. It also occurs for quasi-static loading if the temperature is low. For example polypropylene has a glass transition temperature of 278 K, i.e. 5 °C. At −20 °C it is very brittle. On the other hand T_g for PE is 188 K so that it remains ductile at this low temperature. It is for this reason that in using polymers for underground water pipes and gas pipes polyethylene is used instead of polypropylene.

Brittle fracture usually involves the nucleation of a crack which propagates and spreads through the specimen. It can occur in shear as well as in tension but in what follows we shall deal only with tensile fracture. Consider the case in which a stress is applied to a specimen in which there is a crack of length l orthogonal to the direction of the stress. Then, as we saw in Chapter 8 (p. 226), the crack will grow when the stress reaches a critical value given by

$$\sigma_G = \left(\frac{\gamma E}{l}\right)^{\frac{1}{2}} \tag{13.29}$$

where γ is the fracture surface energy of the polymer. We may readily estimate the value of γ by assuming that fracture simply involves the breaking of C–C bonds across the fracture surface. Suppose that near the fracture surface the relevant part of the polymer chain is normal to the surface. The average area occupied by each H–C–H group is $20 \times 10^{-20}\ \mathrm{m}^2$: thus each unit area of surface is crossed by 5×10^{18} C–C bonds. According to the theory of covalent bonding the rupture of each C–C bond involves an energy of 5.3×10^{-19} J. Thus the energy in producing the *two* unit areas of the fracture surface is $5.3 \times 10^{-19} \times 5 \times 10^{18} \approx 3$ J. Without attempting to distinguish between the total and the free surface energy, we may therefore deduce for the theoretical fracture surface energy

$$\gamma \approx 1.5\ \mathrm{J\ m^{-2}}.$$

However, in fracture experiments the values of γ deduced from equation (13.29) are often 100 or 1000 times larger. This is because (i) severe irreversible viscoelastic flow occurs around the head of the crack and (ii) there is pulling out of chains and fibrils at the tip of the crack usually in the form of crazing as shown in figure 13.15. Both these dissipative processes depend on temperature and on rate of crack propagation: they swamp the free surface energy as estimated above. A material for which the fracture energy is high is said to have high fracture toughness.

The craze material has a thickness of about 2.5 nm near the crack tip, increases to an average thickness of 80 nm as one moves away from the tip and reaches a thickness typically of several hundred nm: at this stage the craze material must break by chain slippage or by rupture. The fact that the density of the craze material is about one-half that of the bulk polymer suggests that the polymer chains pull out of one another by slippage for an appreciable part of the process of craze formation.

Craze formation has been observed in PMMA at 100 K which is nearly 200 K below its T_g. Evidently extensive flow and deformation can occur far below the glass transition temperature. Crazing is most marked in polymers with the greatest distance between chain entanglements. Conversely cross-linking can lead to the complete disappearance of crazing. In that case shear deformation occurs at the crack tip instead of craze formation.

In the presence of certain organic solvents which act as plasticizers crazing occurs at greatly reduced stresses. The solvent thus greatly reduces the fracture energy.

13.19 Cross-linked polymers: some brief comments

These fall into roughly three groups:

(i) The basic material consists of long chains which in the liquid state adopt the intermingled random configuration discussed earlier. The chains are cross-linked by chemical bonding agents or by irradiation. When fully cross-linked the material is hard, glassy and brittle. It cannot be moulded once it has reached this stage.

(ii) The material is as in (i) but the cross-links contain groups that can melt at modest temperatures. When the temperature falls these groups solidify and the polymer is hard and glassy. The material has the merit that at the high temperature it can be moulded.

(iii) Two or more monomers may be mixed and polymerized *in situ* to form a fully branched three-dimensional network.

Broadly speaking cross-linked polymers are like thermoplastics in the glassy state but they have a higher modulus and are stronger in tension. They are brittle at all temperatures.

Figure 13.15. Typical craze formation around the tip of a crack in Perspex. Polymer chains and fibrils are drawn out of each surface as the crack opens.

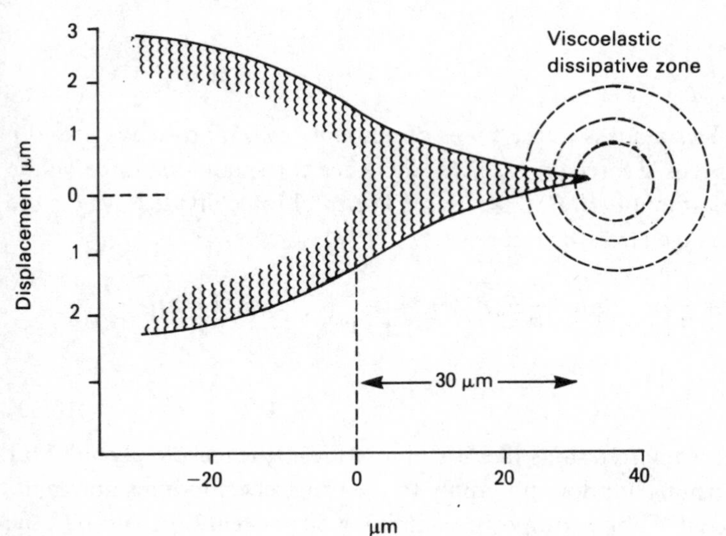

13.20 Polymer composites

It is not the purpose of this chapter to describe the nature and properties of polymer composites as engineering materials. We refer here only to two aspects of physical interest.

Fibres or mats of fibres are often incorporated into polymers as they are moulded to increase their modulus, their strength, their fatigue resistance, and their fracture toughness. Most of the fibres used have diameters between 7 and 15 μm. They are chosen for their high tensile modulus and high tensile strength but are usually relatively brittle (extensibility less than 1 per cent strain). Some typical fibres are carbon, modulus 200–400 GN m^{-2}; highly oriented PE and Kevlar, modulus 150–200 GN m^{-2} and glass, modulus 80–100 GN m^{-2}. These values may be compared with those of typical thermoplastics quoted in Table 13.2, namely 2–4 GN m^{-2}.

(a) *Tensile modulus*

We consider the simplest example: a tensile specimen of uniform cross section A. Uniform continuous fibres are incorporated over the whole length of the specimen. If the volume fraction of the fibre is v then at any section the cross-sectional area of the fibres is Av, of the polymer $A(1-v)$. We write E_f, E_p for the moduli of fibre and polymer respectively. If we apply a tensile force to the composite implying an average stress $\sigma = \text{force}/A$ we assume that the stress distributes itself over the fibre and polymers so that both components of the composite extend by the same amount. If E represents the modulus of the composite the extensional strain is

$$\varepsilon = \frac{\sigma}{E}.$$

This produces a stress in the fibres of amount $(\sigma/E)E_f$ over the area Av implying a tensile force $(\sigma/E)E_f Av$: there is a corresponding force on the polymer of amount $(\sigma/E)E_p A(1-v)$. The total force divided by A is the applied average stress σ. Hence

$$\sigma = \frac{1}{A}\left[\frac{\sigma}{E} E_f Av + \frac{\sigma}{E} E_p A(1-v)\right] = \frac{\sigma}{E}[E_f v + E_p(1-v)]$$

or

$$E = E_f v + E_p(1-v). \tag{13.30}$$

This result is referred to as the rule of mixtures. It is reasonably valid for small extensions but does not apply to the transverse modulus nor to the tensile strength. For a composite containing 50 per cent fibre ($v = 0.5$) the modulus of a typical thermoplastic is increased 20–50-fold. It thus becomes a useful structural material even if the strains are large enough to produce fracture of the relatively brittle fibres. The main assumption of this model is that there is strong adhesion between fibre and polymer.

(b) *Fracture toughness of the composite*

We consider here the simplest case of a crack or notch in a composite and the work done to propagate the crack. If the fibres are normal to the crack and if the adhesion between fibre and polymer is very strong a relatively small opening of the crack will fracture the fibres (figure 13.16(*a*)). Although they increase the modulus they do not add substantially to the work of fracture. On the other hand if the adhesion between fibre and polymer is below some critical value the fibres will be pulled out of the matrix: they may fracture some way back away from the crack but an appreciable length will be drawn out and bridge the crack (figure 13.16(*b*)). In this process work will be expended, the amount of work depending on the length of fibre pulled out and on the shear

Figure 13.16. Crack formation in a fibre–polymer composite. The fibres are assumed to be at right angles to the direction of the crack. (*a*) Strong fibre–polymer adhesion: the fibres fail in a brittle manner after very little extension. (*b*) Moderate fibre–polymer adhesion: the fibres are drawn out of the polymer and dissipate a great deal of frictional energy before finally fracturing.

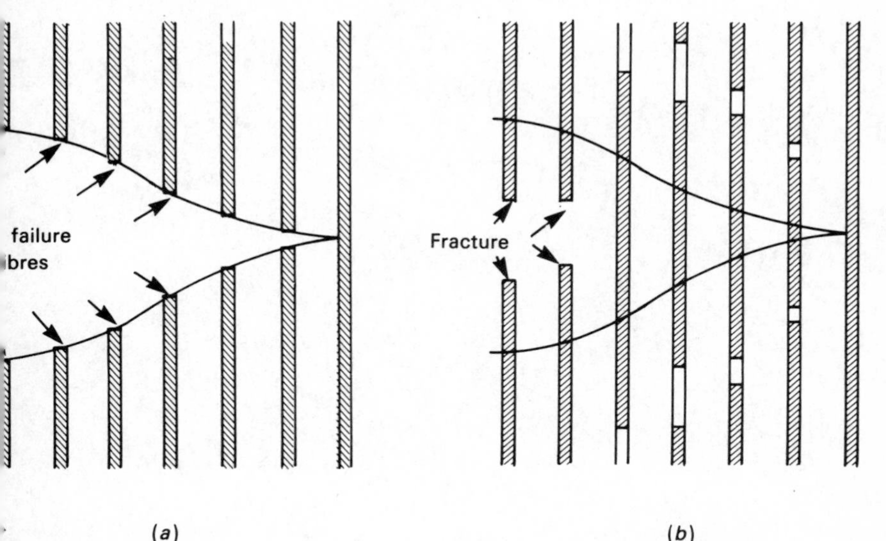

strength of the fibre–polymer interface. The latter can be modified by coating the fibres with materials of appropriate shear strength. In this way the fibres can greatly increase the fracture energy.

We see from this short discussion that strong adhesion is desirable for a maximum increase in modulus: but for high fracture toughness a lower level of adhesion is required. Although the theory of fibre–polymer interaction is well developed, the practical control of fibre–polymer adhesion to produce optimum properties in polymer composites is still something of an art rather than a science.

14

Dielectric properties of matter

In this chapter we shall describe the main dielectric properties of matter and see how far they can be explained in terms of simple atomic models. As we shall see the main mechanism is the polarization of individual atoms and molecules by an electrostatic field.

14.1 Basic dielectric relations

14.1.1 *Dielectric constant (relative permittivity) ε and polarization P*

Suppose we consider a parallel plate condenser carrying a charge σ coulombs for each m^2 of surface. The field in free space between the plates (see figure 14.1(*a*)) is

$$E_v = \sigma/\varepsilon_0, \tag{14.1}$$

where E_v is measured in V m^{-1} and ε_0 is the permittivity of free space and equals 8.85×10^{-12} F m^{-1}. The potential difference between the plates is $\sigma d/\varepsilon_0$ and the capacity C per m^2 of condenser is

$$C = \frac{\text{charge}}{\text{potential}} = \frac{\sigma}{\sigma d/\varepsilon_0} = \frac{\varepsilon_0}{d}. \tag{14.2}$$

We now consider what happens if we fill the intervening space with a dielectric. We find that the capacity or relative permittivity is increased by a factor ε. This is known as the dielectric constant of the material.

For an *isolated* condenser, since σ remains unchanged, this means that the potential difference between the plates is reduced by a factor ε, so that the field is reduced from $E_v = \sigma/\varepsilon_0$ to

$$E = \frac{\sigma}{\varepsilon_0}\frac{1}{\varepsilon}. \tag{14.3}$$

This reduction in field is due to the polarization of the material. The field distorts the atoms and molecules, and orients existing dipoles in such a way as to create a field opposing the applied field. The resultant field is thus reduced.† We define a polarization P as the electrostatic moment per unit volume induced in the dielectric. As we shall see later P is very nearly proportional to the resultant field E. Consider a bar of dielectric of section A, length l. Its total moment due to polarization is

$$P \times \text{volume} = PAl = PA \times l = \text{end charge} \times \text{length}.$$

Thus the moment per unit volume is equivalent to a surface charge P per unit area. We now see that the effective charge density on the plates is $\sigma - P$ (see figure 14.1(b)). The resultant field is then

$$E = \frac{1}{\varepsilon_0}(\sigma - P)$$

or

$$\sigma - P = \varepsilon_0 E. \tag{14.4}$$

Figure 14.1. (a) The field in an isolated condenser is E_v. (b) When filled with a dielectric the material is polarized and the field is reduced to E. (c) Relation between E and E_v.

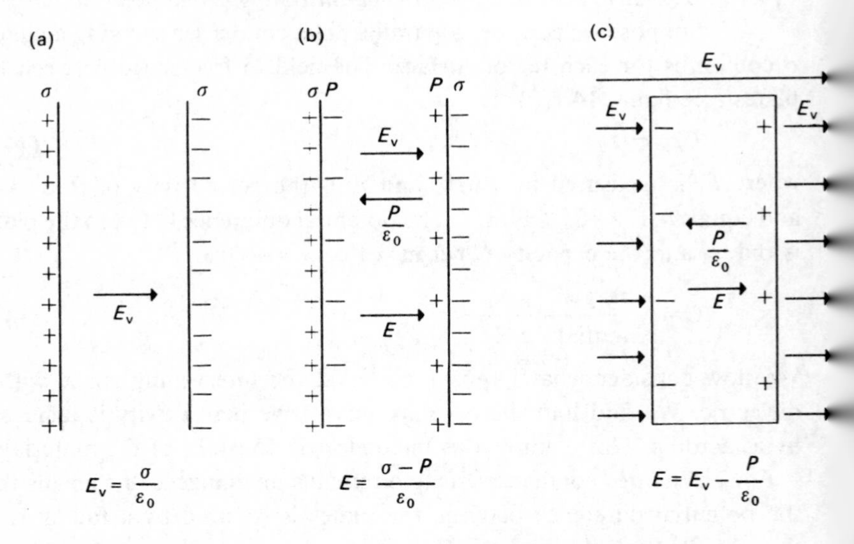

† If we had considered a condenser maintained in a circuit at a constant potential we could show that the insertion of the dielectric involves a further flow of charge such that the total charge, for the same potential, is increased by a factor ε. Here again the free charges on the condenser plates are partly offset by the induced charges in the dielectric.

But from equation (14.3), $\sigma = \varepsilon\varepsilon_0 E$ and equation (14.4) becomes

$$\varepsilon\varepsilon_0 E - P = \varepsilon_0 E,$$

or

$$P = (\varepsilon - 1)\varepsilon_0 E. \tag{14.5}$$

Thus a dielectric which reduces the electrostatic field by a factor ε produces a polarization P which is $(\varepsilon - 1)\varepsilon_0$ times the real field. Equation (14.5) may also be written

$$(\varepsilon - 1) = \frac{P}{\varepsilon_0 E} = \chi, \tag{14.6}$$

where χ is defined as the volume susceptibility of the dielectric. In what follows we shall show how P can be derived in terms of atomic models. Before doing so, however, we refer for completeness to the displacement vector D.

14.1.2 *Displacement vector D*

In passing from vacuum to a dielectric the field changes from E_v to E. Taking the simplest case of a parallel block of dielectric normal to the field we see from figure 14.1(*c*) that

$$E = E_v - P/\varepsilon_0$$

or

$$E_v = E + P/\varepsilon_0 = E\varepsilon.$$

There is a discontinuity in the field. Another quantity called the displacement vector D may now be defined

$$D = \varepsilon E. \tag{14.7}$$

Then in vacuum, $D_v = \varepsilon_v E_v = E_v$ (since $\varepsilon_v = 1$), and in the dielectric

$$D = \varepsilon E = E_v.$$

The quantity D is thus continuous. We shall not use it further.

14.2 **Polarization of gases**

14.2.1 *Mechanisms of polarization*

We first consider the behaviour of gases since the molecules are, generally, so far apart that interactions between them may be neglected. Consequently each gas molecule 'sees' the dielectric field E. This polarizes the molecule in three ways:

(*a*) By distorting the electronic cloud around the nucleus or nuclei of each molecule; we call the polarization due to this mechanism P_e.

(*b*) By stretching or bending bonds in polar molecules; we call this P_a. Usually this is small ($P_a \sim 0.1 P_e$) and can be neglected.

(*c*) By orienting existing molecular dipoles; we call this P_d.

We first consider the electronic polarization P_e. Suppose the molecule is initially non-polar. As a result of electronic displacement the individual molecule (or atom) will acquire a dipole moment p. As we shall see immediately p is proportional to E, so we may write

$$p = \alpha E, \tag{14.8}$$

where α is the molecular (or atomic) polarizability. If there are n molecules per m^3

$$P_e = np = n\alpha E. \tag{14.9}$$

We at once note that the dimensions of P_e are charge/area whilst the dimensions of E are charge/(distance)$^2 \varepsilon_0$, so they are dimensionally identical apart from the $1/\varepsilon_0$ term. Hence $n\alpha$ has dimensions of ε_0. Thus dimensionally

$$[\alpha] = \left[\frac{1}{n}\right] \varepsilon_0 = [L^3]\varepsilon_0.$$

In fact, as we shall see in what follows, L is of the order of the molecular radius.

14.2.2 *The polarization of neutral monatomic molecules*

Consider a neutral atom such as argon or neon, which consists of a central positive charge Ze and an electron cloud of charge $-Ze$ (see figure 14.2(*a*)). We assume that the electron cloud has uniform charge density ρ and extends over a sphere of radius r. If a field E is applied the electron cloud experiences a force EZe in one direction and the nucleus an equal force in the opposite direction. The charges are displaced relative to one another until the attractive force between them exactly equals the force exerted by the field. Let the displacement be x (see figure 14.2(*b*)). The attraction between the nucleus and the electron cloud is the force between Ze on the nucleus and that part of the electronic charge which is within the sphere of radius x.

The latter is $Ze(x/r)^3$. The attractive force (see figure 14.2(*c*)) is then

$$f = \frac{(\text{positive charge}) \times (\text{negative charge})}{4\pi\varepsilon_0 x^2},$$

i.e.

$$f = Ze \times Ze\left(\frac{x}{r}\right)^3 \frac{1}{4\pi\varepsilon_0 x^2} = \frac{Z^2e^2x}{4\pi\varepsilon_0 r^3}. \tag{14.10}$$

This is balanced by the distorting force

$$f = EZe. \tag{14.11}$$

Equating (14.10) and (14.11) we have

$$E = \frac{1}{4\pi\varepsilon_0}\frac{Zex}{r^3}. \tag{14.12}$$

But Zex is the charge times the distance between the charges, i.e. it is the atomic dipole p. Hence

$$p = 4\pi\varepsilon_0 r^3 E. \tag{14.13}$$

A similar result is obtained if the atom is assumed to be a perfectly conducting sphere of radius r. Hence

$$\alpha = p/E = 4\pi\varepsilon_0 r^3.$$

It is interesting to compare the radius of an atom or a molecule, deduced from transport processes (see Table 3.5, p. 75) with the value of $(\alpha/4\pi\varepsilon_0)^{\frac{1}{3}}$ derived from the measured dielectric properties (p/E) of the material. Typical results are given in Table 14.1.

Continuing from equation (14.13) we may now write

$$P_e = np = n4\pi\varepsilon_0 Er^3,$$

Combining this with equation (14.5) we have

$$\varepsilon - 1 = 4\pi nr^3. \tag{14.14}$$

14.2.3 *Dielectric properties and van der Waals critical data*

We now make a wide-ranging and rather unexpected correlation between ε and critical gas data. We note that $4\pi nr^3$ is three times the volume of the molecules in 1 m^3 of gas. The critical volume of the gas is

Figure 14.2. An atom of atomic number Z considered as a central positive charge surrounded by a negative charge of uniform density. The application of a field E displaces the positive relative to the negative charge.

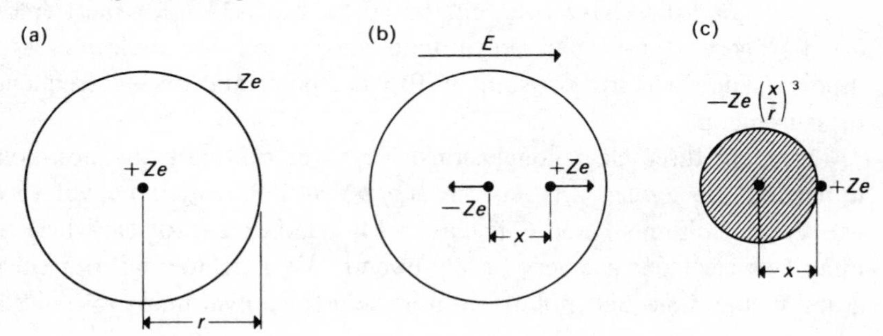

Table 14.1. *Polarizability and molecular radius r (units 10^{-10} m)*

Gas	$\alpha=p/E$ (10^{-40} F m^2)	$(\alpha/4\pi\varepsilon_0)^{\frac{1}{3}}$ (10^{-10} m)	r (10^{-10} m)
He	0.24	0.6	1.1
Ar	1.9	1.2	1.7
H_2	0.95	0.95	1.3
N_2	1.9	1.2	1.8
CO_2	3.1	1.4	2.1

12 times the volume of the molecules. It follows that $4\pi nr^3$ is equal to $\frac{1}{4}(\rho/\rho_c)$, where ρ is the density of the gas and ρ_c the density at the critical point. Hence

$$(\varepsilon-1)=\frac{1}{4}\frac{\rho}{\rho_c}. \tag{14.15}$$

According to Maxwell's theory of electromagnetic waves, at the same wavelength, the refractive index $\mathcal{R}$ is related to the dielectric constant ε and the magnetic permeability μ by the relation

$$\mathcal{R}=(\varepsilon\mu)^{\frac{1}{2}}. \tag{14.16}$$

For dielectrics μ is equal to unity to within one part in 10^6. Since for gases ε is only slightly greater than unity we may write

$$\mathcal{R}=\varepsilon^{\frac{1}{2}}=[1+(\varepsilon-1)]^{\frac{1}{2}}=1+\tfrac{1}{2}(\varepsilon-1), \tag{14.17}$$

so that

$$(\mathcal{R}-1)\approx\tfrac{1}{2}(\varepsilon-1). \tag{14.18}$$

Combining with (14.15) we have

$$\frac{1}{4}\left(\frac{\rho}{\rho_c}\right)=(\varepsilon-1)=2(\mathcal{R}-1). \tag{14.19}$$

These quantities have been compared in Table 14.2 for inert (monatomic) gases, non-polar polyatomic gases, and for molecules with dipoles. The dielectric constant is found from static or low-frequency measurements.

There are three clear conclusions. First, for the inert and non-polar molecules the agreement between $\frac{1}{4}(\rho/\rho_c)$ and the observed values of $(\varepsilon-1)$, see columns 5 and 6, is reasonably good except for He where the inner two electrons are very tightly bound. We conclude that our calculation of the molecular polarization is valid for polyatomic molecules as

Table 14.2. *Dielectric properties of gases at s.t.p.*

1	2	3	4	5	6	7
Type	Gas	ρ (kg m^{-3})	$\rho_c \times 10^{-3}$ (kg m^{-3})	$\frac{1}{4}\frac{\rho}{\rho_c}$ $\times 10^4$	$(\varepsilon-1)$ $\times 10^4$	$2(\mathscr{R}-1)$ $\times 10^4$
Inert	He	0.18	0.069	6.5	0.7	0.7
	Ne	0.90	0.48	4.7	1.3	1.3
	Ar	1.78	0.53	8.4	5.5	5.7
	Xe	5.90	1.15	12.2	15.0	14.0
Non-polar molecules	H_2	0.09	0.031	7.2	2.7	2.8
	N_2	1.25	0.31	10.0	5.8	5.9
	O_2	1.43	0.43	8.3	5.2	5.4
	Cl_2	3.21	0.57	14.1	—	15.4
	CO_2	1.98	0.46	10.8	9.9	9.9
Molecules with dipoles	NH_3	0.77	0.23	8.4	71	7.5
	SO_2	2.93	0.53	13.8	82	13.2
	H_2O	0.60	0.32	4.7	80	5.0

well as for monatomic molecules. Secondly, for all these gases there is very good agreement between $(\varepsilon-1)$ and $2(\mathscr{R}-1)$ – columns 6 and 7. Since the optical refractive index is a measure of the interaction of the molecules with the high-frequency electric field associated with light waves we deduce that this is the same as the electronic polarization produced by a static or low-frequency electric field in these cases.

Thirdly, we note that polar molecules give good agreement between $\frac{1}{4}$ (ρ/ρ_c) and $2(\mathscr{R}-1)$ – columns 5 and 7 – since both are a measure of high-frequency electronic polarization. There is poor agreement between $(\varepsilon-1)$ and $2(\mathscr{R}-1)$ – columns 6 and 7 – since ε is a low-frequency determination and includes the orientation effects of the molecule. By contrast $\mathscr{R}$ involves optical frequencies where only the electronic polarizations can respond. Indeed the difference between $(\varepsilon-1)$ and $2(\mathscr{R}-1)$ for ammonia suggests that the orientation of the dipole contributes 63×10^{-4} to the low-frequency dielectric constant.

14.3 **Polarization of polar molecules**

14.3.1

Apart from the electronic polarization, polar molecules show two other effects:

(*a*) The stretching or bending of bonds. This produces a polarization P_a which is generally very small. We shall not attempt to estimate it quantitatively.

(*b*) An orientation of existing dipoles. The polarization P_d due to this is large and may be as much as $100P_e$.

Materials in which existing dipoles are oriented by the applied field are sometimes referred to as para-electric materials. In what follows we provide a simplified model by means of which we can estimate the magnitude of P_d quantitatively.

The simplest example of a dipole (see figure 14.3) is a diatomic molecule with a charge e at one end and $-e$ at the other and a separation of say 0.1 nm. The permanent dipole has a moment

$$\begin{aligned} p &= \text{charge} \times \text{separation} \\ &= 1.6 \times 10^{-19} \times 10^{-10} \\ &= 1.6 \times 10^{-29}\ \text{C m}. \end{aligned} \tag{14.20}$$

In the absence of a field these dipoles are randomly oriented and show no resultant moment. If an electric field could orient all the molecules we should obtain a polarization per m^3 given by

$$P_d = np.$$

For a gas at s.t.p. n is about 3×10^{25} per m^3. Hence

$$P_d \approx 5 \times 10^{-4}\ \text{C m}^{-2}.$$

The field E_d produced in the dielectric by these dipoles is

$$E_d = P_d/\varepsilon_0 \approx 6 \times 10^7\ \text{V m}^{-1}. \tag{14.21}$$

This would be more than enough to ionize most gases, but experiments show that it is virtually impossible to ionize a gas in this way. The reason is that the calculation ignores thermal motion which vigorously opposes the lining-up of the dipoles. In most practical situations the polarization achieved is very much less than that calculated above.

14.3.2 *Orientation of dipoles in an electric field: the Langevin function (para-electric behaviour)*

We consider the effect of a field E on an assembly of dipoles of moment p. The field attempts to orient the dipoles, thermal motion to maintain disorder.

Figure 14.3. A simple electrostatic dipole.

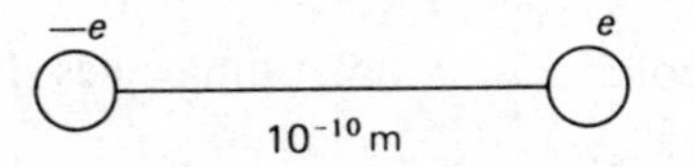

We first calculate the energy of a single dipole in the field if its axis makes an angle θ to the direction of E (see figure 14.4(*a*)). Let the dipole consist of charges q and $-q$ separated by a distance $2a$. Then $p=2aq$. The potential energy is defined as work done on the system by external forces. The forces on the charges are qE and $-qE$ (figure 14.4(*b*)). If both charges were at some arbitrary position, say at O, we should do work $-qEx$ on the positive charge and $qE(-x)$ on the negative charge in moving them to their present positions. The total potential energy ε is

$$\varepsilon=-2qEx=-2qEa\cos\theta$$

or

$$\varepsilon=-Ep\cos\theta. \tag{14.22}$$

We now consider how much space is occupied by orientations between θ and $\theta+\mathrm{d}\theta$. Construct a sphere of arbitrary radius R (figure 14.4(*c*)). The solid angle subtended between θ and $\theta+\mathrm{d}\theta$ is

$$\mathrm{d}\omega=\frac{\text{area of annulus}}{R^2}=\frac{2\pi R\sin\theta R\,\mathrm{d}\theta}{R^2},$$

or

$$\mathrm{d}\omega=2\pi\sin\theta\,\mathrm{d}\theta. \tag{14.23}$$

We now apply the Boltzmann distribution. The number of dipoles with axes between θ and $\theta+\mathrm{d}\theta$ is

$$2\mathrm{d}n=B\exp\left(\frac{-\varepsilon}{kT}\right)\times(\text{element of space})$$

$$=B\exp\left(\frac{pE\cos\theta}{kT}\right)2\pi\sin\theta\,\mathrm{d}\theta. \tag{14.24}$$

Figure 14.4. The orientation of a dipole of moment $2aq$ in a field of strength E.

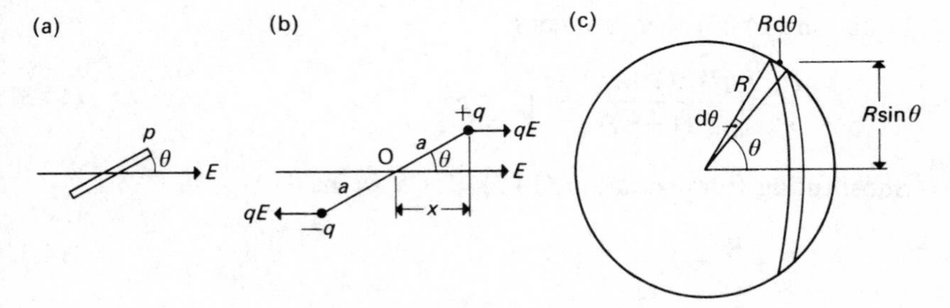

As this is a rather clumsy expression we make the following substitutions:

$$\left.\begin{aligned} x &= \frac{pE}{kT}, \\ z &= \cos\theta \text{ so that } dz = -\sin\theta\, d\theta. \end{aligned}\right\} \tag{14.25}$$

Then equation (14.24) becomes

$$dn = -2\pi B \exp(zx)\, dz. \tag{14.26}$$

We find B by integrating dn for θ from 0 to π, that is for z from $+1$ to -1. This gives n, the total number of dipoles per m^3;

$$n = 2\pi B \int_{-1}^{+1} \exp(zx)\, dz. \tag{14.27}$$

(The change in sign for n follows because we have reversed the order of the limits of integration.) Because of symmetry the resultant polarization normal to E will be zero, and only the polarization parallel to E will be additive. Each dipole contributes a dipole moment parallel to E of amount $p \cos\theta$. Thus the resultant dipole moment per unit volume P_d will be

$$\begin{aligned} P_d &= \int_{\theta=0}^{\theta=\pi} p \cos\theta\, dn \\ &= 2\pi B p \int_{-1}^{+1} z \exp(zx)\, dz. \end{aligned} \tag{14.28}$$

We can express the ratio of P_d to n (using equations (14.27) and (14.28)) as:

$$\frac{P_d}{n} = \frac{p \int z\, e^{zx}\, dz}{\int e^{zx}\, dz}, \tag{14.29}$$

where the integral is from $z = -1$ to $z = +1$.

In most practical situations zx is small so that

$$e^{zx} \approx 1 + zx. \tag{14.30}$$

Equation (14.29) then becomes

$$\frac{P_d}{n} = \left[\frac{p \int z(1+zx)\, dz}{\int (1+zx)\, dz}\right]_{-1}^{+1} = p\frac{x}{3}. \tag{14.31}$$

Substituting from equation (14.25) for x we have

$$P_d = \frac{np^2E}{3kT}. \tag{14.32}$$

We may at once note that for most dielectrics E cannot exceed about $10^7\ \mathrm{V\,m^{-1}}$, otherwise breakdown occurs. For a typical value of $p = 1.6\times10^{-29}$ C m, using $k = 1.4\times10^{-23}\ \mathrm{J\,K^{-1}}$, and $T = 300$ K, we have

$$\frac{pE}{kT} \approx 0.03,$$

so that our assumption implied in equation (14.30) is valid.

Exact integration of equation (14.29) gives the Langevin function

$$P_\mathrm{d} = np\left[\coth\left(\frac{pE}{kT}\right) - \frac{kT}{pE}\right]. \tag{14.33}$$

This function is plotted in figure 14.5 and shows that under normal conditions the linear part of the curve, far removed from saturation, is the only part that is observable. The contribution of P_d to the dielectric constant may now be calculated. We have

$$P_\mathrm{d} = (\varepsilon - 1)_\mathrm{d}\varepsilon_0 E = \frac{np^2E}{3kT}$$

or

$$(\varepsilon - 1)_\mathrm{d} = \frac{np^2}{\varepsilon_0 3kT}. \tag{14.34}$$

For ammonia at s.t.p. we have $n = 3\times10^{25}\ \mathrm{m^{-3}}$, $p \approx 0.5\times10^{-29}$ C m. Then

$$(\varepsilon - 1)_\mathrm{d} = 67\times10^{-4}.$$

Figure 14.5. The Langevin relation for the polarization of dipoles produced by a field E.

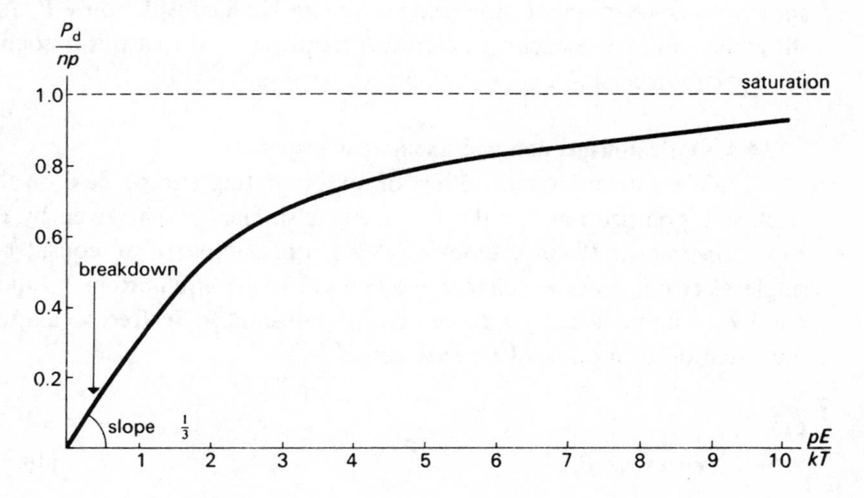

This agrees extremely well with the value quoted in the discussion on p. 369. Measurements of the dielectric constant are often used by the chemist to deduce the value of the molecular dipole.

The orientation polarization can, of course, be increased without exceeding the breakdown of the dielectric by reducing the temperature. This cannot, however, be carried very far since most *polar* substances solidify fairly readily and under these conditions molecular orientation becomes very difficult or impossible.

14.3.3 *Effect of frequency of electric field*

We can at once see the powerful effect of frequency on the polarization. We write

$$P = P_d + P_a + P_e$$

$$= \frac{np^2}{3kT} E + P_a + nr^3 4\pi\varepsilon_0 E.$$

In a static or low-frequency field all these processes will exert their full part. However, the orientation of the whole molecule is relatively slow, and as the frequency increases the orientation lags behind. When the frequency reaches a value of about 10^{12} the dipoles are unable to follow the oscillations of the field: only random orientations remain and these contribute nothing to the resultant polarization. Of the polarization, P, only P_a and P_e remain. At a somewhat higher frequency the stretching of the bonds becomes too sluggish and P_a drops out. Only P_e remains above a frequency of 10^{15}. This is in the optical range. The behaviour is sketched in figure 14.6. However, this ignores the fact that both P_a and P_e may show resonance effects. In the next section we shall consider only P_e and show that resonance occurs at a critical frequency and that this accounts for optical dispersion and anomalous dispersion.

14.4 **Optical dispersion and anomalous dispersion**

We now analyse the effect of an alternating electric field on the electronic polarization P_e; the treatment is similar to that given by G. Joos, *Theoretical Physics*, Blackie (1934). For simplicity we consider a single electron, mass m, charge $-e$, bound to its equilibrium position relative to its nucleus by a force of force-constant k. In free oscillation the equation of motion of the electron is

$$m\ddot{x} + kx = 0,$$

or

$$\ddot{x} + \omega_0^2 x = 0, \tag{14.35}$$

where ω_0 is the natural angular frequency (radians s^{-1}) and $\omega_0^2 = k/m$. The natural frequency ν_0 is given by

$$\nu_0 = \frac{\omega_0}{2\pi} \text{ Hz.} \tag{14.36}$$

We apply a sinusoidal electric field

$$E = E_0 \sin \omega t, \text{ frequency } \nu = \frac{\omega}{2\pi}. \tag{14.37}$$

Ignoring any 'frictional' effects the equation of motion is

$$\ddot{x} + \omega_0^2 x = \frac{-e}{m} E_0 \sin \omega t. \tag{14.38}$$

The steady-state solution is

$$x = \frac{-eE_0 \sin \omega t}{m} \frac{1}{\omega_0^2 - \omega^2}. \tag{14.39}$$

The dipole moment per atom is simply

$$p = -ex. \tag{14.40}$$

Hence

$$p = \frac{e^2 E_0 \sin \omega t}{m(\omega_0^2 - \omega^2)} = \frac{e^2}{4\pi^2 m(\nu_0^2 - \nu^2)} E. \tag{14.41}$$

Figure 14.6. Effect of frequency on polarization ignoring the effect of resonance.

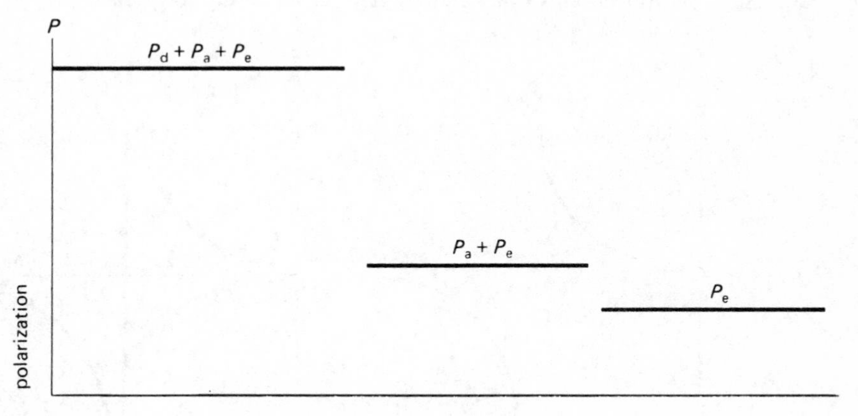

The volume polarization P_e is simply n times this. Using the relationship

$$(\varepsilon - 1) = 2(\mathscr{R} - 1) = \frac{P_e}{\varepsilon_0 E} = \frac{np}{\varepsilon_0 E}$$

and substituting for p from equation (14.41) we obtain

$$\mathscr{R} = 1 + \frac{ne^2}{8\pi^2 \varepsilon_0 m(\nu_0^2 - \nu^2)}. \tag{14.42}$$

This is plotted in figure 14.7(a). It is seen that approaching ν_0 the refractive index rises to plus infinity and just beyond ν_0 it falls to minus infinity. This is the absorption frequency of the atom and is generally of the order of 10^{15}. In practice, radiation resulting from the oscillation involves a certain amount of damping or 'friction', and this prevents infinite values of $\mathscr{R}$ from being reached. The main features are now shown in figure 14.7(b). The refractive index increases with frequency over the region AB; this is the region of 'normal' dispersion. The sudden change in behaviour along CD was, at one stage, considered anomalous and was so called. The above analysis shows, however, that 'anomalous' dispersion is just as normal as 'normal' dispersion.

The frequency ν_0 at which electronic resonance occurs is the same as that involved in the derivation of van der Waals forces. For this reason these forces are sometimes called 'dispersion' forces (see pp. 18–19).

Similarly the frequency at which a resonance effect occurs in P_a is of the order of 10^{12}, so dispersion also occurs in the infrared region. The

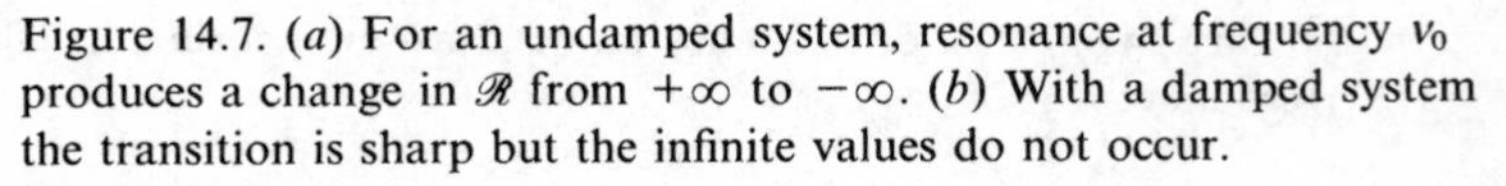
Figure 14.7. (a) For an undamped system, resonance at frequency ν_0 produces a change in $\mathscr{R}$ from $+\infty$ to $-\infty$. (b) With a damped system the transition is sharp but the infinite values do not occur.

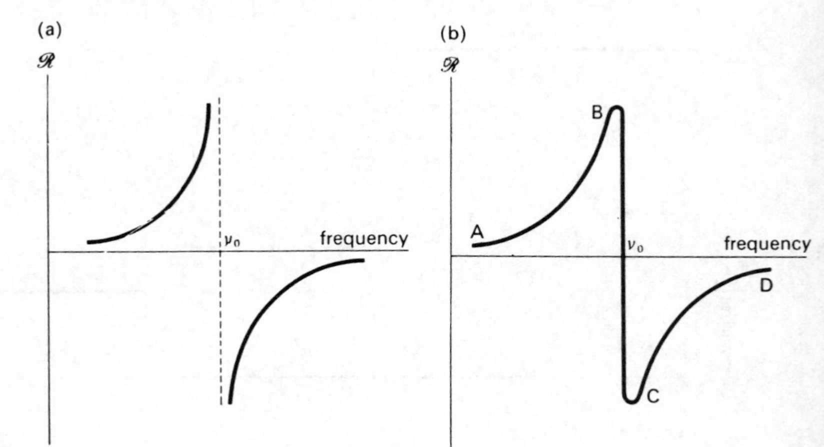

combined behaviour for a dipolar molecule with a single electronic and atomic absorption line is shown schematically in figure 14.8.

14.5 **Dielectric properties of liquids and solids**

14.5.1 *Weak interaction*

We now consider the dielectric behaviour of matter in the condensed state. We first treat the simplest case in which non-polar molecules are involved so that interaction between neighbouring atoms or molecules is weak. In this case we consider only the electronic polarization per atom

$$p=4\pi\varepsilon_0 r^3 E, \tag{14.43}$$

where r is the radius of the individual atom. For simplicity, to obtain an order of magnitude value for the number of atoms n per m^3, assume a simple cubic structure. Then

$$n=\left(\frac{1}{2r}\right)^3. \tag{14.44}$$

The polarization per unit volume, using equations (14.43) and (14.44), is

$$P=np=\frac{\pi\varepsilon_0}{2}E. \tag{14.45}$$

Hence

$$\varepsilon-1=\frac{P}{\varepsilon_0 E}=\frac{\pi}{2}. \tag{14.46}$$

Figure 14.8. Effect of frequency on polarization of a dielectric allowing for resonance effects.

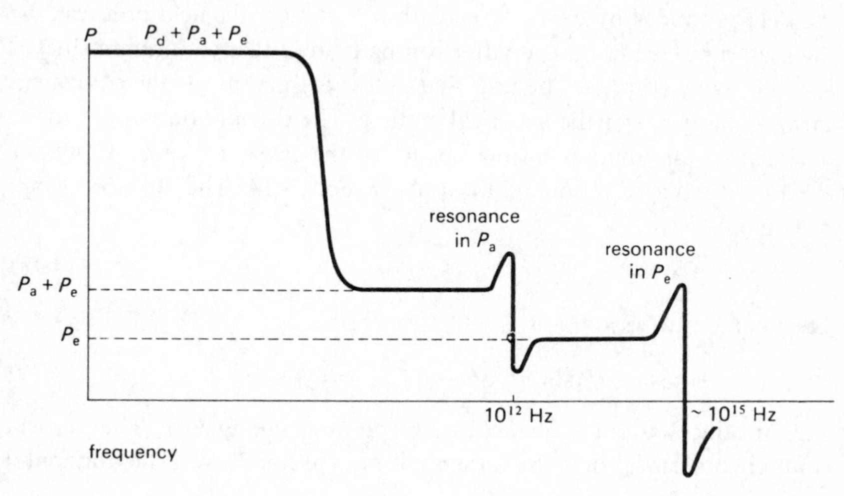

Consequently

$$\varepsilon = 1 + \frac{\pi}{2} \approx 2.6,$$

$$\mathscr{R} = \varepsilon^{\frac{1}{2}} \approx 1.6. \tag{14.47}$$

This analysis shows that for many simple liquids and solids the dielectric constant will be of order 2–3 and the refractive index 1.4–1.7. Clearly in a crystal which has well-defined structural anisotropy, so that the polarizability depends on crystal direction, there will be a corresponding dependence of ε and $\mathscr{R}$ on crystal direction. This is the basic cause of birefringence.

14.5.2 *Strong interaction: the Clausius–Mossotti equation*

When atoms or molecules are far apart as in a gas, they 'see' the average dielectric field E. When, however, they are close together as in a liquid or solid, the atomic (or molecular) dipoles may themselves interact. For example if polarization produces a local dipole of strength $p = 4\pi\varepsilon_0 r^3 E$, the field produced by this dipole at a distance r is of order of magnitude $p/4\pi\varepsilon_0 r^3$, i.e. about E; i.e. the field produced at a neighbour by an induced dipole may be about as large as the field itself. This increases the polarization effect.

We have now to consider how to estimate the resultant field acting on each molecule. The treatment is due to Lorentz. We choose a molecule O and construct a sphere of radius R large compared with the molecular separation (see figure 14.9). The resultant effect at O is the sum of (*a*) the field produced by the material within R and (*b*) the field produced by the material outside R. The latter can be treated as a continuum since it is so far away from O. The field due to this is the same as the equivalent surface charges over the spherical surface. The calculation is as follows.

Construct an annulus on the surface of the cavity making an angle θ, $\theta + \mathrm{d}\theta$ with the direction of the polarization field. The surface charge density is

$$-P \cos \theta. \tag{14.48}$$

The charge on the annulus is

$$-P \cos \theta 2\pi R \sin \theta R \, \mathrm{d}\theta. \tag{14.49}$$

This produces a coulomb field at O pointing along OA. The vertical components cancel out, the horizontal ones are additive. The horizontal component at O, for the annulus, is

$$\left(\frac{1}{4\pi\varepsilon_0}\right)P\cos\theta 2\pi R\sin\theta R\,d\theta\frac{1}{R^2}\cos\theta$$
$$=(1/2\varepsilon_0)P\cos^2\theta\sin\theta\,d\theta. \tag{14.50}$$

For the whole sphere, for θ ranging from 0 to π, the net field E' at O due to the surface charge is

$$E'=\frac{P}{2\varepsilon_0}\int_0^\pi\cos^2\theta\sin\theta\,d\theta=\frac{P}{2\varepsilon_0}\left[\frac{-\cos^3\theta}{3}\right]_0^\pi=\frac{P}{3\varepsilon_0}. \tag{14.51}$$

The resultant field at O is thus $E+E'$ or

$$E+\frac{P}{3\varepsilon_0}. \tag{14.52}$$

We now have to consider the contribution (*a*) due to material within the sphere. Lorentz showed that for a cubic array of dipoles or for a random array of dipoles as in a liquid or a glass, the local field is zero. For such materials we may, therefore, write

$$E_{\text{local}}=E+\frac{P}{3\varepsilon_0}. \tag{14.53}$$

The induced dipole per molecule is

$$p=\alpha E_{\text{local}}, \tag{14.54}$$

Figure 14.9. Charge distribution in a spherical cavity within a dielectric.

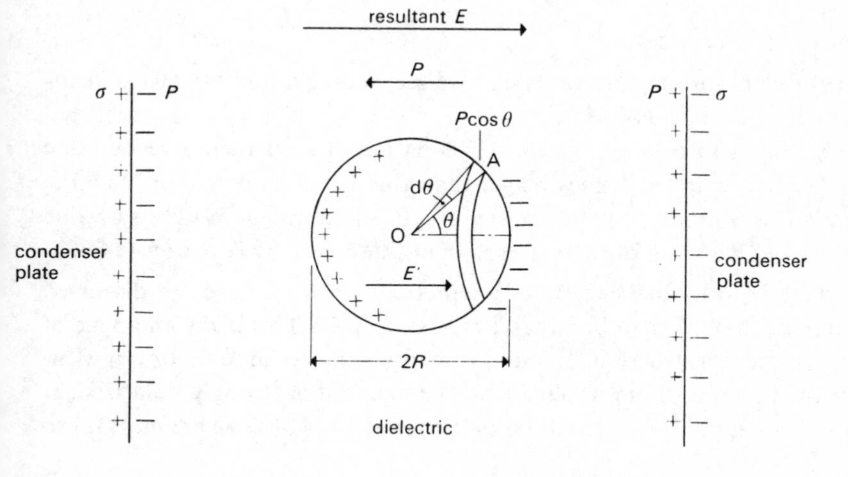

therefore

$$P = np = n\alpha\left(E + \frac{P}{3\varepsilon_0}\right). \tag{14.55}$$

Hence

$$\frac{P}{E} = \frac{n\alpha}{1 - n\alpha/3\varepsilon_0}. \tag{14.56}$$

The basic relation between P and the average field E in the dielectric is

$$(\varepsilon - 1) = \frac{P}{\varepsilon_0 E} = \frac{n\alpha}{\varepsilon_0 - n\alpha/3}. \tag{14.57}$$

Hence

$$(\varepsilon + 2) = (\varepsilon - 1) + 3 = \frac{3\varepsilon_0}{\varepsilon_0 - n\alpha/3}. \tag{14.58}$$

Taking the ratio of equations (14.57) and (14.58), we have finally

$$\frac{\varepsilon - 1}{\varepsilon + 2} = \frac{n\alpha}{3\varepsilon_0}. \tag{14.59}$$

This relation is fairly well obeyed for polar liquids and for dilute solutions of dipolar molecules in a non-polar solvent, as well as for ionic solids of cubic structure. It is known as the Clausius–Mossotti equation.

Before we leave this section we may form some idea of the importance of interaction compared with the results obtained if interaction is ignored. In the latter case, as we saw above,

$$\varepsilon - 1 = \frac{P}{\varepsilon_0 E} = \frac{n\alpha}{\varepsilon_0}. \tag{14.60}$$

Consider three cases:

(*a*) Weak interaction, say $\varepsilon \approx 1$; we see that equation (14.60) is identical with equation (14.59).

(*b*) Slight interaction, say $\varepsilon = 3$; equation (14.59) then gives a value for $(\varepsilon - 1)$, which is $\frac{5}{3}$ times bigger than that given by equation (14.60).

(*c*) Large dipole interaction, say $\varepsilon = 70$; equation (14.59) gives a value for $(\varepsilon - 1)$ which is 24 times larger than that given by equation (14.60).

There is yet a further type of interaction which we have not discussed here, but shall refer to briefly in the next chapter. This is the analogue of ferromagnetism where interaction is intrinsically so large that even without the presence of an applied field the material is strongly polarized. It would correspond to a situation (see equation (14.58)) where $n\alpha \approx 3\varepsilon_0$ so that $\varepsilon \rightarrow \infty$. Such materials are known as ferroelectrics. A ferroelectric material such as barium titanate has a dielectric constant of over 1000.

14.5.3 *Effect of frequency*

With matter in the condensed state frequency has the same effect on the dielectric constant as it has on gases. However, the dipole polarization in liquids is usually more sluggish than in gases – with solids it is often completely absent since rotation of the whole molecular unit within the lattice is difficult or impossible, particularly at temperatures well below the melting point. Results for water and ice are shown in figure 14.10.

14.6 **Summary**

We may summarize the ideas described in this chapter in the following way.

1. The dielectric properties of matter arise from the polarization produced by an electrostatic field. Polarization occurs in three ways:

(*a*) Distortion of electron clouds; all atoms (or molecules) show this effect.

(*b*) With polar molecules stretching or bending of bonds can produce a small contribution to the polarization.

Figure 14.10. Variation of dielectric constant with frequency, for water at 20 °C and for ice at −5 °C and −40 °C. The large dielectric constant at low frequencies at −40 °C shows that molecular rotation is still possible (see p. 236).

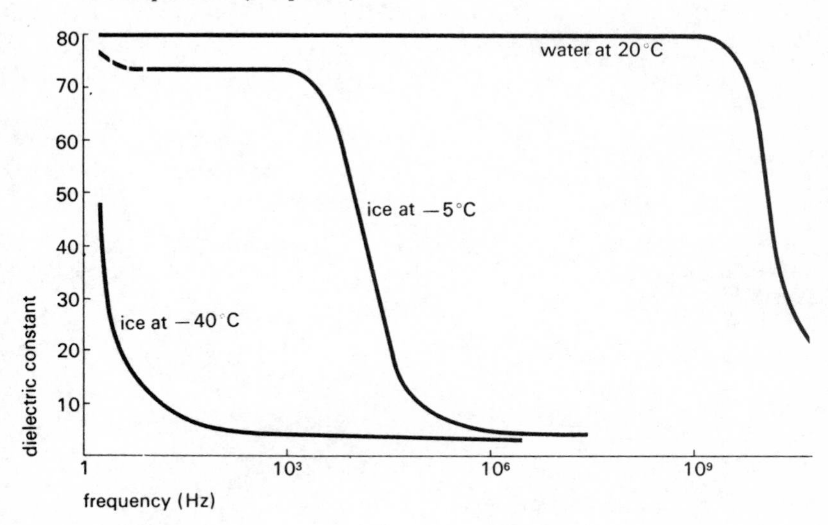

(*c*) With polar molecules orientation of the molecule may produce a very large polarization: the field has to compete with thermal randomization.

2. Temperature has no effect on (*a*) since this involves electronic properties. Increasing temperature resists the polarizing influence of the field for both (*b*) and (*c*). It is particularly marked with (*c*). However, the molecular orientation cannot be increased indefinitely by lowering the temperature since orientation of the molecular dipoles may become impossible when the material is solid.

3. Frequency of the electrostatic field has a marked effect on all types of polarization. Orientation becomes increasingly difficult as the frequency is raised and becomes impossible at frequencies above about 10^{12} Hz. The stretching or bending of bonds also becomes unable to follow the field at frequencies above about 10^{14} Hz, and only the electronic polarization remains up to a frequency of 10^{15}–10^{16} Hz. This is in the optical range and is therefore the only part of the dielectric mechanism which plays a part in the refractive index of the material. There is usually a resonance effect at frequencies in this region and this accounts for optical dispersion. The bond stretching (or bending) mechanism also shows a resonance effect but this generally occurs in the infrared.

4. With matter in the condensed phase the behaviour is complicated by neighbour–neighbour interaction. This can greatly increase the effective polarization of the material.

15

Magnetic properties of matter

15.1 Introduction

We know that there are certain metallic bodies which attract one another and which, if suitably suspended will tend to point in the north–south direction. These materials are magnetic. We also know that if a current is passed through a coil of wire, the coil itself behaves like a magnet and will attract a bar magnet or another similar coil. If we insert certain 'magnetic' materials into the coils the interaction between the coils will be changed. The coil also possesses a certain self-inductance and if we insert similar 'magnetic' materials into the coil its self-inductance will be increased.

These are some of the basic observations involved in magnetism and the purpose of this chapter is to discuss the magnetic properties of matter in terms of atomic and electronic mechanisms. There are many resemblances to dielectric effects the main differences being that

(*a*) there is no such thing as an isolated magnetic pole;
(*b*) there is no magnetic equivalent of a condenser;
(*c*) there is no dielectric analogue of a solenoid.

The most fundamental approach to magnetism is due to Einstein. He considered the electric field generated by a moving charge and the force it would exert on another charge moving parallel to it with the same velocity. Because of relativity effects the space ahead of the moving charge is crowded and this compression of space in which the electric field operates produces an additional force on the other charge over and above that calculated from the classical coulombic law. This extra force turns out to be identical with the magnetic force exerted between two parallel wires carrying an electric current.

Thus magnetism is a second-order relativistic electrostatic effect – it is not a new type of force. We shall not discuss this in detail any further.

15.2 The magnetic equations

We first define the magnetic field B as that type of field which produces a force on a conductor carrying a current. If a field B (teslas) acts at an angle θ on a short length of conductor $\mathrm{d}l$ (metres) carrying a current I (amperes) it is found that the conductor experiences a force F (newtons) in a direction at right angles to a plane containing B and $\mathrm{d}l$ (figure 15.1(*a*)) of magnitude

$$F = BI\,\mathrm{d}l\sin\theta. \tag{15.1}$$

Consider now a coil of irregular shape in the xy plane, the field B pointing in the x direction (figure 15.1(*b*)). There is a force on the element at P of amount

$$BI\,\mathrm{d}l\sin\theta = BI\,\mathrm{d}y \tag{15.2}$$

out of the paper and an equal force at Q into the paper. These two forces exert a couple

$$BI\,\mathrm{d}y\,x_1 = BI\,\mathrm{d}A, \tag{15.3}$$

where x_1 is the distance PQ and $\mathrm{d}A$ is the area of the strip between P and Q. For the whole coil the couple is then BIA, where A is the area of the coil. Thus the coil behaves as though it had a magnetic moment IA. We

Figure 15.1. (*a*) A field B acting on a current element I produces a force normal to the plane containing B and I. (*b*) Calculation of moment exerted by a field B on a coil carrying current I.

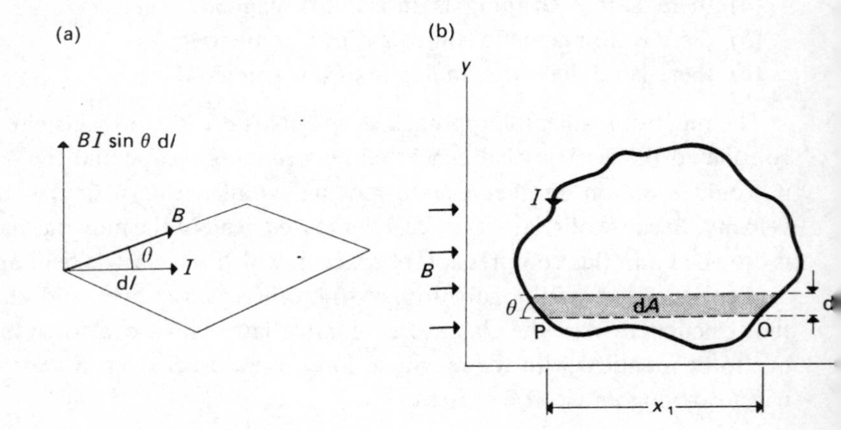

may thus define magnetic moment m of a coil as

$$m = IA \tag{15.4}$$

and add to this that when lying in a coplanar field B it experiences a couple

$$G = BIA = Bm. \tag{15.5}$$

Thus a coil carrying a current behaves in a magnetic field like a magnet. It is for this reason that ever since A. M. Ampère, it has been considered that the magnetism of atoms is due to circulating currents within the atom; consequently every 'north' pole has associated with it an equal 'south' pole. For this reason it is not possible to envisage the existence of isolated magnetic poles in the way that there are isolated charges in electrostatics. The earlier view regarded the circulating currents as arising from electrons travelling in orbits; later work has included the electron spin which, from this point of view, can also be regarded as a minute circulating current.

Let us now consider the behaviour of a long solenoid containing n_0 turns per unit length. The solenoid is assumed to be so long that the free ends do not produce any disturbing effects at the centre. If we were to place a small test coil at the centre of the solenoid we should find that the solenoid produces a couple on the coil from which we should deduce that the magnetic field has a value

$$B = \mu_0 I n_0, \tag{15.6}$$

where μ_0 = permeability of free space = $4\pi \times 10^{-7}$ H m^{-1}.

Suppose the solenoid was of length l and has a cross-section of area A. We could compare its magnetic properties with a permanent magnet of length l, cross-sectional area A, possessing a uniform magnetic moment M per unit volume. The total moment of the magnet is

$$M \times \text{volume} = MAl. \tag{15.7}$$

For the solenoid each turn produces a magnetic moment (see equation (15.4))

$$m = IA,$$

so that the moment of the whole solenoid is

$$m \times n_0 l = IAn_0 l. \tag{15.8}$$

Comparing equations (15.7) and (15.8) we see that the coil and the magnet have identical magnetic properties if

$$MAl = IAn_0 l,$$

i.e.

$$M = In_0. \tag{15.9}$$

Since the field in the solenoid (see equation (15.6)) is $\mu_0 In_0$, it follows that the equivalent field inside the magnet is also $\mu_0 In_0$. Consequently, from equation (15.9)

$$B = \mu_0 In_0 = \mu_0 M. \tag{15.10}$$

We now consider the effect of filling the solenoid with 'magnetic' material. We assume that the solenoid is so long that the end poles have a negligible 'demagnetizing' effect. Alternatively this could be achieved by forming the solenoid in a toroidal shape so that there are no free ends. The field in the solenoid in air is (see equation (15.6))

$$B = \mu_0 In_0.$$

Self-induction experiments show that B is increased by the presence of the 'magnetic' material by a factor μ known as the magnetic permeability; which for weakly magnetic material is a constant (for strongly magnetic materials, μ is itself a function of B). Thus within the solenoid

$$B = \mu\mu_0 In_0.$$

But we may also regard B as being the sum of two parts: one part is due to the solenoid itself ($\mu_0 In_0$) and the other part is due to the magnetism M produced in the magnetic material ($\mu_0 M$), thus we may write:

$$B = \mu\mu_0 In_0 = \underbrace{\mu_0 In_0}_{\text{due to solenoid}} + \underbrace{\mu_0 M}_{\text{due to magnetic material}}. \tag{15.11}$$

We now define In_0 as the magnetizing field H of the solenoid (amperes per metre)

$$H = In_0. \tag{15.12}$$

Then equation (15.11) can be written as

$$B = \mu\mu_0 H = \mu_0 H + \mu_0 M, \tag{15.13}$$

$$\mu - 1 = M/H = \chi, \tag{15.14}$$

where χ is the volume susceptibility of the magnetic material.

If there are n atomic or molecular or electronic units per unit volume each of which acquires magnetic moment m we have

$$M = nm.$$

Then equation (15.14) becomes

$$\mu - 1 = \frac{nm}{H} = n\alpha, \tag{15.15}$$

where α is defined as the magnetic polarizability per unit (atom, molecule, electron).

In experimental magnetism we apply a magnetizing field H and find the magnetic moment M produced in the material under examination. In what follows we shall show how we may deduce M or χ or α in terms of atomic mechanisms. Before doing so we may draw on the parallelism between a solenoid and a magnet to emphasize a further point of difference between magnetic and dielectric properties. If electric charges are placed on a dielectric the electrostatic field is always from positive to negative whether it is observed inside or outside the dielectric. With a permanent magnet, however, the situation is different. Outside the material the field is in the direction away from the 'north' pole and towards the 'south' pole: within the material the field is from south to north. The explanation for this is quite obvious if we replace the permanent magnet by an equivalent solenoid. Perhaps more relevant is the idea that every elementary magnetic unit within the magnet, down to the individual atom or even the spinning electron, can be regarded as equivalent to a minute circulating current.

15.3 Diamagnetism: Langevin's treatment

All atoms possess a diamagnetic component. This is because an applied field always modifies the circulating 'electronic' current in such a way that it opposes the field. We consider the simplest case, an electron of charge e, mass m_e moving with velocity v in a circular orbit of radius r. If τ is the time for the electron to describe one revolution, e behaves like a current I of magnitude

$$I = \frac{e}{\tau} = \frac{ev}{2\pi r}. \tag{15.16}$$

The magnetic moment of the orbiting electron is

$$m = \pi r^2 I = \frac{evr}{2}. \tag{15.17}$$

Consider now the effect of a field of induction B which is normal to the orbit. The flux ϕ in the circuit is

$$\phi = \pi r^2 B. \tag{15.18}$$

If the field varies with time an e.m.f. will be induced in the circuit of magnitude

$$\text{e.m.f.} = -\frac{d\phi}{dt} = -\pi r^2 \frac{dB}{dt}. \qquad (15.19)$$

This e.m.f. acts around the current loop and is equivalent to an electrostatic field E, where

$$\begin{aligned}\text{e.m.f.} &= 2\pi r E \\ &= -\pi r^2 \frac{dB}{dt}.\end{aligned} \qquad (15.20)$$

Hence

$$E = -\frac{r}{2}\frac{dB}{dt}. \qquad (15.21)$$

This field exerts a force eE on the electron and this in turn produces a change dv in its velocity:

$$eE = m_e \frac{dv}{dt} = -\frac{er}{2}\frac{dB}{dt}, \qquad (15.22)$$

$$\frac{dv}{dt} = -\frac{er}{2m_e}\frac{dB}{dt}. \qquad (15.23)$$

Consequently in the time that B changes by ΔB, v changes by Δv where

$$\Delta v = -\frac{er}{2m_e}\Delta B. \qquad (15.24)$$

According to equation (15.17) a change in v produces a corresponding change in the magnetic moment of the orbit:

$$\Delta m = \frac{er}{2}\Delta v.$$

Combining with equation (15.24),

$$\Delta m = -\frac{e^2 r^2}{4m_e}\Delta B. \qquad (15.25)$$

We see that Δm is always in a sense which opposes ΔB. Consider, for example, two identical orbits with opposite orbital momenta. Each orbit has a magnetic moment m but the net moment is zero. If a field ΔB is applied one orbit is slowed down and the other is speeded up: the net change is additive for the two orbits (see figure 15.2).

There is another point of interest. If the orbital velocity is reduced by Δv, the centrifugal force on the electron is reduced by $\Delta(m_e v^2/r) =$

$2m_e v\Delta v/r$. From equation (15.24) this equals $ve\Delta B$. But the field ΔB itself exerts a radial force on the electron of amount: velocity × charge × $\Delta B = ve\Delta B$. This exactly compensates for the reduced centrifugal force: consequently if the initial orbit is stable under some specified force between nucleus and electron the diamagnetic mechanism does not introduce any change.

For diamagnetic materials $B \approx \mu_0 H$ to 1 part in 10^5. Equation (15.25) then becomes

$$\Delta m = \frac{-e^2r^2}{4m_e}\mu_0 \Delta H. \tag{15.26}$$

By definition the magnetic susceptibility per electron is simply $\Delta m/\Delta H$. If there are n atoms per unit volume the susceptibility per unit volume is

$$\chi = n\frac{\Delta m}{\Delta H} = -\frac{n}{4}\sum \mu_0 \frac{e^2r^2}{m_e}, \tag{15.27}$$

where the summation is for all the orbital electrons in each atom.

If the orbits are not normal to B each orbit undergoes precession and χ is slightly decreased. Similarly if the orbits are not circular a lower value of χ is obtained. A better value for χ is

$$\chi = -\frac{n}{6}\sum \mu_0 \frac{e^2r^2}{m_e}. \tag{15.28}$$

We may at once derive an order-of-magnitude value for χ. If one mole of a typical metal occupies 10^{-5} m^3, $n_0 = 6 \times 10^{28}$. If there are say 10 orbital electrons and r is of order 10^{-10} m we have, per m^3

$$\chi = -(2 \text{ to } 3) \times 10^{-6}\ \mathrm{J\,T^{-2}}. \tag{15.29}$$

Figure 15.2. Sketch showing nature of diamagnetism; (*a*) when no field is applied the net amount is zero, (*b*) when field ΔB is applied there is an additive change in the magnetic moments of both orbits.

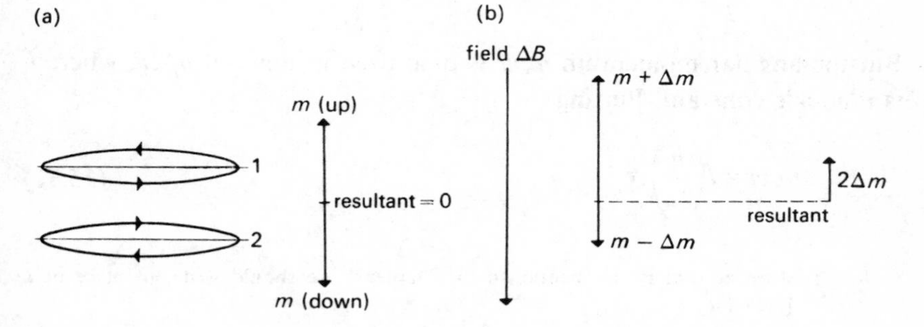

Consequently $\mu_0 M$ is quite small compared with B. This value of χ is typical for most solids and liquids. In the gaseous state the main change is in n_0. At s.t.p. n_0 is about 10^3 times smaller than in the condensed state so that for gases at s.t.p.

$$\chi \approx 10^{-9}.$$

Note that χ is independent of temperature since the internal structure of the atom is virtually unchanged by normal temperatures.

In the model described above, diamagnetism is attributed to the influence of the magnetic field on the orbital momentum of the electrons. What then happens to the s-electrons, which possess zero orbital momentum? A detailed quantum theory treatment shows that s-electrons possess zero angular momentum only in the absence of a magnetic field. Upon the application of a magnetic field they acquire a small amount of orbital momentum. This produces an equivalent amount of diamagnetism to that calculated above. Thus it follows that all the electrons contribute to diamagnetism.

15.4 **Paramagnetism: the Langevin function**

If an atom possesses a permanent magnetic dipole because the electrons have either orbital momentum or spin, the dipole will orient in an applied magnetic field and produce a magnetization in the same direction as the applied field. Materials which are made of such atoms are attracted (weakly) by a magnetic field and are known as paramagnetic substances. As in the case of dielectric dipoles the orientation by the applied field is opposed by thermal disorder.

We first consider the order of magnitude of the permanent dipole – arising from an orbiting electron. For a circular orbit of radius r and electron velocity v we have (see equation (15.17)).

$$m = \frac{evr}{2}. \qquad (15.30)$$

But the angular momentum $m_e vr$ is quantized in units of $h/2\pi$, where h is Planck's constant. Putting

$$m_e vr = l\left(\frac{h}{2\pi}\right),\dagger \qquad (15.31)$$

† More accurately, as mentioned in Chapter 1 we should write, in place of l, $[l(l+1)]^{\frac{1}{2}}$.

where l is an integer, and substituting in equation (15.30) we have

$$m = l\left(\frac{eh}{4\pi m_e}\right). \tag{15.32}$$

The quantity in the bracket is known as the Bohr magneton. As a first approximation we expect to find certain atoms in which the angular momenta do not cancel out, so that one orbit (or one spin) is left unpaired. The magnetic moment of such an atom will be one Bohr magneton, the magnitude of which is

$$\frac{eh}{4\pi m_e} \approx 10^{-23}\ \mathrm{J\,T^{-1}}. \tag{15.33}$$

Since orbits, and spins, tend to occur in pairs of opposite momenta (or spins) we should not expect to find atoms with magnetic moments much greater than this. This has an important consequence which greatly eases further analysis. As we shall see below, at low temperatures and moderate fields H, it is possible to orient all the dipoles and produce magnetic saturation. Nevertheless, even at this stage, the magnetic moment M per unit volume is small compared with H. Consequently the difference between B and $\mu_0 H$, or between H and the true field seen by the individual atom, is negligible and can be ignored. In what follows, therefore, we consider simply the effect of the field B.

Following the analysis for para-electric behaviour given in the previous chapter, we expect to find for small and moderate fields a volume susceptibility

$$\chi = \frac{M}{H} \approx \frac{\mu_0 M}{B} = \frac{\mu_0 n m^2}{3kT}. \tag{15.34}$$

For a solid occupying 10^{-5} m^3 per mole, $n = 6 \times 10^{28}$. Assuming m is one magneton we find that, at 300 K, χ has a value, per m^3

$$\chi \approx 6 \times 10^{-4}\ \mathrm{J\,T^{-2}}.$$

This is 200–300 times greater than the diamagnetic susceptibility. Consequently the diamagnetic susceptibility, which all atoms possess, is swamped by the paramagnetic susceptibility if the atom contains an unpaired orbit or spin.

For strong fields or low temperatures we obtain the corresponding function

$$\chi = \frac{nm}{H}\left(\coth\frac{mB}{kT} - \frac{kT}{mB}\right). \tag{15.35}$$

This is drawn in figure 15.3. Although this curve is identical with that obtained for para-electrics there are two basic differences between paramagnetic and para-electric materials. The first concerns saturation. Low temperatures clearly favour saturation in both cases. However, most dielectrics possessing electrostatic dipoles become solids at low temperature and orientation becomes difficult or impossible; orientation is thus frozen-out. With paramagnetics we are only concerned with the orientation of orbits (or spins) and this can occur at the very lowest temperatures. Thus one can greatly enhance paramagnetism by operating at very low temperatures; one cannot in general do this with para-electrics.

As an example of saturation with paramagnetics we note from figure 15.3 that this occurs when $mB/kT > 5$, i.e. when

$$B \approx \mu_0 H > \frac{5kT}{m}. \tag{15.36}$$

If $m = 10^{-23}$ J T^{-1} and $T = 1$ K, we obtain saturation when

$$H > 6 \times 10^6 \text{ A m}^{-1}.$$

Such a magnetizing field can be easily obtained and, with it, saturation can be achieved at temperatures of about 1 K. It should be noted that this argument is valid for gases where the individual molecules are far apart (n relatively small) and do not interact with one another. It also applies to ionic salts where each ion behaves like an isolated dipole but in such materials the number of dipoles per unit volume will not be large.

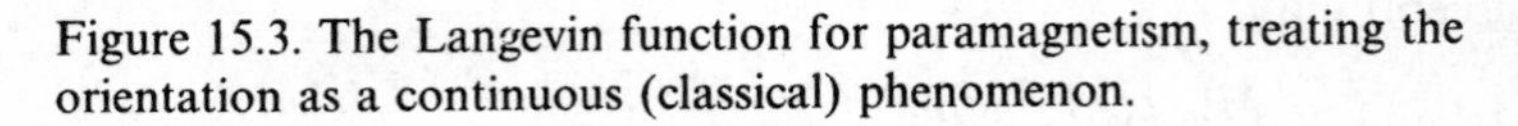

Figure 15.3. The Langevin function for paramagnetism, treating the orientation as a continuous (classical) phenomenon.

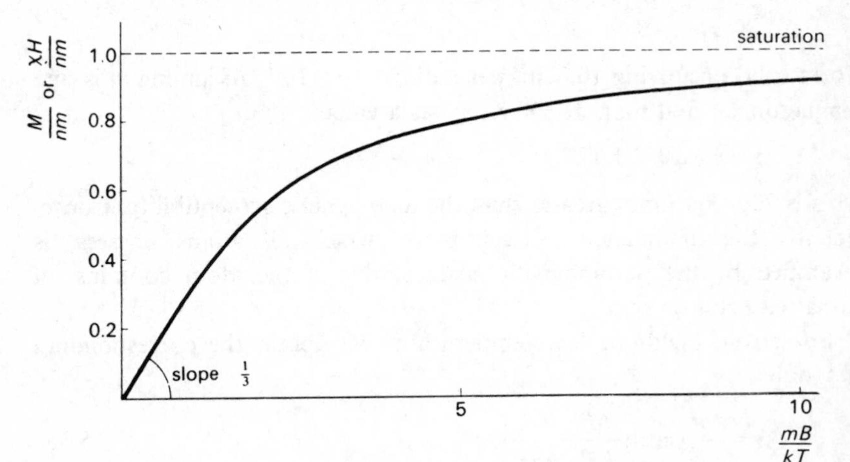

For example in chrome alum, $CrK(SO_4)_2 \cdot 12H_2O$, the Cr^{3+} ions have permanent dipoles and so contribute to the paramagnetism, but the K^+ and SO_4^- ions show only diamagnetic effects. The number of Cr^{3+} ions per unit volume is relatively small. Thus for both gases and ionic salts, although saturation can be obtained as indicated above with reasonable magnetizing fields by working at low temperatures the overall magnetization will not be as large as might at first sight be expected because n will not be very large.

On the other hand with paramagnetic metals, where n is large, the argument given above is not valid. The metal may be considered to consist of positive ions in a sea of electrons. The positive ions generally consist of closed electronic shells or an even number of electrons with paired spins. They are diamagnetic. The valency electrons (which provide electrical conductivity) have spins which can be aligned by an applied magnetizing field (see section 15.6 below). These therefore contribute to the paramagnetism of the metal. The field also affects the translational motion of these electrons and it can be shown that this produces a diamagnetic contribution which is just about one-third of the paramagnetism. We shall not deal with this but continue discussion of the paramagnetic effect. Since this is due to the free valency electrons in the metal, the number able to take part in this process is not the total number of free electrons n but, as pointed out on p. 250, only those at the top of the Fermi level E_F, i.e. only a fraction of order kT/E_F. Consequently the quantity n per unit volume in equation (15.34) should be replaced by nkT/E_F. The volume susceptibility then becomes

$$\chi = \mu_0 \frac{nkT}{E_F} \frac{m^2}{3kT} = \frac{\mu_0 n m^2}{3E_F}. \tag{15.37}$$

We see that the susceptibility is independent of temperature, a result in good agreement with the observed behaviour of paramagnetic *metals*.

We have discussed the possibility of saturation with paramagnetic salts and gases. The second difference between paramagnetic and dielectric behaviour concerns the nature of the polarizing field. Consider the typical dense solid discussed after equation (15.34): $n = 6 \times 10^{28}$ per m^3, m is of the order of 1 Bohr magneton. Suppose that at very low temperatures (~ 1 K) all the dipoles are aligned by the application of a magnetizing field $H = 6 \times 10^6$ A m^{-1} as suggested by equation (15.36). Then the magnetic moment per unit volume $M = nm = 6 \times 10^5$. Thus M is still a good deal smaller than H and we do not need to distinguish between $B = \mu_0 H$ and $B = \mu_0 H + \mu_0 M$ (see equation (15.13)). Thus we are not faced with the difficult problem that occurs in dielectric studies of attempting to allow for the effect of the polarization itself on the field producing the polarization. Only in ferromagnetism does this become important. In that case, as we shall see below, the interaction due to the magnetic polarization of the material is so strong that it completely dominates the behaviour, i.e. the resultant field is dominated by M rather than by H.

The difference between the minute volume susceptibility of paramagnetics as compared with para-electrics is due to the difference in size of the unit dipole. In electrostatics p has a value of about 1.6×10^{-29} C m: in paramagnetism m is of the order of the Bohr magneton and has a value of about 10^{-23} J T^{-1}. We have:

$$\text{in para-electricity } \chi = \frac{np^2}{\varepsilon_0 3kT}, \quad \text{in paramagnetism } \chi = \frac{\mu_0 nm^2}{3kT}.$$

Thus for materials with the same values of n the ratio $\chi_{\text{mag}}/\chi_{\text{diel}}$ is equal to

$$\frac{\mu_0 \varepsilon_0 m^2}{p^2} \approx 10^{-5}.$$

Note that the quantity $\mu_0 \varepsilon_0$ in the numerator is equal to the reciprocal of the square of the velocity of light. This emphasizes the basic connection between electrostatic and electromagnetic phenomena deduced by Clerk Maxwell and explained by Einstein in terms of relativistic effects (see p. 383). It is for the latter reason that magnetic phenomena are 'second-order' effects compared with electrostatic phenomena. Thus for matter in the condensed state $(\mu - 1)$ is of the order 10^{-4} in magnetism while $(\varepsilon - 1)$ is of order 1 to 10 in dielectrics.

15.5 **Ferromagnetism**

15.5.1 *Theory of ferromagnetism*

Certain materials such as iron, nickel and cobalt show very high magnetization M (typically $M \approx 10^6$ A m^{-1}) for fairly modest fields even at room temperature. They reach magnetic saturation very easily. With an 'ordinary' specimen of iron this occurs for magnetizing fields of the order of a few 10^{-1} A m^{-1} (see figure 15.4); with good single crystals in favourable orientations for a field of order only 10^{-4} A m^{-1}, so under these conditions χ is about 1000.

A typical value of M at saturation is about 10^6 A m^{-1}. If one mole occupies 10^{-5} m^3 the number of atoms per m^3 is 6×10^{28}. At saturation, if these give a moment of 10^6 A m^{-1} each atom must contribute a magnetic moment of $10^6/(6 \times 10^{28}) \approx 1.6 \times 10^{-23}$, i.e. about 1 Bohr magneton. This suggests that a saturated ferromagnetic consists of completely aligned dipoles. In this sense they are like paramagnetics, but with paramagnetics this result can only be achieved by using enormous fields or by operating at extremely low temperatures. With ferromagnetics alignment occurs with relatively weak fields.

It was this conclusion that led A. Weiss to suggest that these materials are spontaneously magnetized over fairly large domains and that these domains are easily aligned when an external field is applied. There is a cooperative interaction between the atoms so that an additional field λM acts on each atom. This resembles the Lorentz field in dielectrics $\frac{4}{3}\pi P$, but the quantity λ is immensely larger, of the order 10^4. The resultant field acting on each atom and producing magnetization may then be written

$$B=\mu_0(H+\lambda M). \tag{15.38}$$

The Langevin equation for the alignment of the dipoles becomes

$$M=mn\left(\coth x-\frac{1}{x}\right), \tag{15.39}$$

where now

$$x=\frac{m}{kT}(H+\lambda M)\mu_0. \tag{15.40}$$

The behaviour of the Langevin equation as a function of H is best seen by rewriting equation (15.39) as

$$\frac{M}{nm}=\coth x-\frac{1}{x}, \tag{15.41}$$

Figure 15.4. Magnetic moment M produced in a typical magnetic material for magnetizing fields of order 10^{-1} A m^{-1}.

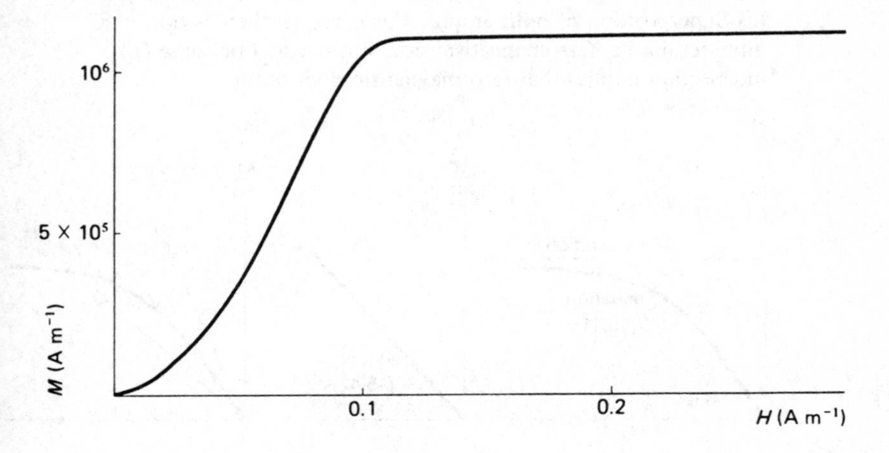

and equation (15.40) as

$$\frac{M}{nm}=\frac{xkT}{m^2\lambda n\mu_0}-\frac{H}{\lambda nm}. \tag{15.42}$$

Consider the behaviour when $H=0$. This corresponds to the spontaneous magnetization of ferromagnetic materials in the absence of an applied field. We plot equation (15.41) as M/nm against x and obtain the typical curve of figure 15.5(*a*). The slope near the origin is $\frac{1}{3}$. In figure 15.5(*b*) we have plotted equation (15.42) as M/nm against x, for $H=0$. We obtain a straight line of slope $kT/m^2\lambda n\mu_0$. Clearly if this slope is greater than $\frac{1}{3}$ no solution is possible and there is no magnetization (figure 15.5(*c*), curve (i)). If the slope is less than $\frac{1}{3}$, magnetization and saturation are easily achieved (figure 15.5(*c*), curve (ii)).

It is at once evident that there is a critical temperature above which no ferromagnetization will occur. This is known as the Curie temperature T_c and is given by

$$\frac{kT_c}{m^2\lambda n\mu_0}=\frac{1}{3}. \tag{15.43}$$

If the saturation value of M is written as M_0 $(M_0=nm)$ the drop-off of M with increasing temperature is shown in figure 15.6 as a plot of M/M_0 against T/T_c. Above T_0 only paramagnetism and diamagnetism remain. In what follows we shall ignore the diamagnetism and calculate in a very simple way the paramagnetism above T_c. If the fields are not too strong

Figure 15.5. Graphical method of explaining ferromagnetism. (*a*) Graph of equation (15.41). (*b*) Graph of equation (15.42). (*c*) Superposition of both graphs. For curve (i) there is no intersection, i.e. ferromagnetism does not occur. For curve (ii) intersection implies that ferromagnetism does occur.

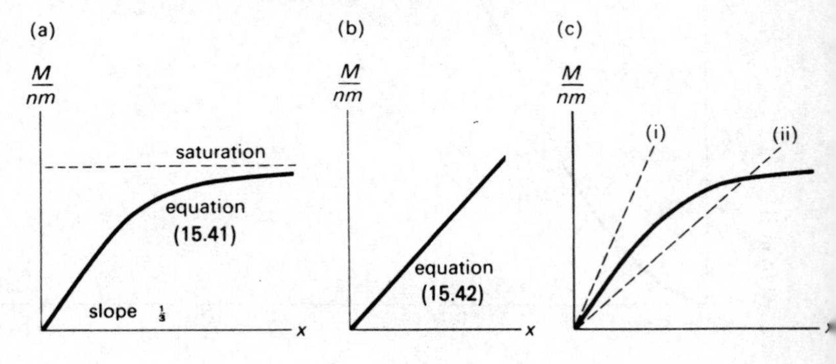

the Langevin equation (15.41) may be written

$$\frac{M}{nm}=\frac{x}{3}. \tag{15.44}$$

Similarly, using equation (15.43), equation (15.42) may be written

$$\frac{M}{nm}=\frac{xT}{3T_c}-\frac{H}{\lambda nm}. \tag{15.45}$$

Substituting for x from equation (15.44) we obtain

$$\frac{M}{nm}\left(\frac{T}{T_c}-1\right)=\frac{H}{\lambda nm}. \tag{15.46}$$

Consequently the paramagnetic susceptibility becomes

$$\chi=\frac{M}{H}=\frac{1}{\lambda}\left(\frac{T_c}{T-T_c}\right). \tag{15.47}$$

From equation (15.43) this becomes

$$\chi=\frac{nm^2\mu_0}{3k(T-T_c)}. \tag{15.48}$$

We see at once that λ has disappeared explicitly although, of course, it is implicit in the value of T_c. Equation (15.48) is well obeyed and gives good quantitative agreement for most ferromagnetic materials on the assumption that each atom contributes of the order of 1 or 2 Bohr magnetons.

Figure 15.6. Drop in magnetic moment of a ferromagnetic material as temperature is raised. Above the Curie temperature T_c the material shows only diamagnetism and paramagnetism.

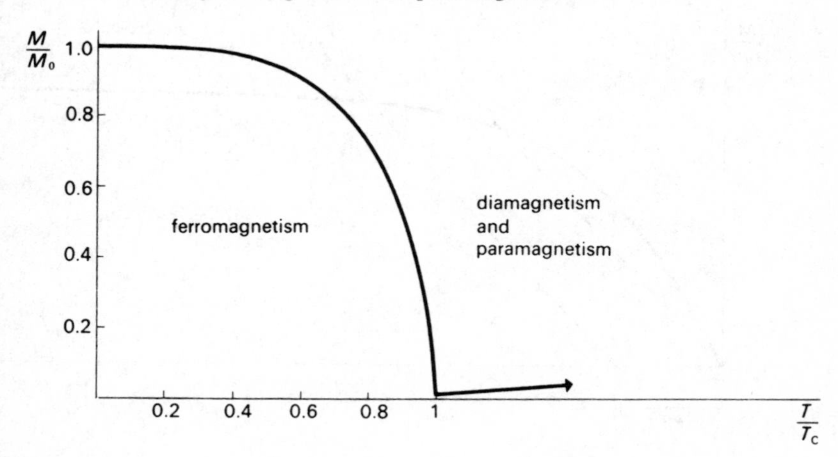

Finally we may note that in the ferromagnetic range an applied field has little effect compared with the internal field. Even for a field $H \approx 1\ \mathrm{A\,m^{-1}}$ the interval $mH\mu_0/kT$ shown in figure 15.7 is small and has very little effect on the resultant magnetization.

15.5.2 *The origin of the internal field: ferromagnetism and antiferromagnetism*

The internal field B_0 at saturation in a ferromagnetic material is given by

$$B_0 = \lambda M\mu_0 = \lambda mn\mu_0 .$$

For typical values $\lambda = 5000$, $m = 1$ Bohr magneton ($10^{-23}\ \mathrm{J\,T^{-1}}$). $n = 6 \times 10^{28}$ atom $\mathrm{m^{-3}}$, $\mu_0 = 4\pi \times 10^{-7}\ \mathrm{Hm^{-1}}$ we obtain $B_0 = 4000$ T. This is an extremely strong field, very much greater than the ordinary field acting on a dipole due to the magnetization of its neighbours (for this λ would be of order unity as in the dielectric case).

It is not possible to explain the origin of this large internal field in simple terms. Indeed it introduces very difficult quantum mechanical problems. We may, however, describe it in the following terms. The large value of λ arises from the 'exchange' interaction between the atomic dipoles. This interaction depends on the ratio of the orbital radius r to the distance a between neighbouring atoms. Over a critical range (a/r about 3) the magnetic exchange interaction is large and positive. This implies that for atoms with large dipoles and of the correct a/r ratio strong positive interaction will occur, λ will be large and the material will

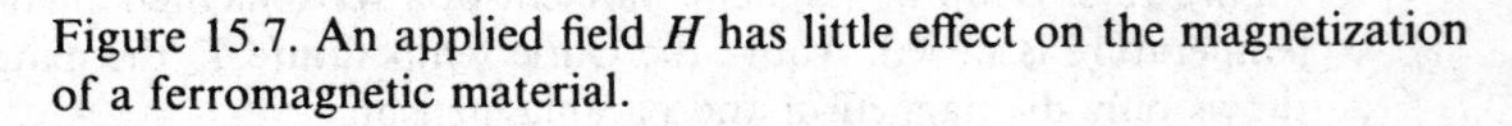
Figure 15.7. An applied field H has little effect on the magnetization of a ferromagnetic material.

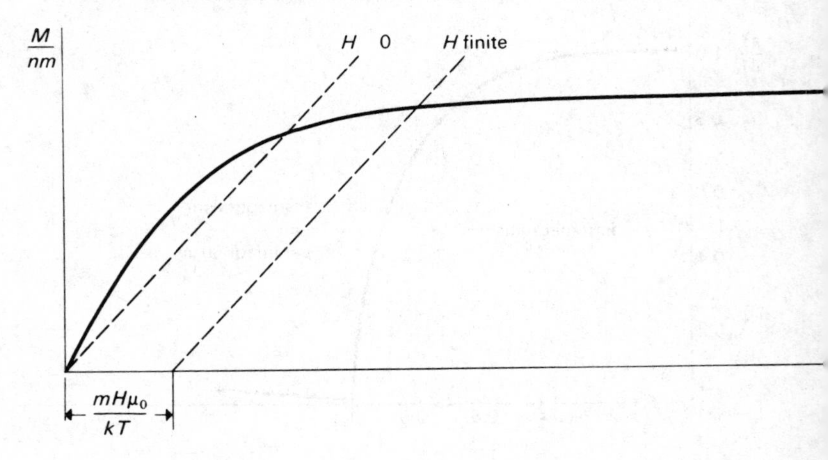

be ferromagnetic (Fe, Co, Ni). This is shown schematically in figure 15.8. It is also seen that for small values of a/r the interaction can be negative. This favours an antiparallel orientation of neighbouring spins, and materials showing this property (MnF_2, MnO, CoO) are known as anti-ferromagnetic substances: at low temperatures the interaction is effective and an applied field produces little magnetization so that the susceptibility is small. As the temperature is raised the anti-ferromagnetic interaction becomes less effective and the susceptibility of the material increases. Above a critical temperature known as the Néel temperature the interaction is completely swamped by thermal motion. The material ceases to be anti-ferromagnetic and becomes paramagnetic in the, more or less, classical sense. The detailed behaviour is very complicated and we shall not discuss it further.

15.6 Quantum treatment of magnetic properties

The classical treatment of diamagnetism is satisfactory from two points of view. First, the concept of induced currents producing fields which oppose the applied field seems reasonable and tractable. Secondly, it gives the right answer for the diamagnetic susceptibility. From all other points of view it is wrong.

Detailed considerations, first described by N. Bohr in 1911, show that if classical mechanics is applied *rigorously* to an orbiting electron the paramagnetic and diamagnetic terms cancel one another exactly so that there is no net induced magnetization. The reason is briefly as follows. Assuming that an electron orbiting around a nucleus is a stable classical

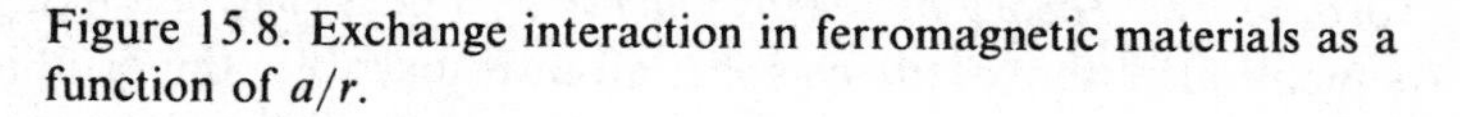

Figure 15.8. Exchange interaction in ferromagnetic materials as a function of a/r.

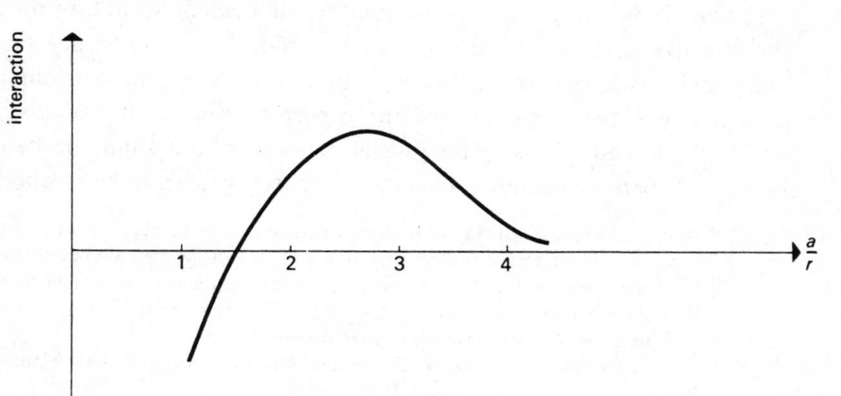

system, the radius of a given orbit can have a continuous range of values from 0 to ∞ rather than the one particular size assumed in the Langevin theory. Further, whereas the Langevin theory applies the Boltzmann energy distribution to the external degrees of freedom of the orbiting electron, i.e. to the orientation of the orbits in paramagnetism, it does not do so to the internal degrees of freedom, i.e. to the orbital energies of the electron in its treatment of diamagnetism. This is logically inconsistent. If the Boltzmann distribution is applied to both orientation energies and orbital energies the paramagnetic and diamagnetic parts exactly compensate one another. The Langevin treatment gives the correct answer because it assumes a finite radius and an orbital energy independent of temperature; in a sense it incorporates hidden quantum conditions although in fact Langevin's theory (1905) preceded Bohr's quantized atom (1913) by eight years.

Classical electromagnetic concepts cannot really be applied to a classical atom since a non-quantized orbiting electron constitutes an unstable system (see below). The only situation for which such concepts can be applied meaningfully is to a free 'electron-gas', for example the electron plasma in a metal. Here again, the effect of a magnetic field is to produce diamagnetic and paramagnetic effects which exactly cancel one another. The reason is simply that in classical electromagnetism a field can deflect an electron but it cannot do work on it, so that although the electrons are deflected into helical paths their energy distribution remains unaltered.† This implies that the magnetic field produces no energy change in the system, i.e. the system behaves as though it has no net magnetic property. A discussion of this issue is given in a very readable and interesting form in *The Theory of Electric and Magnetic Susceptibilities*.‡

As mentioned above, the classical atom is unstable; the orbiting electrons would spiral into the nucleus radiating all their energy in the process. The effective lifetime of such an atom would be infinitesimally small. Atoms exist as stable entities precisely because energy states (or angular momenta) are quantized. To produce a perturbation of the energies by the classical methods is quite invalid. However, quantum mechanics shows that there is a genuine diamagnetic effect which can be calculated

† Of course a magnetic field exerts a force on a conductor through which a current is passing. If the conductor moves, work is done, but the energy does not come from the magnetic field. The energy is provided by the electric current flowing through the conductor in overcoming the back e.m.f. generated in the conductor itself as a result of its movement in the magnetic field.

‡ J. H. van Vleck, *The Theory of Electric and Magnetic Susceptibilities* (Clarendon Press, 1932).

rigorously. The analysis is far too advanced to be given here. It gives the same answer as the classical treatment which we have given in this chapter in section 15.2. For this reason, as in most other elementary texts, we have persisted in giving the simple treatment: the answer is right though the physics is not entirely satisfactory.

We have taken over from our chapter on dielectric properties the classical treatment of paramagnetism. With para-electrics quantum effects play a trivial part and the classical treatment is perfectly satisfactory. This is not so with paramagnetics. The reason is that the individual dipole can take up only a discrete number of orientations. If, for example, an atomic system has a resultant angular momentum represented by a vector of length $\frac{3}{2}$, the only orientations which the system can acquire in a magnetic field correspond to components of $\frac{3}{2}$, $\frac{1}{2}$, $-\frac{1}{2}$ and $-\frac{3}{2}$ in the direction of the field. In classical physics the orientations would cover the whole continuous range such that the components have values from $+\frac{3}{2}$ to $-\frac{3}{2}$. The effect on paramagnetism is not very marked but it is worth considering the simplest possible case when the dipole of the individual atom is due to a single electron spin. The only orientations allowed are those which are exactly parallel and antiparallel to the applied field. If there are n atoms per m^3, and of these n_+ are those with orientations along the field and n_- those with orientations opposing the field, we have by the Boltzmann principle

$$n_+ = A \exp\left(\frac{mB}{kT}\right) \tag{15.49}$$

$$n_- = A \exp\left(\frac{-mB}{kT}\right). \tag{15.50}$$

Since $n_+ + n_- = n$ we find that

$$A = n\left[\exp\left(\frac{mB}{kT}\right) + \exp\left(-\frac{mB}{kT}\right)\right]^{-1}. \tag{15.51}$$

The net moment per unit volume is then

$$M = (n_+ - n_-)m \tag{15.52}$$

$$= nm \frac{\left[\exp\left(\frac{mB}{kT}\right) - \exp\left(-\frac{mB}{kT}\right)\right]}{\left[\exp\left(\frac{mB}{kT}\right) + \exp\left(\frac{mB}{kT}\right)\right]}$$

$$= nm \tanh\left(\frac{mB}{kT}\right). \tag{15.53}$$

In form this curve is very similar to the Langevin function. However, there is a quantitative difference in the region of small fields. Since for small values of x, $\tanh x \approx x$ we have from equation (15.53) for small values of H,

$$M \approx nm\left(\frac{mB}{kT}\right) = \frac{nm^2B}{kT}.$$

The susceptibility is then

$$\chi = \frac{M}{H} = \frac{nm^2}{kT}\frac{B}{H} = \frac{nm^2\mu\mu_0 H}{kTH} \approx \frac{\mu_0 nm^2}{kT}$$

since μ is very nearly equal to unity. Comparison with equation (15.34) shows that this is just three times the value deduced from the Langevin function. Similar corrections may be applied to the discussion of ferromagnetism and the Curie temperature given above. However the correction factor of 3 will only apply rigorously for a single-electron-spin system. With most materials the situation is usually more complicated and in fact the correction factor becomes less than 3.

15.7 Ferromagnetic domains

We consider here briefly the magnetization of a single crystal of a ferromagnetic material such as iron. With iron the 6 cube edge directions [100] are the directions of easy magnetization. If a magnetic field H is applied along the edge direction the domains are aligned easily and saturation is achieved for a very small field. If the field is applied along, say, the face diagonal direction [110] the easy orientation directions are at 45°

Figure 15.9. Magnetization of a single crystal of iron along a cube direction [100] and along a face diagonal direction [110].

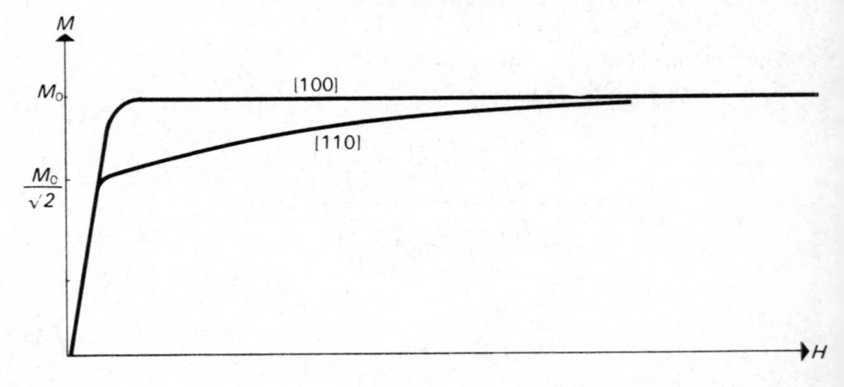

to the field. Consequently for small fields the net magnetization in the direction of H is $1/\sqrt{2}$ of what it would be if the field were applied along the [100] direction. Experiments show that this is so (see figure 15.9). If the field is increased further magnetization can occur only by twisting the domain directions out of their easy orientations. This is a far more difficult process and relatively strong fields must be used to achieve it.

15.8 Magnetic hysteresis

When a magnetic field is applied to a magnetic material, the magnetization generally increases as the field is increased. If the field is reduced, the magnetization diminishes but does not follow the original curve – it 'lags behind'. This effect is known as magnetic hysteresis.

For the following simple description of magnetic hysteresis I am indebted to Dr Shoenberg. Consider the behaviour of a material which has a single easy direction of magnetization. In the unmagnetized state as many domains are magnetized to the left as to the right (see figure 15.10(*a*)). If a field is applied to the right the domains all become magnetized to the right and the material is saturated (figure 15.10(*b*)). The work done is used in moving the domain walls surrounding the right-pointing domains, until they swallow up the left-pointing domains. If the field is now removed no further change occurs and the material shows 100 per cent remanence (see figure 15.10(*c*)). The magnetization can be reduced only by applying a strong field to the left.

In general, metal specimens contain internal stresses. If the stresses are small they merely decide which of the easy directions of magnetization

Figure 15.10. Magnetization of material showing a single direction of easy magnetization. (*a*) Initial unmagnetized state. (*b*) Effect of applying a weak field to the right is to orient all the domains in that direction. This gives magnetic saturation; on removing the field the magnetization remains unchanged so that the remanence is almost equal to the saturation magnetization. (*c*) Magnetization curve showing that the remanence is very large and that demagnetization can be achieved only by reversing the sign of the field.

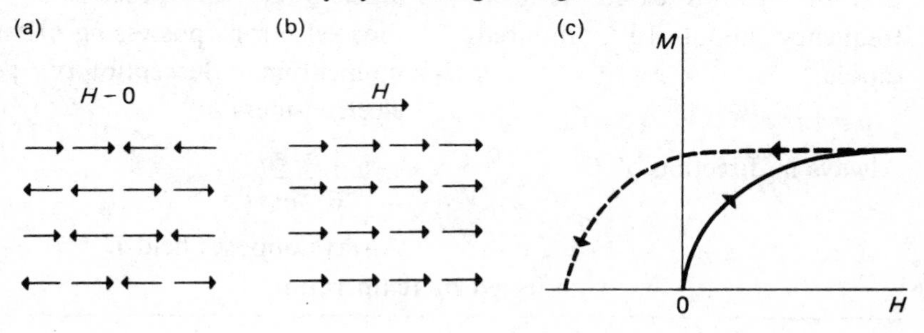

are easiest of all. If the stresses are large they may completely dominate the behaviour; in that case the easy domain directions are determined by the stresses rather than by crystal orientation. Consider for simplicity a specimen in which, as a result of local stresses, the domain directions are either horizontal, vertical or at 45° to these directions (figure 15.11(*a*)). If a strong horizontal field is applied all the domains line up with the field and the material achieves saturation (figure 15.11(*b*)). If the field is now removed the domains return to the original orientations they had in figure 15.11(*a*), but with the sense imposed on them by the applied field. This is shown in figure 15.11(*c*). Under these conditions the material has a remanance of about half the saturation magnetization.

Figure 15.11. Magnetization of a material showing a limited number of directions of easy magnetization (vertical, horizontal and at 45°). (*a*) Initial unmagnetized state, (*b*) a horizontal field *H* aligns all the domains, (*c*) on removing the field the domains revert to their original directions but the sense is that of the field *H*. Under these conditions the remanence is about half the saturation magnetization.

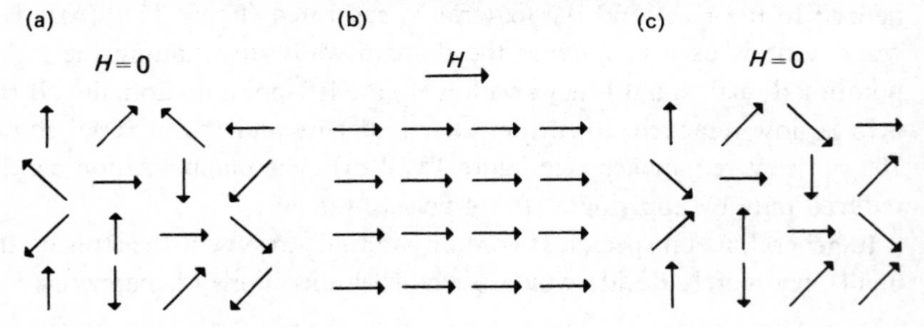

15.9 Summary: comparison of dielectric and magnetic properties

We now summarize in note form the main points of resemblance and difference in dielectric and magnetic properties.

Dielectric	Magnetic
Electronic polarization. Distortion of electron cloud. Critical frequency about 10^{15}. Induced dipole: $p=4\pi\varepsilon_0 r^3 E$ Always in direction of E	Diamagnetism. Deceleration or acceleration or precession of those electrons possessing orbital momentum. Susceptibility per electron-mass m_e $-\frac{1}{6}\frac{\mu_0 e^2 r^2}{m_e}$ Always opposes field H.
Unaffected by temperature	

Dielectric (cont.)	Magnetic (cont.)
Atomic polarization; stretching or bending of bonds if molecule has permanent dipole. Critical frequency about 10^{12}	Not applicable
Para-electric: orientation of molecular dipole p by applied field E, opposed by thermal motion. Typical value of $p = 1.6 \times 10^{-29}$ C m. For small E, susceptibility per molecule is: $\frac{n}{\varepsilon_0}\frac{p^2}{3kT}$	Paramagnetic: orientation of atomic dipole by applied field H, opposed by thermal motion. Typical value of $m = 10^{-23}$ J T^{-1}. For small H, susceptibility per atom is: $\frac{\mu_0 n m^2}{3kT}$ (classical) $\frac{\mu_0 n m^2}{kT}$ (quantum mechanical)
Para-electric polarization is of order 10^4 larger than paramagnetic. Hence need to consider real field rather than applied field. Very markedly dependent on temperature. Full behaviour described by Langevin function.	
Para-electric behaviour cannot be easily enhanced by low temperatures because dielectric solidifies and molecules cannot be oriented.	Paramagnetic effects can be greatly enhanced by low temperatures since only orbits or spins have to be oriented.
Ferroelectric Cooperative phenomena in which internal field is augmented by some type of specific interaction. Large oriented domains of permanent dipoles. Above some critical temperature (Curie temperature) the interaction is ineffective, and the materials cease to be ferroelectric or ferromagnetic.	*Ferromagnetic*

Appendix

Values of some physical constants

Quantity	Usual symbol	Value
Charge on electron	e	1.602×10^{-19} C
Rest-mass of electron	m_e	9.109×10^{-31} kg
Electron charge/mass	e/m_e	1.759×10^{11} C kg^{-1}
Rest-mass of proton	M	1.673×10^{-27} kg
Ratio: proton/electron mass	M/m_e	1836
Boltzmann constant	k	1.381×10^{-23} J K^{-1}
Avogadro's number	N_A	6.022×10^{23} mol^{-1}
Gas constant	R	8.315 J mol^{-1} K^{-1}
Mechanical equivalent of heat	—	4.186 J cal^{-1}
Volume of 1 mole of gas at s.t.p.	V	22.41×10^{-3} m^3
Faraday constant	F	96 490 C mol^{-1}
Planck's constant	h	6.626×10^{-34} J s
Planck's constant/2π	$\hbar$	1.055×10^{-34} J s
Bohr magneton $[eh/4\pi m_e)$	μ_B	9.274×10^{-24} J T^{-1}
Bohr radius	a_0	0.529×10^{-10} m
Gravitational constant	G	6.673×10^{-11} N m^2 kg^{-2}
Energy equivalent of 1 eV		1.602×10^{-19} J
Velocity of light	c	299 800 km s^{-1}
Permeability of free space	μ_0	$4\pi \times 10^{-7}$ H m^{-1}
Permittivity of free space	$\varepsilon_0 = (\mu_0 c^2)^{-1}$	8.854×10^{-12} F m^{-1}

The values in the above table for N_A, R, J and F are 'physical' constants based on the number of atoms in 0.012 kg of ^{12}C. The equivalent constants used by chemists are sometimes based on a different molar convention and the values quoted may differ by up to one part in a thousand. The calorie in this table is the 15 °C calorie defined on p. 33.

Index